AF581667

(Mass Spectrometric) Non Target Screening–Techniques and Strategies

(Mass Spectrometric) Non Target Screening–Techniques and Strategies

Editor

Thomas Letzel

MDPI • Basel • Beijing • Wuhan • Barcelona • Belgrade • Manchester • Tokyo • Cluj • Tianjin

Editor
Thomas Letzel
Consultancy and Education
AFIN-TS GmbH
Augsburg
Germany

Editorial Office
MDPI
St. Alban-Anlage 66
4052 Basel, Switzerland

This is a reprint of articles from the Special Issue published online in the open access journal *Molecules* (ISSN 1420-3049) (available at: www.mdpi.com/journal/molecules/special_issues/non_target_screen).

For citation purposes, cite each article independently as indicated on the article page online and as indicated below:

LastName, A.A.; LastName, B.B.; LastName, C.C. Article Title. *Journal Name* **Year**, *Volume Number*, Page Range.

ISBN 978-3-0365-6125-7 (Hbk)
ISBN 978-3-0365-6640-5 (PDF)

Contents

About the Editor

Thomas Letzel

PD Dr. Thomas Letzel was trained as a chemical laboratory technician, has studied chemistry in Munich and worked on his Ph.D. thesis in the field of environmental analysis at the Technical University of Munich. After a post-doctoral stay at the VU Amsterdam, he was head of the analytical research group (AFG) at the Technical University of Munich (TUM) from 2003 to 2018 and held a teaching professorship at the Ludwig-Maximilians University in Munich from 2014 to 2017. Since 2018 he is also one of the executive directors at the start-up company 'Analytisches Forschungsinstitut für Non-Target Screening' (AFIN-TS GmbH). In AFIN-TS he continues basic and advanced research in analytical chemistry and simultaneously consults, supports and teaches common instrumental analysis, such as LC-MS, GC-MS, etc., as well as HILIC, RPLC-HILIC, SFC, and non-target screening solutions (also for commercial partners). In the past 20 years, he has devoted himself to teaching and imparting knowledge about chromatographic and mass spectrometric analysis, as well as applications in various research areas, such as environmental and food analysis, as well as pharmaceutical analysis and bioanalytics. So far, he has published 84 peer-reviewed research articles, 24 review articles, six books and more than 80 open-view articles. With www.chemnixblog.de, www.analyticsplus.org and others he actively operates educational online platforms.

Preface to "(Mass Spectrometric) Non Target Screening–Techniques and Strategies"

Mass spectrometric non-target screening is a preferably comprehensive and untargeted (predominantly organic molecules detecting) approach combining (robust) analytical measurements with adapted data evaluation concepts, systematic compound identification workflows, and statistical data interpretation. It is well suited for the identification of new, unexpected, and/or unknown organic compounds, as well as monitoring 'molecular fingerprints'and profiling 'process-relevant'molecules via statistical methods.

Non-target screening (NTS) provides special benefits for the investigation of complex samples in various disciplines. Even well-characterized samples can contain relevant and yet not detected compounds. Depending on the nature of the sample and the aim of the analytics, the presence of such unknown compounds can be of concern and relevance. Non-target screening is a universal technique that can be applied in many areas, such as environmental, forensic, food, metabolomics and other analysis. Since several meetings in 2016 (and later) regarding that topic, and many books and publications, the NTS topic has gained a new level of knowledge and awareness. A lot of this knowledge was presented on face-to-face, hybrid, and digital conferences and workshops, such as the 'International Conference on Non-Target Screening' and the 'Gas Chromatography meets Non-Target Screening' (details regarding the programs and others see on https://afin-ts.de). Further, many of which presenting there also published in these years in this special NTS issue.

Concluding, in the years 2019–2022, 14 articles were published and presented in this Special Issue, whereby it contains 4 peer-reviewed review articles and 10 peer-reviewed research articles. The focus of the review articles is on semi-quantitative NTS (Malm et al.), on GC-IMS (Capitain and Weller), on GC-API-MS (Li et al.) and on liquid biomarkers (for Equine anti-doping (Tou et al.). Two research articles deal with analytical data processing concepts (Rubino and Yang et al., respectively) and eight research articles present strategies in various disciplines. Hunynh et al. presented sample preparation methods for NTS in urban waters, Panse et al. water-solved molecules after UV photo dissociation, whereas Morimoto et al. studied the aqueous matrix tea and Stegmueller et al. human urine (after coffee consumption). Further metabolomics studies are presented by Satria at al. (in Lingzhi), Dionisio et al. (in Maize roots), and Xue et al. (in Lettuce as well as Maize). Last but not least, Alvarez et al. presented a glycomics-driven glycoproteomics analytical platform for unknown inhibitors.

All in all, this Special Issue is reflecting new NTS developments in the recent years and is at the same time a good starting point for the screening need of the forthcoming ones. If you are interested to publish your resent developments open-access do not hesitate to contact me for a publication in the next Special Issue.

Thomas Letzel
Editor

MDPI

Review

Guide to Semi-Quantitative Non-Targeted Screening Using LC/ESI/HRMS

Louise Malm [1,†], Emma Palm [1,†], Amina Souihi [1], Merle Plassmann [2], Jaanus Liigand [3] and Anneli Kruve [1,*]

1 Department of Materials and Environmental Chemistry, Stockholm University, Svante Arrhenius väg 16, 114 18 Stockholm, Sweden; loma5202@student.su.se (L.M.); emma.palm@mmk.su.se (E.P.); amina.souihi@mmk.su.se (A.S.)

2 Department of Environmental Science, Stockholm University, Svante Arrhenius väg 8, 114 18 Stockholm, Sweden; Merle.Plassmann@aces.su.se

3 Quantem Analytics, 510 08 Tartu, Estonia; jaanus@quantem.co

* Correspondence: anneli.kruve@su.se

† These authors contributed equally to this paper.

Abstract: Non-targeted screening (NTS) with reversed phase liquid chromatography electrospray ionization high resolution mass spectrometry (LC/ESI/HRMS) is increasingly employed as an alternative to targeted analysis; however, it is not possible to quantify all compounds found in a sample with analytical standards. As an alternative, semi-quantification strategies are, or at least should be, used to estimate the concentrations of the unknown compounds before final decision making. All steps in the analytical chain, from sample preparation to ionization conditions and data processing can influence the signals obtained, and thus the estimated concentrations. Therefore, each step needs to be considered carefully. Generally, less is more when it comes to choosing sample preparation as well as chromatographic and ionization conditions in NTS. By combining the positive and negative ionization mode, the performance of NTS can be improved, since different compounds ionize better in one or the other mode. Furthermore, NTS gives opportunities for retrospective analysis. In this tutorial, strategies for semi-quantification are described, sources potentially decreasing the signals are identified and possibilities to improve NTS are discussed. Additionally, examples of retrospective analysis are presented. Finally, we present a checklist for carrying out semi-quantitative NTS.

Keywords: Ionization; quantification; decision making; NTS strategies

Citation: Malm, L.; Palm, E.; Souihi, A.; Plassmann, M.; Liigand, J.; Kruve, A. Guide to Semi-Quantitative Non-Targeted Screening Using LC/ESI/HRMS. *Molecules* **2021**, *26*, 3524. https://doi.org/10.3390/molecules26123524

Academic Editor: Thomas Letzel

Received: 11 May 2021
Accepted: 4 June 2021
Published: 9 June 2021

1. Different Strategies

A general workflow for non-targeted screening (NTS), as described by Hollender et al. [1], includes representative sampling followed by enrichment suitable for the sample matrix. E.g., for water matrices, solid-phase extraction (SPE) is ordinarily used; however, in order to not lose compounds of interest, the stationary phase used should be chosen with care [1,2]. For separation and analysis of the sample, liquid chromatography electrospray ionization high resolution mass spectrometry (LC/ESI/HRMS) is utilized. Commonly, the chromatographic separation is performed in reversed phase, and the HRMS is run in full scan mode. After collecting the data, the next step is peak detection and grouping of peaks related to the same molecular structure, and also comparison of sample peaks with peaks from compounds present in blanks [1–3]. This step can often be semi-automated, using either the vendor, third-party or open-source software (e.g., Thermo Scientific™ Compound Discoverer™ software, [4] envipy [5] from EAWAG or MZmine [6]). Still, even after this data pre-processing step, there are usually too many peaks left for all to be confidently (to level 1 and 2 according to the Schymanski scale) identified, [7] thus the peaks are prioritized before the ones with highest priority are fully identified with analytical standards [1,3]. Every so often, we simply prioritize the compound for which we have information—e.g., if standards are available, if they are toxic or known to be very abundant or persistent. It might also be of interest to prioritize compounds with many transformation products (TPs), as the TPs can sometimes be more toxic [8] or more abundant [8,9] than the parent compound. To accurately prioritize the peaks, it is desired to know the concentrations of the compounds, however, the ionization efficiency (*IE*) of different compounds varies tremendously in ESI, making it inappropriate to only compare the intensities in the spectrum for prioritizing. Instead, semi-quantitative approaches should be included in the NTS workflow before prioritizing the peaks.

As mentioned, *IE* varies significantly in the ESI source, depending on the properties of the compound itself but also the properties of the sample matrix and the eluent used [10,11]. While the ionization efficiency is influenced by many factors two general rules to consider are: (1) in positive ionization mode stronger bases and in negative ionization mode stronger acids, and (2) more hydrophobic compounds, have higher ionization efficiency. Additionally, the increased amount of organic solvent will increase the *IE*, since the evaporation rate of the droplet will increase [11]. Therefore, compounds eluting later from reversed phase chromatography tend to have higher ionization efficiency values. Moreover, the pH of the mobile phase in combination with the acidity/basicity of the compound also have great influence of the *IE*: compounds with acidic moieties usually ionize more efficiently in basic environments in negative ESI mode (ESI-), while compounds with basic moieties usually have higher *IE* in acidic mobile phase in positive ESI mode (ESI+) [12,13]. Many additional factors impacting ionization efficiency have been studied previously relating to compound structure, mobile phase, and instrument parameters.

A measurement that is closely related to the ionization efficiency of a compound is the response factor (RF), the ratio of the detected peak area and the concentration of the compound (Equation (1)). This ratio is important in some of the semi-quantification strategies presented below.

$$RF = \frac{peak\ area}{concentration} \tag{1}$$

As the main focus of this tutorial is on the practical implementation of semi-quantification, we will not go too deep into details regarding ionization efficiency and kindly direct the interested reader towards excellent reviews/studies published previously, e.g., for compound properties see Kruve [2] and Cech and Enke [12]; for mobile phase properties Kostiainen and Kauppila [14] and Kiontke et al. [15]; for source parameters Page et al. [16] Pieke et al. [17] and Espinosa et al. [18].

1.1. Structurally Similar Compounds

One approach for semi-quantification is to use standards with structural similarities to the unknown compound(s). This approach however is not true non-target screening but rather suspect screening, since a tentative structure of the unknown, or suspect, is needed. The assumption is that the response of the structurally similar compound will be similar to that of the suspect, and therefore, the RF of the similar standard can be used to estimate the concentration of the unknown (Equation (2)).

$$c_{suspect\ compound} = \frac{peak\ area_{suspect\ compound}}{RF_{similar\ compound}} \tag{2}$$

An online tool [19,20] is available from the University of Athens to search for structurally similar compounds based on 2D similarities, e.g., number of similar functional groups and distance between functional groups [21]. The suspects are compared with compounds from NORMAN SusDat Database, [22] with similarity down to 5%. The similarity is measured as Tanimoto similarity of substructure-based fingerprints. However, the similarity metric does not yet take into consideration if the properties are relevant or not for ionization efficiency in electrospray.

One special class of compounds are transformation products. Many compounds, e.g., pesticides and pharmaceuticals, degrade naturally in the environment, giving rise to transformation products [3,9,23]. TPs can also be formed by chemical degradation, e.g., via oxidation processes in wastewater treatment plants [3,24]. Although TPs can be more persistent, [9] abundant [3,9] and/or toxic [8,24] than the parent compound, they are rarely included in environmental targeted analysis, e.g., in water protective plans. This is primarily because of the unknown toxicity [9] and/or structure [2] of many TPs. However, it is possible to semi-quantify transformation products based on the parent compounds RF, in an analogous approach as the similar compound approach described above (Equation (3)).

$$c_{TP} = \frac{peak\ area_{TP}}{RF_{parent\ compound}} \tag{3}$$

This strategy is based on the presumed structural similarity between TPs and their parent compounds; however, sometimes the TPs have lost functional groups that largely affect the ionization efficiency. As an example, we can look at atrazine and some its known degradation products, to see how the structure changes, see Table 1. We can compare how the similarity score, obtained from the University of Athens online tool, [19,20] changes with the structure. As seen, the similarity decreases drastically as the structural differences increase. Surprisingly, the hydrophobicity of the compound does not seem to influence the overall similarity, as the log*P* varies quite a lot between the parent and TPs while the similarity scores remain high. This, however, makes the approaches utilizing structurally similar compounds prone to higher errors, as the hydrophobicity is known to have a high impact on the ionization efficiency [11,25]. The decrease in response factors with the decrease in similarities further indicates that such approaches, although easy to use, might give higher errors than alternative approaches.

Table 1. Structurally similar compounds do not necessarily mean that they have similar chromatographic properties. Here we see atrazine and some known TPs with corresponding score on the similarity to atrazine (parent) obtained from the online tool from University of Athens [19] and the RF measured in our laboratory. Chemicalize was used for prediction of pK_a and log*P*, July 2020, https://chemicalize.com/ (accessed on 9 June 2021) developed by ChemAxon (http://www.chemaxon.com) (accessed on 9 June 2021) [26].

Compound	Similarity Score	pK_a	log*P*	RF
Atrazine 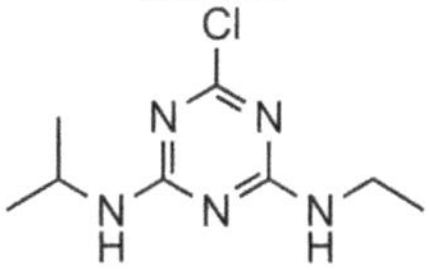	100%	4.2	2.2	7.3×10^{17}
Atrazine-2-hydroxy 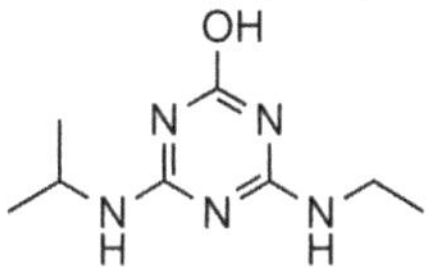	58.3%	3.6	−3.1	1.0×10^{17}
Atrazine-desethyl 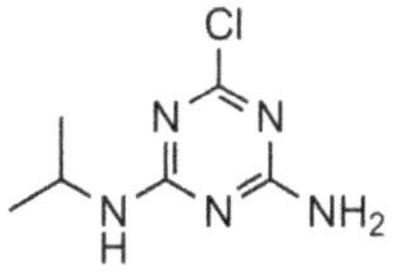	53.9%	4.4	0.8	4.1×10^{16}

Table 1. *Cont.*

Compound	Similarity Score	pK_a	log$P$	RF
Atrazine-desethyl-2-hydroxy 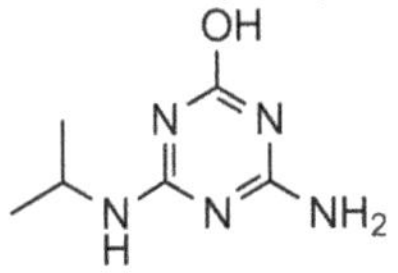	38.9%	3.0	−3.5	3.0×10^{16}
Atrazine-desisopropyl 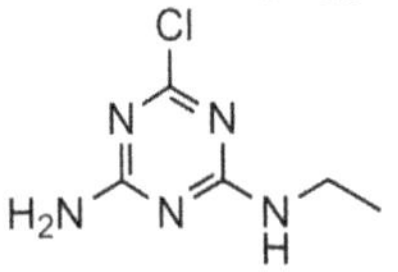	44.6%	4.4	0.4	2.7×10^{16}
Atrazine-desisopropyl-2-hydroxy 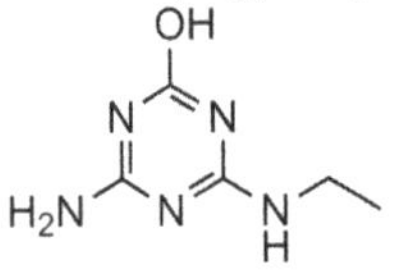	31.5%	3.1	−3.9	4.6×10^{16}
Atrazine-desethyl-desisopropyl 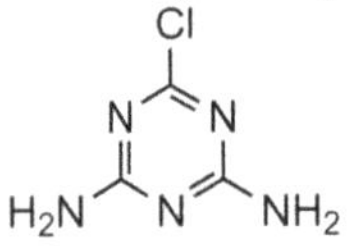	19.8%	4.6	−0.2	4.9×10^{15}
Atrazine-desethyl-desisopropyl-2-hydroxy 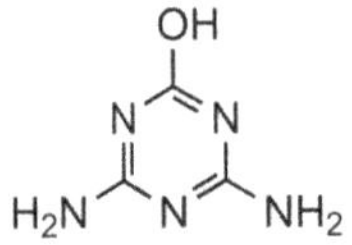	13.4%	3.1	−4.5	4.0×10^{14}

Additional limitations in the transformation product quantification with parent compounds are that sometimes, the parent compound and transformation products cannot be detected with the same analysis mode. For example, chlorothalonil is not detected in LC/ESI/HRMS while the corresponding TPs are well detectable. Furthermore, while flufenacet is usually analyzed in ESI+, its transformation products flufenacet-OXA and flufenacet-ESA are analyzed in negative ionization mode [9].

Theoretically, this approach works for all compounds that have either structurally similar compounds or a parent compound. Although the parent compound-TP approach is usually applied in environmental analysis, it can also be used to estimate concentrations of metabolites in, e.g., biological samples [27].

1.2. Close Eluting Compounds

Another strategy for semi-quantification of unknowns is to use the RF of the compound with known concentration (internal standard) that is eluting closest to the unknown compound in the chromatogram. This strategy was proposed by Pieke et al. [17] and is based on the assumption that compounds that elute close in time in reversed phase LC, i.e., compounds that share similarities in chromatographic properties, will also have similar RF. The factors influencing retention in reversed phase LC are similar to those influencing *IE*, e.g., the polarity of the compound and its acid/base properties. Thus, the concentration of the unknown compound can be estimated (Equation (4)).

$$c_{unknown\ compound} = \frac{peak\ area_{unknown\ compound}}{RF_{closest\ eluting\ compound}} \quad (4)$$

How many internal standards are needed depends on the sample, i.e., on how many unknown compounds that it contains. However, a good rule of thumb is to have both early, middle, and late eluting standards with known concentration, since we do not know beforehand where our unknown compounds will elute. Because of the large variation in *IE*, which strongly affects the RF, it has also been proposed to include internal standards that covers a wide range of response factors [17]. This strategy is in theory applicable for any class of compounds that can be analyzed with LC/MS, in both ESI+ and ESI-. However, to our knowledge it has not yet been validated for a wide range of compounds.

For this approach, the chromatographic conditions also play an important role. In some cases, the closest eluting compound may not be the same at different pH conditions as the compound's retention time changes with protonation and deprotonation. We evaluated the impact of chromatographic conditions for a set of 81 compounds with mobile phases of three different pHs. For two compounds, metsulfuron-methyl and valsartan, drastic changes in retention time were observed. This also changes the closest eluting compound and thus the accuracy of the response factor prediction. For example, metsulfuron-methyl and its closest eluting compound prometryn have response factor ratio of 0.692 at pH 2.7 while at pH 5.0 the closest eluting compound is 2-napthoic acid and the response factor ratio is 857, as shown in Table 2. This variation indicates that retention time and ionization efficiency are influenced by somewhat different factors.

Table 2. The closest eluting compound is in some cases not the same in different mobile phases, as is the case for valsartan and metsulfuron-methyl. Here we see how the retention time change causes the change of closest eluting compound and how this can affect the accuracy of the response factor prediction in ESI+.

Compound	pH	Neighbour	$RT_{compound}$ (min)	$RT_{neighbour}$ (min)	$RF_{compound}$	$RF_{neighbour}$	$\frac{RF_{compound}}{RF_{neighbour}}$
Valsartan	2.7	Estrone	12.58	12.58	5.99×10^{11}	1.11×10^{12}	0.54
	5.0	Chlorthalidone	7.30	7.41	1.52×10^{11}	9.42×10^{10}	1.61
	8.0	Sulfamethazine	5.23	5.27	7.93×10^{11}	4.77×10^{12}	0.166
Metsulfuron-methyl	2.7	Prometryn	10.32	10.26	4.31×10^{12}	6.23×10^{12}	0.692
	5.0	2-napthoic acid	9.62	9.62	2.17×10^{12}	2.53×10^{9}	857
	8.0	Gabapentin	3.78	3.75	5.80×10^{12}	2.52×10^{12}	2.30

1.3. Predicting Ionization Efficiency

The last approach for semi-quantification of suspects/unknowns is to use machine learning to train models to predict the *IE* of the compounds. Factors that influence the ionization efficiency has been widely studied by many researchers, [12,28,29] and there are also many papers describing models to predict the *IE* [25,30,31]. However, most earlier studies and models have focused on either ESI+ or ESI-, and on one class of compounds. A comprehensive comparison of previous *IE* studies has been done by Liigand et al. [32] Recently, we have presented an *IE* prediction model that works for positive and negative ionization mode [10]. Data from 106 eluent compositions with 353 compounds in positive mode, and 33 eluent compositions with 101 compounds in negative mode, was used. The model utilizes random forest regression to predict the ionization efficiency based on 2D PaDEL descriptors (450 in ESI+ and 145 in ESI-) together with eluent descriptors. Because the ionization efficiency can differ so drastically between different instruments, all *IE* measurements used for the model development were performed relative to tetraethylammonium and benzoic acid for ESI+ and ESI-, respectively. Furthermore, all *IE* values were unified by performing linear regression between the measurements from two different instruments, to transform the predictions to instrument specific values, and thereby making the results comparable between different instruments [10].

The obtained *IE* predictions are related to a predicted RF (Equation (5)), which is used to calculate the concentration of the suspect compound (Equation (6)).

$$logRF_{pred} = slope \times logIE_{pred} + intercept \tag{5}$$

$$c_{suspect\ compound} = \frac{peak\ area_{suspect\ compound}}{10^{logRF_{pred}}} \tag{6}$$

A commercial online tool [33] provided by Quantem Analytics is available to semi-quantify suspect compounds based on the *IE* prediction model. It requires information about analysis conditions (ionization mode, eluent composition, gradient program) and a csv file with SMILES, retention time and signal obtained, as well as at least 5 compounds with known concentration. These compounds with known concentration are used to calibrate the predictions made by the tool.

Currently, the *IE* prediction model described here is only applicable for compounds that form $[M+H]^+$, $[M]^+$, and $[M-H]^-$ ions, as such compounds was used to train the model. Therefore, some compounds, e.g., biomolecules like sugars, are not suitable to semi-quantify with this approach, as they predominantly form adducts [10]. Additionally, any machine learning model can only be applied to the compounds structurally similar to the ones used in training the models. Therefore, the application range depends on the concrete model and its basis. The *IE* prediction model published by Liigand et al. [10] was trained and tested on a large range of compound classes, e.g., benzenoids, organo-heterocyclic compounds, organic acids, etc., and therefore covers a wide chemical space. It has also been validated on pesticides and mycotoxins.

The second semi-quantification approach is the only true non-targeted approach, as no tentative structure of the suspect compound is required as opposed to the other strategies. However, the use of high resolving instruments has improved the mass accuracy obtained, and thereby made it easier to assign a tentative structure based on the exact mass and MS/MS fragmentation spectra [1,10,34].

Most of the semi-quantification strategies described here are relatively straight forward and easy to use. Therefore, we suggest investigating if combinations of the approaches could improve the results and decrease the errors related to the predicted concentrations.

2. Sources Decreasing the Signal of the Molecular Ion

Any of the described approaches can only account for the factors it has been based on. Quantification models that are trained to account for the differences in the ionization efficiency of different compounds, for example, will not be able to account for the factors in other parts of the analysis that also influence the signal of the compounds. In non-targeted LC/HRMS many sources for signal variation occur, starting from the very beginning of the workflow to the end. In sample preparation, different compounds may have very-different recoveries while chromatography may cause peak artifacts. Additionally, the softness of ionization conditions and ion transportation may influence the intensity of the signals observed in the final spectrum. Finally, the automatic integration has to be accurate to obtain reliable results.

2.1. Sample Preparation

The relatively low concentrations of compounds in the samples of interest often require preparing the samples before analysis. Many samples, including biological tissues, soil, or other heterogeneous non-liquid samples, also require sample preparation to obtain a liquid sample. The methods used for preparing water samples prior to non-targeted LC/HRMS analysis may involve sample filtration, [35] vacuum-assisted evaporative concentration, [9] SPE, [36] liquid-liquid extraction (LLE), [37,38] or similar. While filtration can be sufficient for preparation of drinking water, more complicated water samples such as wastewater are often extracted with SPE or LLE. LLE is mostly used for preparing samples for GC/MS analysis but can also be used for preparing water samples for LC/HRMS [37]. In LC/ESI/HRMS, SPE is the most widely used sample preparation strategy, whereby very-different protocols are used. Hydrophilic-lipophilic balance, cation exchange, anion exchange, reversed-phase or mixed-mode SPE sorbents are the most-commonly used, either alone [39,40] or in combination [36].

In sample preparation, some compounds will be extracted while other compounds will be removed due to the selectivity of the SPE procedure. This will determine which compounds can be detected with the NTS methods and directly influences the scope of the methods. However, extraction also influences quantification. Even if extracted, the recovery of different compounds varies. Schultze et al. [41] found reversed-phase SPE to yield superior recoveries and detection limits over anion- and cation exchange SPE for water samples. However, only 159 of 251 studied analytes yielded recoveries within the acceptable range of 60 to 123%. Even more so, lowest recoveries that still enabled analyte detection were close to 1%. From the quantification point of view, a recovery of 50% will end up with a quantification error of a factor of 2, while recovery of 1% results in an error of a factor of 100. These effects are much more dramatic in comparison to targeted methods, as no isotopically labelled internal standards are available for correcting with the poor recovery.

Therefore, we suggest (1) avoiding extraction where possible, and (2) further research on extraction efficiency. In some cases, it is possible to avoid sample preparation. For example, the detection limits of contemporary LC/ESI/HRMS devices often reach ng/L levels, so many environmental contaminants can be analyzed directly from the drinking and surface water samples. In our experiences, the matrix effect caused by remaining "matrix" components is negligible compared to the losses in extraction efficiency. This can be also seen in Figure 1, where we have compared the quantification accuracy for 30 micropollutants with a generic hydrophilic-lipophilic balance SPE vs. direct injection. It is clear that many compounds have poor recoveries with SPE; therefore, the estimated concentrations may be an order of magnitude lower than the actual concentrations.

However, in some cases the sample preparation has to be used. In such cases a generic sample preparation, proven in the literature, is suggested [42]. It is advantageous to use a sample preparation method for which recoveries for many different compounds have been previously determined and are publicly available. This will allow to evaluate the suitability of the extraction method in the context of quantification. A word of caution is needed here, although informative, the recoveries determined for test compounds do not have to apply for the compounds detected with non-targeted analysis. Especially if the physicochemical properties are significantly different.

Another possibility is to account for the differences in extraction efficiency with modeling approaches. The current problems here are that very few datasets of recoveries are available, and the extraction conditions used in SPE vary tremendously. This includes both the choice of SPE stationary phase, but also solvent. In contrast to electrospray ionization efficiency, the choice of solvents can impact the extraction recoveries of different compound groups very differently, while in electrospray many effects are common to all compounds and can be generalized. To overcome the sparsity of data needed for efficient modeling, both standardization of the sample preparation methods as well as common efforts in collecting data are required.

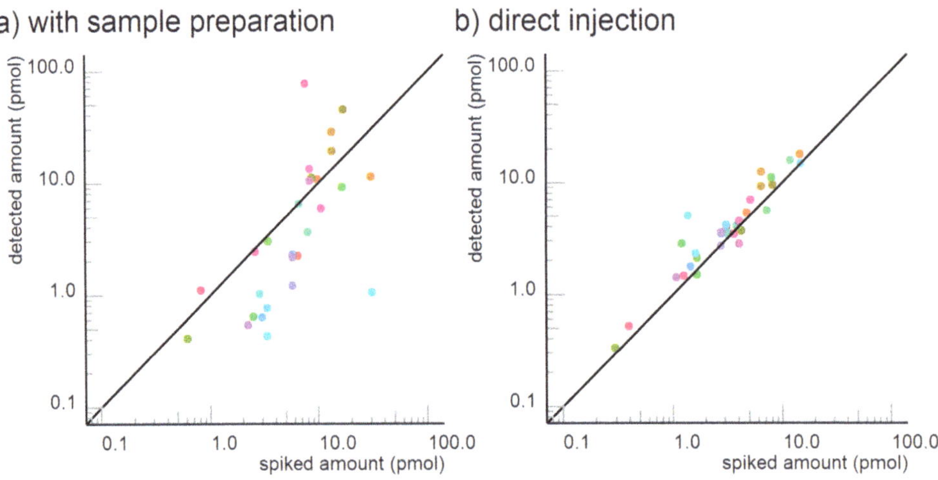

Figure 1. Sample preparation can have a devastating effect on the quantification accuracy as seen here: (**a**) quantification results for contaminants analysis with SPE extraction and (**b**) without SPE extraction, but with a high injection volume. Color corresponds to compound, and both analyzes come from the average of three replica.

2.2. Chromatography

In non-targeted LC/HRMS, the quantification of unknown compounds is highly affected by the chromatographic conditions [13,14]. For example, the mobile phase pH influences the sensitivity of positive ionization mode; acidic conditions show better sensitivity towards basic compounds because their pK_a values are higher than the pH, and protons are abundant. The sensitivity of analysis decreases at higher pH because the protonation is less efficient, see Figure 2a. However, sometimes the acidic mobile phase can decrease retention and symmetric peak shape [43]. In basic conditions, the "wrong-way-round" ionization could improve the sensitivity towards basic compounds using ammonium formate [44]. In addition, a recent study reported that basic mobile phase significantly improved the ionization of both basic and acidic pharmaceuticals in positive mode [43].

The ionization efficiency of a compound is also be affected by acids and buffers. It is widely discussed that trifluoroacetic acid causes ionization suppression in electrospray due to formation of gas phase ion pairs [45,46]. In addition, formate and acetate have higher ability to form ion-pairs compared to fluoride and therefore they lead to lower ionization efficiencies compared to buffer containing fluoride [47]. Ojakivi et al. [48] has observed that acids provide higher ionization efficiencies compared to buffers containing ammonium ions at same pH. This was explained by the fact that ammonium ions have a higher tendency to occupy the charged surface of droplet.

Additionally, the organic solvent and its content may influence the ionization of compounds in reversed phase chromatography. The sensitivity generally increases with higher amount of organic modifier. Polar compounds, which are less retained by the column, elute in more water rich mobile phase and therefore their ionization may be low. Although, at equal organic modifier concentration, methanol and acetonitrile yield indistinguishable ionization efficiency values, [49] the organic solvent and its content may influence the ionization of compounds in reversed phase chromatography. Methanol has a lower elution strength compared to acetonitrile, which allows compounds to elute later at a higher organic solvent content; therefore, the ionization of compounds is improved [14]. Moreover, methanol might improve the peak shape and symmetry of basic compounds [14]. In Figure 2b, we have compared the sensitivity of 27 compounds analyzed with methanol vs. acetonitrile as organic modifier. For this dataset no one organic modifier had significant advantages over another. The differences in sensitivity are generally small, less than 3×.

However, one compound shows 10 × higher sensitivity with acetonitrile. This comparison also accounts for the different retention times with the two modifiers. In case of non-targeted analysis, one still needs to understand the chemical space they are most interested in. Although acetonitrile is a widely used organic modifier in reversed phase LC, it may not be suitable in all circumstances. E.g., Colizza et al. [50] have observed that in case of cyclic peroxides, already small amount of acetonitrile in the eluent composition results in strong ion suppression.

The use of binary mobile phase may also influence the quantification of compounds due to peak artifacts in reversed phase LC [51]. The injection of sample can introduce disturbance of the equilibrium state in the column between the mobile and stationary phases. The analytes flow through the column with different velocity than the original equilibrium. Thus, the chromatographic system sets a new equilibrium via a relaxation process, which leads to additional signals in the chromatogram called system or solvent peak [52]. The quantification and identification of unknown compounds become more complicated due to the presence of peak artifacts in the chromatogram.

In general, we have observed that compounds with higher ionization efficiency have lower prediction errors in semi-quantification. This means that it is favorable to use a mobile phase with a pH that aids the ionization in the mode one is using. Thus, in positive mode, mobile phase with lower pH will likely give lower prediction error and mobile phase with higher pH is more likely to give lower prediction errors in negative mode.

We strongly encourage community to use generic standardized methods to collect standardized datasets for future model development of any kind/phenomenon in the benefits of the community and society. In a recent quality assurance paper, [53] every participant in an interlaboratory comparison used slightly different gradient, which makes it unnecessarily complicated to use these datasets for generalization training, i.e., retention time models. For example, Domingo-Almenara et al. [54] have done tremendous work and collected reversed phase retention times for more than 80,000 compounds. However, to use their data in other laboratories, the gradient and solvent needs to match between laboratories. This becomes less feasible when the database has been collected with niche mobile phases or exotic gradients, or if laboratories intend to use very specific chromatographic conditions. Therefore, standardization of chromatographic conditions is highly needed. While comparing the gradient programs used when collecting MassBank data, many different gradients have been used. However, generally 0.1% formic acid is used as water phase and the gradient starts from 1 to 13% of water phase. To generalize these methods, we recommend using 0.1% formic acid in ESI positive mode and ammonium hydroxide or ammonium formate pH = 8.0 in ESI negative mode. In both cases a linear gradient from 5 to 100% of organic modifier over 15 min for a 10 cm C18 column with 3 μm particle size has proven generic. Currently more ionization efficiency data have been collected with acetonitrile than with methanol, with the modeling results for acetonitrile having higher confidence. As a result, we suggest using acetonitrile as the organic modifier in reversed phase chromatography if semi-quantification is intended.

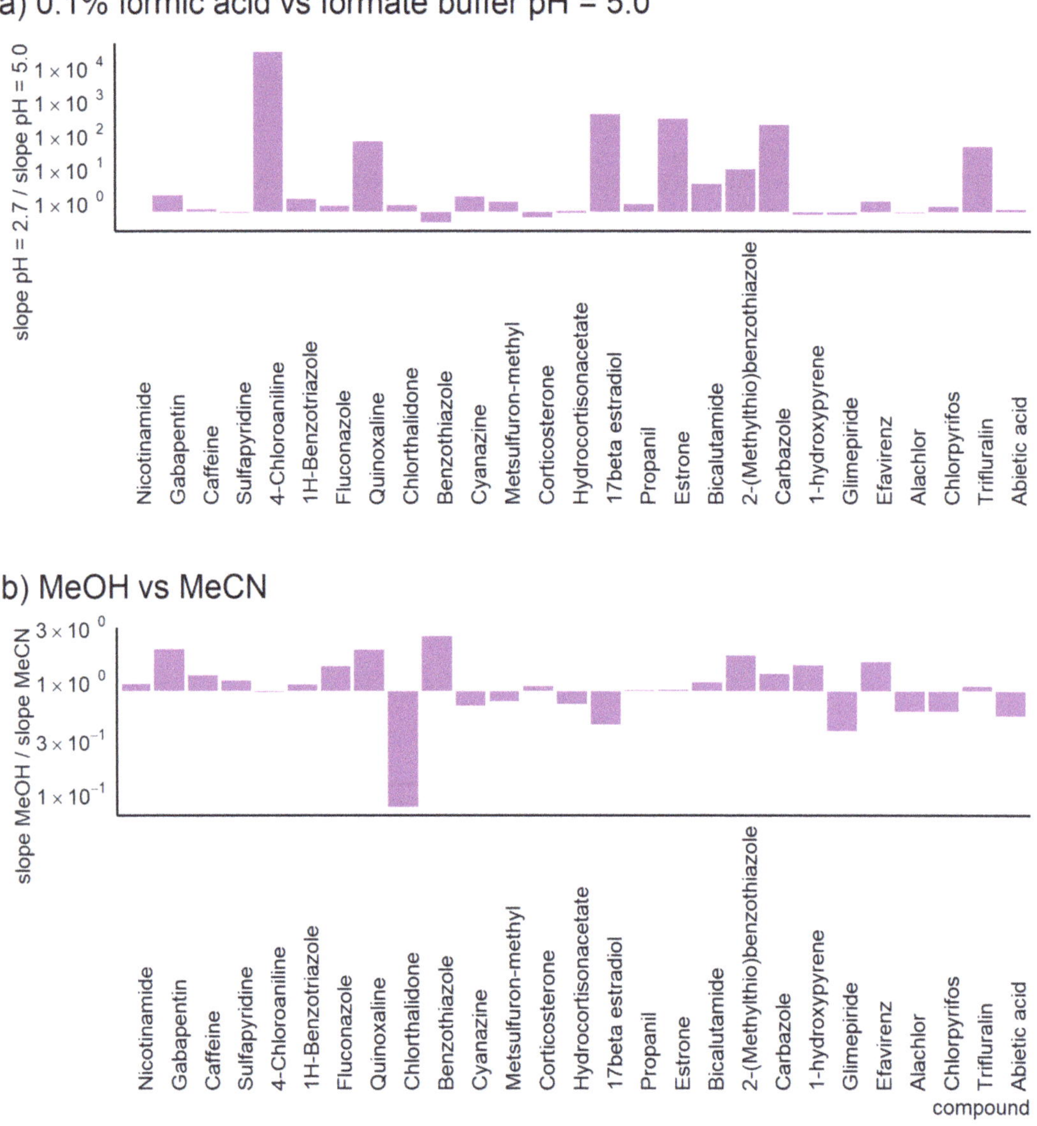

Figure 2. The comparison of sensitivity obtained with (**a**) 0.1% formic acid vs. 10 mM formate buffer at pH = 5.0 and (**b**) using methanol or acetonitrile as organic solvent. In all cases reversed phase chromatography with the same gradient program was used (linear gradient from 5% to 100% of organic modifier over 15 min).

2.3. Ionization Conditions and Ion Transport

Different ionization sources have different softness. Some sources cause extensive fragmentation while other sources facilitate the formation of the molecular ion only. It is suggested that fragmentation occurs after the ejection of the molecular ions from charged droplets to the gas phase. This fragmentation can be brought about in the source, but also while entering the vacuum region of the mass spectrometer. Due to the fact that fragmentation yields are strongly different from source to source and compound to compound, all gas phase ions formed from the compound of interest needs to be accounted for.

In addition to ESI source design and conditions, ion optics and/or use of ion mobility before the mass analyzer may also play a role [55]. Seo et al. [56] observed that protonated *ortho*-aminobenzoic acid is very fragile and proposed that collision with N_2 in the ion mobility cell or source region provides enough energy for efficient fragmentation. We have a similar observation for *ortho*-aminobenzoic acid when using Agilent instrument with ionFunnel ion focusing. An abundant peak at m/z = 120, corresponding to water loss, is observed in the mass spectrum, but only a small or no molecular ion peak.

One remaining challenge in NTS is identifying which fragments have been formed from specific compounds. Most data treatment software and packages do not enable automatic grouping of in-source fragments together with the molecular ion. This is also understandable, as it is essentially impossible to automatically identify which ions simply have very close retention times, and which belong to the same compound. The option here is to analyze fragmentation spectra. The fragments formed in-source and fragments formed from the same compound in MS/MS are the same. Therefore, knowing the fragment ions from the data independent or data dependent acquisition experiments will allow identifying the in-source fragments too. These possibilities, however, have not yet been automated in current software.

2.4. Data Processing

Software used in non-target screening mostly does a good job when it comes to integrating peaks. In some software, the integration algorithm can be chosen by the user depending on the chromatography (e.g., MZmine [6]), while in others, the software itself decides which the best algorithm to use is (e.g., Compound Discoverer [4]). However, many purely non-targeted processing software do not allow changing or deleting peak integrations after processing. And in some instances, like peak splitting or long tailing/fronting, only parts of the peaks might be integrated. Additionally, the alignment between samples can for some compounds with unstable retention times be difficult. Both a wrong integration and a bad alignment can result in a feature being listed twice with different retention times. This means that one always needs to check the integrations of those substances being used during semi-quantification. A solution for substances with bad integration is then to switch to a quantification software to integrate those peaks and, if necessary, do the integration manually.

3. Improving the Performance

When estimating the concentrations, it is often believed that the accuracy of the estimation will depend on two things: (1) which features are being used, and (2) what kind of model is selected. In many ways this is true. Firstly, the features must be able to describe how well the analytes will ionize in the source, and secondly, the model must be suitable for the type of data that we have. However, making a better model is not the only way to improve the prediction accuracy. In fact, decisions can be made already in the analysis step to get more accurate estimations of the concentrations.

3.1. Increasing the Accuracy by Combining Positive and Negative Mode

Generally, when doing an experiment, be it targeted or non-targeted, one of the first things to decide is how the analytes will be detected. In case of non-targeted analysis with LC/ESI/HRMS, the decision shifts to whether the analysis should be carried out in positive or negative mode. For ESI+, the main advantage is that it can detect more compounds, [57] which is especially beneficial in NTS as the compounds present in the sample are unknown. ESI- on the other hand has the advantage of lower noise levels, [58] and in some cases also has higher sensitivity, which allows for lower detection limits [57]. Different compounds will also ionize to different extent in the different modes depending on several factors. E.g., acids generally ionize better in negative ESI mode, while bases ionize better in positive ESI mode due to their ability to become deprotonated and protonated [58]. For non-target analysis, however, we might not know if the compounds are acids or bases. So how do we choose which mode to use?

One possible answer to that question is that we do not choose one mode, but instead run our analysis with alternating positive and negative mode. Even better would be to run the analysis once in positive mode with an acidic eluent, [13] and once in negative mode with a basic eluent, [59] as different eluents are beneficial in the different modes. Doing the analysis this way has multiple advantages: one major being the simultaneous detection of compound that ionize only in positive or negative mode. This means that more compounds in general are detected, giving a more complete picture of what is present in the sample. Another advantage is the confidence in the quantification results. For the compounds that ionize in both positive and negative mode, the ionization efficiency in both modes can be modeled [10] and used for estimating of the concentration. These two estimated concentrations can then be compared, and if they are similar, it will give more confidence to the accuracy of the models' predictions. In addition to being able to compare the two obtained concentrations, both predictions could also be combined.

We have tested four approaches to obtain predicted concentrations using the model developed by Liigand et al. [10] on 39 compounds that are detected in both positive and negative mode at pH = 2.7 and 10, respectively. The alternatives for quantification are to use the concentrations obtained from either (1) positive or (2) negative mode, (3) using the mean of the concentrations from the two modes, or (4) selecting the concentration from the mode with the highest peak area/sensitivity.

The predictions were done by first predicting the ionization efficiency for each compound in the two modes, followed by calculating the predicted concentrations in the two separate modes (equations 4 and 5). The predicted concentrations for the two other approaches, the mean and the choice based on peak area, were calculated with Equations (7) and (8).

$$c_{mean} = \frac{c_{positive} + c_{negative}}{2} \tag{7}$$

$$c_{choice} = \begin{cases} c_{positive}, & \text{if } peak\ area_{positive} > peak\ area_{negative} \\ c_{negative}, & \text{otherwise} \end{cases} \tag{8}$$

The prediction errors were calculated by comparing the predicted concentrations with the known concentrations (Equation (9)).

$$prediction\ error = \begin{cases} \frac{c_{predicted}}{c_{known}}, & \text{if } c_{predicted} > c_{known} \\ \frac{c_{known}}{c_{predicted}}, & \text{otherwise.} \end{cases} \tag{9}$$

The obtained errors for both the individual concentrations, and the combined approaches were then compared and a statistically significance (Kruskal-Wallis test which gave a p-value of 1.073×10^{-5}. This suggests that at least one of the approaches gives significantly different errors from the other approaches. This was then further investigated with a Dunn's test which showed that all approaches differ in their prediction accuracy at the 95% confidence level) was observed; therefore, mean and median errors were calculated for all approaches. The lowest mean errors were obtained for the mean and choice approaches, 3.6 and 4.6, respectively. These are significantly lower than the mean errors for the concentrations obtained from the individual modes, which were 40.0 for ESI+ and 12.8 for ESI-. Similarly, the median errors were also lower for the approaches using the combined concentrations, although they were fairly low for the individual modes too. In positive mode, the median error was 3.3 and in negative mode it was 4.0. For the combined modes the corresponding values were 2.3 and 2.9 for the mean and the choice approach, respectively. In Figure 3a,b, the approaches of using positive or negative mode individually can be seen to yield much higher errors for some of the compounds, compared to the combined approaches.

The reason for this improvement in the prediction accuracy can be explained from how ensemble models improve prediction accuracy. For the mean approach, when the prediction accuracy is poor in one mode, this can be partially compensated for by a better prediction accuracy in the other mode. For the choice approach on the other hand, we have observed that the prediction accuracy tends to be poorer for compounds with a low response factor. Consequently, by always relying on the mode with the highest peak area, and thereby the highest response factor, some of the worst errors should be avoided. This is visualized in Figure 3c, where compounds with overpredicted concentrations in positive mode will generally be underpredicted in negative mode and vice versa.

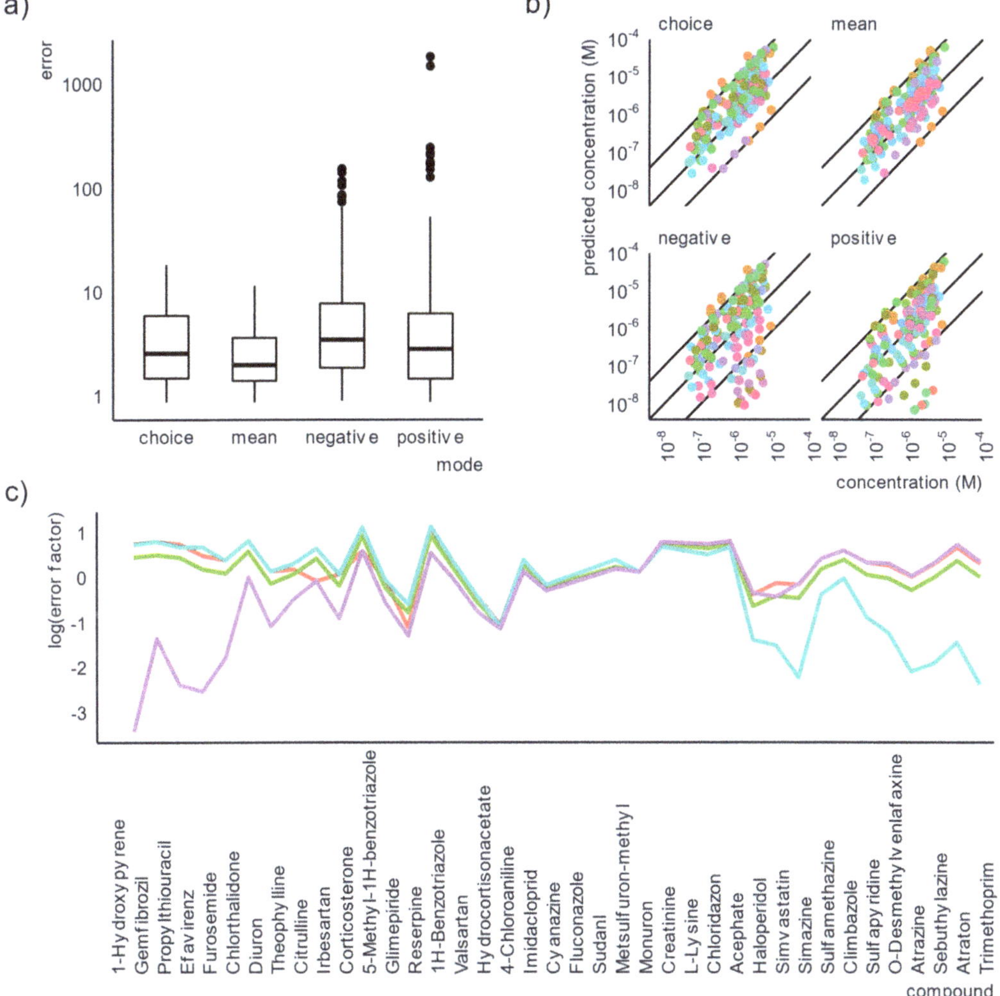

Figure 3. Running the analysis in both ESI+ and ESI− and combining the results appear to yield lower prediction errors. Here we see (**a**) boxplot of the errors of the concentrations predicted with the four different approaches, (**b**) scatter plot of the known concentration vs. the predicted concentration, colors are corresponding to different compounds. The outer lines show an error of a factor of 10 and the middle line shows the ideal prediction, and (**c**) plot of the error for each of the compounds in each mode. The purple line shows positive mode, the blue line shows negative mode, the red line shows the mean approach, and the green line shows the choice approach.

3.2. *Dilution of the Sample*

To do semi-quantification of the compounds in a sample, what is really done is an estimation of how well the compound will ionize in the mass spectrometer. Or, in other words, an estimation of the slope of its hypothetical calibration curve. This hypothetical slope is then used to calculate the concentration of the analyte. However, semi-quantification approaches do not tell the linear range of the hypothetical calibration curve. Therefore, if the measured concentration does not fall within the linear range of the calibration curve, we risk underestimating the concentration.

One way to remedy this is to analyze dilutions of the sample and verify linearity. Instead of only analyzing the same sample in two or three replicas, different dilutions should also be analyzed, e.g., 1×, 2× and 5×. Later, the semi-quantification approach would be applied independently for each of the dilutions and corrected for the dilution made. The comparison of these concentrations reveals which compounds in the sample have concentrations that fall outside the linear range of the calibration curve. The concentrations of these compounds should be estimated from one of the more diluted samples. For example, if the 1× dilution yields a significantly lower predicted concentration we can suspect exceeding upper limit of linearity. However, if the concentrations predicted for 2× and 5× dilutions match, these results can be averaged to obtain a final prediction.

To illustrate this, we used data from Wang et al. [60] and calculated the prediction errors of the concentration for pyridaben in a solvent matrix which had concentrations falling outside the linear range. The calibration curve and prediction errors are shown in Figure 4. The prediction errors were found to be between 2.1× and 2.9× for the dilutions in the linear range, whereas the concentration above the linear range had prediction errors of a factor of 5.4× and 11.1×. Briefly, by making sure the semi-quantification is performed in the linear range, the prediction error can be reduced.

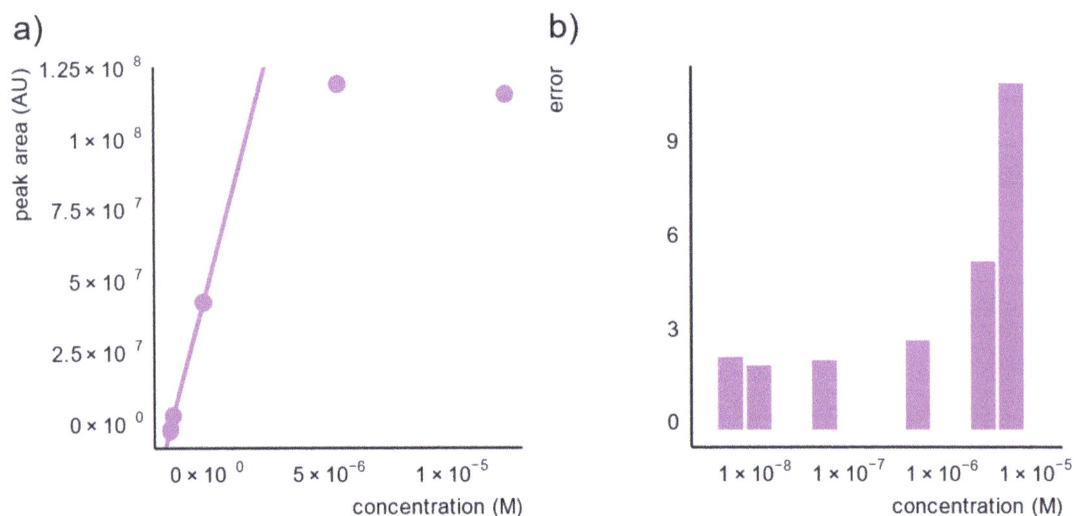

Figure 4. Performing semi-quantification on diluted samples within the linear range reduces the prediction errors, as seen here: (**a**) the calibration curve for pyridaben, and (**b**) bar-plot of the errors for each point in the calibration curve.

3.3. *Quality Control*

Quality control (QC) measures are as important in non-targeted screening as they are in targeted analysis and can shed light on human errors, malfunctioning of instrument, or severe contamination of the instrument. Our suggestions are to use the full potential of such measures to ensure high quality of the semi-quantification.

Firstly, technical replicates are a very-good measure to pinpoint human errors while handling samples. We suggest analyzing samples in 2, or if possible 3, replica. This can be advantageously combined with the analysis of different dilutions of the samples, described in more detail in Section 3.2. However, technical replicates can unfortunately not correct for systematic effects, such as poor recovery in sample preparation, or overlooking fragments formed in-source. Therefore, it must be stressed that agreement between technical replicates does not necessarily translate to high accuracy in semi-quantification, but rather assures that no fatal errors have occurred during sample processing or analysis.

Secondly, analysis of dirty samples over a long period of time can make the ion optics of the instrument very dirty, and thereby reduce the sensitivity of the instrument [61]. Therefore, the same sample analyzed at the beginning and at the end of the sequence may yield significantly different peak areas. A possibility to overcome this is by spiking the samples with the compounds used for transferring predicted ionization efficiency values to the instrument specific response factors. The response factors of these calibration compounds will be affected by the cleanliness of the instrument in the same manner as the unknown compounds, thus changes in instrumental signal are automatically accounted for.

Additionally, QC samples, i.e., analysis of samples spiked with known concentrations of analytes, can be used to evaluate the stability of the instrument, both the within-day and day-to-day performance, and can also account for drift in signal. [62,63] This is especially important during long analyzes, as signal have been shown to drift over ±25% during a 12 h run [64]. For more details on quality control in non-targeted screening see recent reviews by Knolhoff et al. [65] and Schultze et al. [66].

4. Opportunities Beyond Instrumental Analysis

Contributing to non-targeted analysis does not only mean needing to acquire new and expensive instrumentation and run new samples. Retrospective analysis as well as data repositories give opportunities to contribute without running any new experiments.

Community is fortunately already starting to store collected data for re-analyzes in the future with more advance tools and more knowledge. NORMAN digital sample freezing platform [67] is one of these platforms giving the possibility to run new quantification and/or identification experiments on already collected samples. The NORMAN community have set a nice example to collect and record as comprehensive data and metadata as possible, making it available for the next research ideas. Other communities that use non-targeted analysis, e.g., metabolomics, also have open data repositories (e.g., Metabolights [68], Metabolomics Workbench [69], GNPS MassIVE [70]) for collecting raw data files.

Based on the data from such platforms, semi-quantitation makes it possible to discover pollutants in environmental samples, or biomarkers in biological samples, that have previously flown under the radar in methods based solely on signal. Revealing the concentration of these compounds can prove their importance, even if their peak area is small, or allow comparison with samples collected and measured later.

As mass spectrometric signal varies somewhat between days, and a lot between laboratories, semi-quantification open ways to better evaluate time series behavior in samples, or directly compare results of large cohorts measured by multiple labs. Furthermore, for the wider community, it is not only important to know what is in the sample, but also how much of it there is. Thus, semi-quantification combined with already collected data from repositories can reveal a lot of new, useful information. A general scheme of how to perform retrospective semi-quantification is presented in Figure 5.

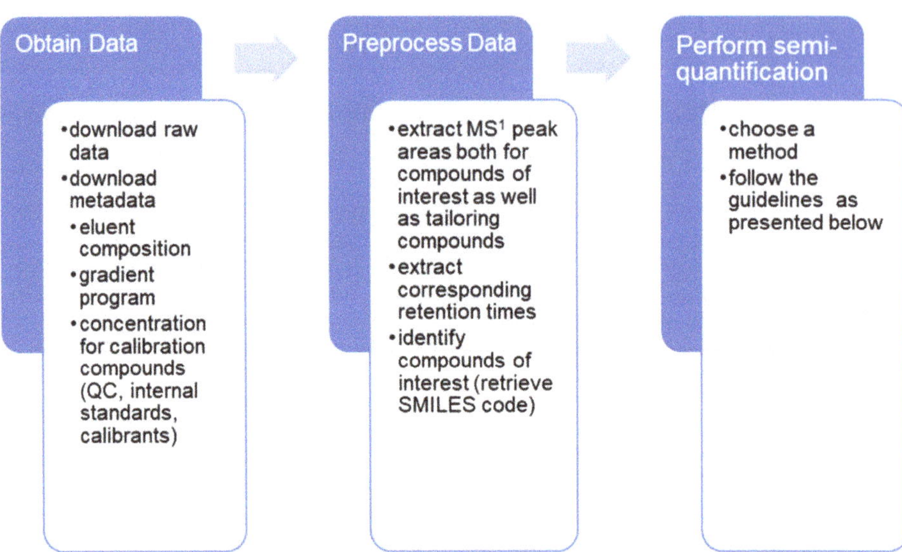

Figure 5. Performing retrospective analysis generally requires three main steps: acquire the data, data pre-processing and applying semi-quantification.

Our experiences show that retrospective analysis is fully feasible if the necessary data are available. We have applied retrospective analysis on green tea samples originally measured at University of North Carolina, Greensborough, USA, in 2016 [71]. Semi-quantification of these samples was performed by us in the summer of 2018 at University of Tartu, Estonia. The retrospective semi-quantification became feasible because they measured reference standard, NIST green tea, together with the samples in the previous study. Reference standard with known concentrations is a great tool to use for calibrating the predictions to laboratory, matrix and method specific response factors. We were able to show that, using solely predicted ionization efficiencies for concentration estimation for polyphenols and similar natural products in green tea, the average error of predicted concentration was 1.9×. Moreover, in the former study, they detected and identified 5 additional compounds, for which they were lacking analytical standards. In our retrospective study, we could provide concentration estimates for those compounds as well.

5. Carrying Out Semi-Quantitative NTS in Practice

Generally, steps taken during the non-targeted analysis directly influence the accuracy of the semi-quantification. We suggest generic chromatographic method alongside a minimal sample preparation method to not lose compounds. However, doing analysis in both positive and negative mode as well as analyzing dilutions together with replica can significantly improve the accuracy. A step-by-step summary with notes is brought in Table 3.

Table 3. An overview of the steps in NTS together with the possible impact on semi-quantification accuracy and measures to improve this accuracy.

NTS Step	Suggested Procedure	Reasoning
Sample Preparation	Avoid extensive sample preparation where possible.	If you analyze fairly simple liquids like water samples, urine, beverages, etc., direct injection of the sample is suggested over sample clean-up to avoid losses of analyte. Matrix effect can be evaluated by comparing the results from different dilutions, see below.
Standards	Prepare a solution with a set of compounds with known concentrations and analyze this set together with your samples. A suggested set of compounds could be tetrahexylammonium salt, haloperidol, diphenyl phthalate, tetraethylammonium salt, phenylalanine, dimethyl phthalate, progesterone, alanine, uracil, and saccharin for positive ESI mode. In negative mode we suggest 4-aminobutyric acid, sorbic acid, vanillin, benzoic acid, salicylic acid, p-nitrophenol, 3-nitrobenzenesulphonamide, perfluorobutyric acid, tetradecanoic acid, 3,5-diiodosalicylic acid, perfluorooctanesulfonic acid [32].	Make sure that these compounds cover a wide ionization efficiencies range and elute over the full chromatographic run. These compounds will be used to transfer the ionization efficiency predictions to your instrument scale.
Sample Analysis	Analyze samples on at least two dilutions.	Additional to running duplicates or triplicates, you can also analyze your samples at different dilutions or with different injection volumes. This will allow you to assure that all measurements are in the linear range as well as account for possible matrix effect. Matrix effect [72] is known to be less severe for more diluted samples.
Chromatography	Use generic chromatographic parameters. Avoid exotic additives and organic solvents. To generalize these methods, we recommend using 0.1% formic acid in ESI positive mode and ammonium hydroxide or ammonium formiate pH = 8.0 in ESI negative mode. In both cases a linear gradient from 5 to 100% of acetonitrile over 15 min for a 10 cm C18 column with 3 µm particle size has proven generic.	The predictions of any model are applicable only to the conditions used in the training/validation of the model. Therefore, rare LC conditions are likely not to be covered by the quantification model used.
Ionization Conditions	Choose soft ionization conditions, use the default parameters of the vendor as guide. If possible, do not alter these parameters too much. Run in both positive and negative ESI mode.	Soft conditions are likely to cause less fragmentation. The units and range of values of source parameters depend on the vendor, so exact parameters cannot be transferred between instruments. However, using vendor recommendations across instruments yields similar relative ionization efficiency values, and therefore, semi-quantification results [73]. Running analysis in both positive and negative mode enables combining and comparing results for polyfunctional compounds ionizing in both modes.
MS Parameters	Use a wide scan range, e.g., from 100 to 1000 Da.	Wide scan range will enable pinpointing fragments formed in the ionization source.

Table 3. *Cont.*

NTS Step	Suggested Procedure	Reasoning
Data Processing	Combine the signal of the precursor ion and fragments together. Check that the integrations is acceptable.	Fragmentation occurs separately from ionization and is not accounted in the prediction algorithms. Poor integration of tailing or split peaks may significantly decrease semi-quantification accuracy.

6. Conclusions

To aid in decision making, non-targeted LC/ESI/HRMS analysis together with methods for semi-quantifications are increasingly used. In this tutorial, we have presented appropriate NTS workflows and ready-to-use semi-quantification approaches. To avoid high errors related to the estimated concentrations, our overall recommendation is to use as general methods as possible throughout the entire analysis.

We suggest doing as little sample preparation as possible, both to prevent losing compounds in the process but also to avoid poor recoveries of compounds. If possible, the best sample preparation is simply no sample preparation, however, if it is needed it is best to use generic methods. During analysis, it is recommended to run the samples in both positive and negative ESI mode to increase the number of detected compounds. If it is feasible, best results are achieved when using different mobile phases for ESI+ and ESI-: suggested mobile phases are presented in Section 2.2 and in Table 3, together with a generic gradient. We also propose to run the same sample at different dilutions, at least two but preferably more, to determine that the analysis is in the linear range by comparing the estimated concentrations with the dilutions. Before applying semi-quantification, it is important to account for fragmentation occurring in the ion source, especially if the sample might contain compounds that easily fragment, e.g., phthalates, carboxylic acids. In our experiences, we have seen that semi-quantification strategies based on ionization efficiency predictions give more accurate results. Therefore, we encourage the use and further development of such approaches. However, in the case of no tentative structures of the unknown compounds, only one semi-quantification strategy presented here is applicable, namely the closest eluting compound approach.

In conclusion, we cannot stress enough the need of using generic methods and parameters, as suggested in Table 3, as well as the benefits of unifying the methods across communities, when doing NTS.

Author Contributions: Methodology, L.M. and E.P.; validation, L.M. and E.P.; formal analysis, L.M., E.P., J.L., A.S., and A.K.; data curation, E.P.; writing—original draft preparation, L.M., E.P., A.S., M.P., J.L. and A.K.; writing—review and editing, L.M., E.P., A.S., M.P., J.L. and A.K.; visualization, L.M. and E.P.; supervision, A.K.; project administration, A.K.; funding acquisition, A.K. All authors have read and agreed to the published version of the manuscript.

Funding: The funding has been generously provided by Swedish Research Council for Sustainable Development grant 2020-01511.

Institutional Review Board Statement: Not applicable.

Informed Consent Statement: Not applicable.

Data Availability Statement: The data were available from the authors. The code is available from: https://github.com/kruvelab/NTS-semi-quant_review/ (accessed on 9 June 2021).

Conflicts of Interest: The authors declare no conflict of interest.

References

1. Hollender, J.; Schymanski, E.L.; Singer, H.P.; Ferguson, P.L. Nontarget Screening with High Resolution Mass Spectrometry in the Environment: Ready to Go? *Environ. Sci. Technol.* **2017**, *51*, 11505–11512. [CrossRef]
2. Kruve, A. Semi-quantitative Non-target Analysis of Water with Liquid Chromatography/High-resolution Mass Spectrometry: How Far Are We? *Rapid Commun. Mass Spectrom.* **2018**, *33*, 54–63. [CrossRef]
3. Bletsou, A.A.; Jeon, J.; Hollender, J.; Archontaki, E.; Thomaidis, N.S. Targeted and Non-Targeted Liquid Chromatography-Mass Spectrometric Workflows for Identification of Transformation Products of Emerging Pollutants in the Aquatic Environment. *TrAC Trends Anal. Chem.* **2015**, *66*, 32–44. [CrossRef]
4. *Compound Discoverer Software*; Thermo Fisher Scientific: Courtaboeuf Cedex, France.
5. Schmitt, U.; Loos, M.; Singer, H. EAWAG. Available online: https://www.eawag.ch/en/department/uchem/projects/envipy/ (accessed on 10 May 2021).
6. Pluskal, T.; Castillo, S.; Villar-Briones, A.; Orešič, M. MZmine 2: Modular Framework for Processing, Visualizing, and Analyzing Mass Spectrometry-Based Molecular Profile Data. *BMC Bioinform.* **2010**, *11*, 395. [CrossRef] [PubMed]
7. Schymanski, E.L.; Jeon, J.; Gulde, R.; Fenner, K.; Ruff, M.; Singer, H.P.; Hollender, J. Identifying Small Molecules via High Resolution Mass Spectrometry: Communicating Confidence. *Environ. Sci. Technol.* **2014**, *48*, 2097–2098. [CrossRef]
8. Escher, B.I.; Fenner, K. Recent Advances in Environmental Risk Assessment of Transformation Products. *Environ. Sci. Technol.* **2011**, *45*, 3835–3847. [CrossRef] [PubMed]
9. Kiefer, K.; Müller, A.; Singer, H.; Hollender, J. New Relevant Pesticide Transformation Products in Groundwater Detected Using Target and Suspect Screening for Agricultural and Urban Micropollutants with LC-HRMS. *Water Res.* **2019**, *165*, 114972. [CrossRef] [PubMed]
10. Liigand, J.; Wang, T.; Kellogg, J.; Smedsgaard, J.; Cech, N.; Kruve, A. Quantification for Non-Targeted LC/MS Screening without Standard Substances. *Sci. Rep.* **2020**, *10*, 5808. [CrossRef]
11. Kruve, A. Strategies for Drawing Quantitative Conclusions from Nontargeted Liquid Chromatography–High-Resolution Mass Spectrometry Analysis. *Anal. Chem.* **2020**, *92*, 4691–4699. [CrossRef]
12. Cech, N.B.; Enke, C.G. Practical Implications of Some Recent Studies in Electrospray Ionization Fundamentals. *Mass Spectrom. Rev.* **2001**, *20*, 362–387. [CrossRef]
13. Liigand, J.; Kruve, A.; Leito, I.; Girod, M.; Antoine, R. Effect of Mobile Phase on Electrospray Ionization Efficiency. *J. Am. Soc. Mass Spectrom.* **2014**, *25*, 1853–1861. [CrossRef]
14. Kostiainen, R.; Kauppila, T.J. Effect of Eluent on the Ionization Process in Liquid Chromatography–Mass Spectrometry. *J. Chromatogr. A* **2009**, *1216*, 685–699. [CrossRef]
15. Kiontke, A.; Oliveira-Birkmeier, A.; Opitz, A.; Birkemeyer, C. Electrospray Ionization Efficiency Is Dependent on Different Molecular Descriptors with Respect to Solvent PH and Instrumental Configuration. *PLoS ONE* **2016**, *11*, e0167502. [CrossRef] [PubMed]
16. Page, J.S.; Kelly, R.T.; Tang, K.; Smith, R.D. Ionization and Transmission Efficiency in an Electrospray Ionization—Mass Spectrometry Interface. *J. Am. Soc. Mass Spectrom.* **2007**, *18*, 1582–1590. [CrossRef]
17. Pieke, E.N.; Granby, K.; Trier, X.; Smedsgaard, J. A Framework to Estimate Concentrations of Potentially Unknown Substances by Semi-Quantification in Liquid Chromatography Electrospray Ionization Mass Spectrometry. *Anal. Chim. Acta* **2017**, *975*, 30–41. [CrossRef] [PubMed]
18. Espinosa, M.S.; Folguera, L.; Magallanes, J.F.; Babay, P.A. Exploring Analyte Response in an ESI-MS System with Different Chemometric Tools. *Chemom. Intell. Lab. Syst.* **2015**, *146*, 120–127. [CrossRef]
19. Aalizadeh, R.; Nika, M.-C.; Thomaidis, N.S. Development and Application of Retention Time Prediction Models in the Suspect and Non-Target Screening of Emerging Contaminants. *J. Hazard. Mater.* **2019**, *363*, 277–285. [CrossRef]
20. Similar Compound Finder. Available online: http://dsfp.chem.uoa.gr/semiquantification/ (accessed on 10 May 2021).
21. Kruve, A.; Aalizadeh, R.; Malm, L.; Alygizakis, N.; Thomaidis, N.S. Interlaboratory Comparison on Strategies for Semi-Quantitative Non-Targeted LC-ESI-HRMS. Available online: https://www.norman-network.net/sites/default/files/files/QA-QC%20Issues/Invitation%20letter%20JPA%202020%20semi-quant%20inter%20lab%20%28002%29.pdf (accessed on 21 July 2020).
22. NORMAN Network; Aalizadeh, R.; Alygizakis, N.; Schymanski, E.; Slobodnik, J.; Fischer, S.; Cirka, L. S0 | SUSDAT | Merged NORMAN Suspect List: SusDat (Version NORMAN-SLE-S0.0.3.2) [Data set]. *Zenodo* **2021**. [CrossRef]
23. Kruve, A.; Kiefer, K.; Hollender, J. Benchmarking of the Quantification Approaches for the Non-Targeted Screening of Micropollutants and Their Transformation Products in Groundwater. *Anal. Bioanal. Chem.* **2021**, *413*, 1549–1559. [CrossRef]
24. Richardson, S.D.; Kimura, S.Y. Water Analysis: Emerging Contaminants and Current Issues. *Anal. Chem.* **2020**, *92*, 473–505. [CrossRef]
25. Chalcraft, K.R.; Lee, R.; Mills, C.; Britz-McKibbin, P. Virtual Quantification of Metabolites by Capillary Electrophoresis-Electrospray Ionization-Mass Spectrometry: Predicting Ionization Efficiency Without Chemical Standards. *Anal. Chem.* **2009**, *81*, 2506–2515. [CrossRef]
26. ChemAxon. Available online: https://chemicalize.com/ (accessed on 7 July 2020).
27. Dahal, U.P.; Jones, J.P.; Davis, J.A.; Rock, D.A. Small Molecule Quantification by Liquid Chromatography-Mass Spectrometry for Metabolites of Drugs and Drug Candidates. *Drug Metab. Dispos.* **2011**, *39*, 2355–2360. [CrossRef] [PubMed]

28. Oss, M.; Kruve, A.; Herodes, K.; Leito, I. Electrospray Ionization Efficiency Scale of Organic Compounds. *Anal. Chem.* **2010**, *82*, 2865–2872. [CrossRef]
29. Huffman, B.A.; Poltash, M.L.; Hughey, C.A. Effect of Polar Protic and Polar Aprotic Solvents on Negative-Ion Electrospray Ionization and Chromatographic Separation of Small Acidic Molecules. *Anal. Chem.* **2012**, *84*, 9942–9950. [CrossRef]
30. Panagopoulos Abrahamsson, D.; Park, J.-S.; Singh, R.R.; Sirota, M.; Woodruff, T.J. Applications of Machine Learning to In Silico Quantification of Chemicals without Analytical Standards. *J. Chem. Inf. Modeling* **2020**, *60*, 2718–2727. [CrossRef]
31. Mayhew, A.W.; Topping, D.O.; Hamilton, J.F. New Approach Combining Molecular Fingerprints and Machine Learning to Estimate Relative Ionization Efficiency in Electrospray Ionization. *ACS Omega* **2020**, *5*, 9510–9516. [CrossRef] [PubMed]
32. Liigand, P.; Liigand, J.; Kaupmees, K.; Kruve, A. 30 Years of Research on ESI/MS Response: Trends, Contradictions and Applications. *Anal. Chim. Acta* **2021**, *1152*, 238117. [CrossRef]
33. Quantem Analytics. Available online: https://app.quantem.co/ (accessed on 10 May 2021).
34. Dührkop, K.; Nothias, L.-F.; Fleischauer, M.; Reher, R.; Ludwig, M.; Hoffmann, M.A.; Petras, D.; Gerwick, W.H.; Rousu, J.; Dorrestein, P.C.; et al. Systematic Classification of Unknown Metabolites Using High-Resolution Fragmentation Mass Spectra. *Nat. Biotechnol.* **2021**, *39*, 462–471. [CrossRef] [PubMed]
35. Brunner, A.M.; Bertelkamp, C.; Dingemans, M.M.L.; Kolkman, A.; Wols, B.; Harmsen, D.; Siegers, W.; Martijn, B.J.; Oorthuizen, W.A.; ter Laak, T.L. Integration of Target Analyses, Non-Target Screening and Effect-Based Monitoring to Assess OMP Related Water Quality Changes in Drinking Water Treatment. *Sci. Total Environ.* **2020**, *705*, 135779. [CrossRef]
36. Schymanski, E.L.; Singer, H.P.; Longrée, P.; Loos, M.; Ruff, M.; Stravs, M.A.; Ripollés Vidal, C.; Hollender, J. Strategies to Characterize Polar Organic Contamination in Wastewater: Exploring the Capability of High Resolution Mass Spectrometry. *Environ. Sci. Technol.* **2014**, *48*, 1811–1818. [CrossRef]
37. Sørensen, L.; McCormack, P.; Altin, D.; Robson, W.J.; Booth, A.M.; Faksness, L.-G.; Rowland, S.J.; Størseth, T.R. Establishing a Link between Composition and Toxicity of Offshore Produced Waters Using Comprehensive Analysis Techniques—A Way Forward for Discharge Monitoring? *Sci. Total Environ.* **2019**, *694*, 133682. [CrossRef]
38. Blum, K.M.; Andersson, P.L.; Renman, G.; Ahrens, L.; Gros, M.; Wiberg, K.; Haglund, P. Non-Target Screening and Prioritization of Potentially Persistent, Bioaccumulating and Toxic Domestic Wastewater Contaminants and Their Removal in on-Site and Large-Scale Sewage Treatment Plants. *Sci. Total Environ.* **2017**, *575*, 265–275. [CrossRef]
39. Baz-Lomba, J.A.; Salvatore, S.; Gracia-Lor, E.; Bade, R.; Castiglioni, S.; Castrignanò, E.; Causanilles, A.; Hernandez, F.; Kasprzyk-Hordern, B.; Kinyua, J.; et al. Comparison of Pharmaceutical, Illicit Drug, Alcohol, Nicotine and Caffeine Levels in Wastewater with Sale, Seizure and Consumption Data for 8 European Cities. *BMC Public Health* **2016**, *16*, 1035. [CrossRef]
40. González-Mariño, I.; Gracia-Lor, E.; Rousis, N.I.; Castrignanò, E.; Thomas, K.V.; Quintana, J.B.; Kasprzyk-Hordern, B.; Zuccato, E.; Castiglioni, S. Wastewater-Based Epidemiology To Monitor Synthetic Cathinones Use in Different European Countries. *Environ. Sci. Technol.* **2016**, *50*, 10089–10096. [CrossRef] [PubMed]
41. Schulze, T.; Ahel, M.; Ahlheim, J.; Aït-Aïssa, S.; Brion, F.; Di Paolo, C.; Froment, J.; Hidasi, A.O.; Hollender, J.; Hollert, H.; et al. Assessment of a Novel Device for Onsite Integrative Large-Volume Solid Phase Extraction of Water Samples to Enable a Comprehensive Chemical and Effect-Based Analysis. *Sci. Total Environ.* **2017**, *581–582*, 350–358. [CrossRef] [PubMed]
42. Dzuman, Z.; Zachariasova, M.; Veprikova, Z.; Godula, M.; Hajslova, J. Multi-Analyte High Performance Liquid Chromatography Coupled to High Resolution Tandem Mass Spectrometry Method for Control of Pesticide Residues, Mycotoxins, and Pyrrolizidine Alkaloids. *Anal. Chim. Acta* **2015**, *863*, 29–40. [CrossRef]
43. Kafeenah, H.I.S.; Osman, R.; Bakar, N.K.A. Effect of Mobile Phase PH on the Electrospray Ionization Efficiency and Qualitative Analysis of Pharmaceuticals in ESI + LC-MS/MS. *J. Chromatogr. Sci.* **2019**, *57*, 847–854. [CrossRef]
44. Zhou, S.; Cook, K.D. Protonation in Electrospray Mass Spectrometry: Wrong-Way-Round or Right-Way-Round? *J. Am. Soc. Mass Spectrom.* **2000**, *11*, 961–966. [CrossRef]
45. Shou, W.; Naidong, W. Simple Means to Alleviate Sensitivity Loss by Trifluoroacetic Acid (TFA) Mobile Phases in the Hydrophilic Interaction Chromatography–Electrospray Tandem Mass Spectrometric (HILIC–ESI/MS/MS) Bioanalysis of Basic Compounds. *J. Chromatogr. B* **2005**, *825*, 186–192. [CrossRef]
46. Mallet, C.R.; Lu, Z.; Mazzeo, J.R. A Study of Ion Suppression Effects in Electrospray Ionization from Mobile Phase Additives and Solid-Phase Extracts. *Rapid Commun. Mass Spectrom.* **2004**, *18*, 49–58. [CrossRef]
47. Snyder, L.R.; Kirkland, J.J.; Glajch, J.L. *Practical HPLC Method Development*, 2nd ed.; Wiley: New York, NY, USA, 1997; ISBN 978-0-471-00703-6.
48. Ojakivi, M.; Liigand, J.; Kruve, A. Modifying the Acidity of Charged Droplets. *ChemistrySelect* **2018**, *3*, 335–338. [CrossRef]
49. Rebane, R.; Kruve, A.; Liigand, J.; Liigand, P.; Gornischeff, A.; Leito, I. Ionization Efficiency Ladders as Tools for Choosing Ionization Mode and Solvent in Liquid Chromatography/Mass Spectrometry. *Rapid Commun. Mass Spectrom.* **2019**, *33*, 1834–1843. [CrossRef]
50. Colizza, K.; Mahoney, K.E.; Yevdokimov, A.V.; Smith, J.L.; Oxley, J.C. Acetonitrile Ion Suppression in Atmospheric Pressure Ionization Mass Spectrometry. *J. Am. Soc. Mass Spectrom.* **2016**, *27*, 1796–1804. [CrossRef]
51. Buszewski, B.; Bocian, S.; Felinger, A. Artifacts in Liquid-Phase Separations–System, Solvent, and Impurity Peaks. *Chem. Rev.* **2012**, *112*, 2629–2641. [CrossRef] [PubMed]
52. Srbek, J.; Coufal, P.; Bosáková, Z.; Tesařová, E. System Peaks and Their Positive and Negative Aspects in Chromatographic Techniques. *J. Sep. Sci.* **2005**, *28*, 1263–1270. [CrossRef]

53. Oberacher, H.; Sasse, M.; Antignac, J.-P.; Guitton, Y.; Debrauwer, L.; Jamin, E.L.; Schulze, T.; Krauss, M.; Covaci, A.; Caballero-Casero, N.; et al. A European Proposal for Quality Control and Quality Assurance of Tandem Mass Spectral Libraries. *Environ. Sci. Eur.* **2020**, *32*, 43. [CrossRef]
54. Domingo-Almenara, X.; Guijas, C.; Billings, E.; Montenegro-Burke, J.R.; Uritboonthai, W.; Aisporna, A.E.; Chen, E.; Benton, H.P.; Siuzdak, G. The METLIN Small Molecule Dataset for Machine Learning-Based Retention Time Prediction. *Nat. Commun.* **2019**, *10*, 5811. [CrossRef]
55. Minkus, S.; Bieber, S.; Moser, S.; Letzel, T. Optimization of Electrospray Ionization Parameters in a RPLC-HILIC-MS/MS Coupling by Design of Experiment. Available online: http://afin-ts.de/literature/?lang=en (accessed on 11 May 2021).
56. Seo, J.; Warnke, S.; Gewinner, S.; Schöllkopf, W.; Bowers, M.T.; Pagel, K.; von Helden, G. The Impact of Environment and Resonance Effects on the Site of Protonation of Aminobenzoic Acid Derivatives. *Phys. Chem. Chem. Phys.* **2016**, *18*, 25474–25482. [CrossRef]
57. Liigand, P.; Kaupmees, K.; Haav, K.; Liigand, J.; Leito, I.; Girod, M.; Antoine, R.; Kruve, A. Think Negative: Finding the Best Electrospray Ionization/MS Mode for Your Analyte. *Anal. Chem.* **2017**, *89*, 5665–5668. [CrossRef] [PubMed]
58. Cole, R.B. *Electrospray and MALDI Mass Spectrometry Fundamentals, Instrumentation, Practicalities, and Biological Applications*; Wiley: Hoboken, NJ, USA, 2011; ISBN 978-1-118-21155-7.
59. Kruve, A.; Kaupmees, K. Predicting ESI/MS Signal Change for Anions in Different Solvents. *Anal. Chem.* **2017**, *89*, 5079–5086. [CrossRef]
60. Wang, T.; Liigand, J.; Frandsen, H.L.; Smedsgaard, J.; Kruve, A. Standard Substances Free Quantification Makes LC/ESI/MS Non-Targeted Screening of Pesticides in Cereals Comparable between Labs. *Food Chem.* **2020**, *318*, 126460. [CrossRef]
61. Lagerwerf, F.M.; van Dongen, W.D.; Steenvoorden, R.J.J.M.; Honing, M.; Jonkman, J.H.G. Exploring the Boundaries of Bioanalytical Quantitative LC–MS–MS. *TrAC Trends Anal. Chem.* **2000**, *19*, 418–427. [CrossRef]
62. Kirwan, J.A.; Broadhurst, D.I.; Davidson, R.L.; Viant, M.R. Characterising and Correcting Batch Variation in an Automated Direct Infusion Mass Spectrometry (DIMS) Metabolomics Workflow. *Anal. Bioanal. Chem.* **2013**, *405*, 5147–5157. [CrossRef] [PubMed]
63. Brunius, C.; Shi, L.; Landberg, R. Large-Scale Untargeted LC-MS Metabolomics Data Correction Using between-Batch Feature Alignment and Cluster-Based within-Batch Signal Intensity Drift Correction. *Metabolomics* **2016**, *12*, 173. [CrossRef] [PubMed]
64. Jiang, F.; Liu, Q.; Li, Q.; Zhang, S.; Qu, X.; Zhu, J.; Zhong, G.; Huang, M. Signal Drift in Liquid Chromatography Tandem Mass Spectrometry and Its Internal Standard Calibration Strategy for Quantitative Analysis. *Anal. Chem.* **2020**, *92*, 7690–7698. [CrossRef] [PubMed]
65. Knolhoff, A.M.; Premo, J.H.; Fisher, C.M. A Proposed Quality Control Standard Mixture and Its Uses for Evaluating Nontargeted and Suspect Screening LC/HR-MS Method Performance. *Anal. Chem.* **2021**, *93*, 1596–1603. [CrossRef]
66. Schulze, B.; Jeon, Y.; Kaserzon, S.; Heffernan, A.L.; Dewapriya, P.; O'Brien, J.; Gomez Ramos, M.J.; Ghorbani Gorji, S.; Mueller, J.F.; Thomas, K.V.; et al. An Assessment of Quality Assurance/Quality Control Efforts in High Resolution Mass Spectrometry Non-Target Workflows for Analysis of Environmental Samples. *TrAC Trends Anal. Chem.* **2020**, *133*, 116063. [CrossRef]
67. Alygizakis, N.A.; Oswald, P.; Thomaidis, N.S.; Schymanski, E.L.; Aalizadeh, R.; Schulze, T.; Oswaldova, M.; Slobodnik, J. NORMAN Digital Sample Freezing Platform: A European Virtual Platform to Exchange Liquid Chromatography High Resolution-Mass Spectrometry Data and Screen Suspects in "Digitally Frozen" Environmental Samples. *TrAC Trends Anal. Chem.* **2019**, *115*, 129–137. [CrossRef]
68. Haug, K.; Cochrane, K.; Nainala, V.C.; Williams, M.; Chang, J.; Jayaseelan, K.V.; O'Donovan, C. MetaboLights: A Resource Evolving in Response to the Needs of Its Scientific Community. *Nucleic Acids Res.* **2019**, gkz1019. [CrossRef]
69. Sud, M.; Fahy, E.; Cotter, D.; Azam, K.; Vadivelu, I.; Burant, C.; Edison, A.; Fiehn, O.; Higashi, R.; Nair, K.S.; et al. Metabolomics Workbench: An International Repository for Metabolomics Data and Metadata, Metabolite Standards, Protocols, Tutorials and Training, and Analysis Tools. *Nucleic Acids Res.* **2016**, *44*, D463–D470. [CrossRef]
70. Wang, M.; Carver, J.J.; Phelan, V.V.; Sanchez, L.M.; Garg, N.; Peng, Y.; Nguyen, D.D.; Watrous, J.; Kapono, C.A.; Luzzatto-Knaan, T.; et al. Sharing and Community Curation of Mass Spectrometry Data with Global Natural Products Social Molecular Networking. *Nat. Biotechnol.* **2016**, *34*, 828–837. [CrossRef]
71. Kellogg, J.J.; Graf, T.N.; Paine, M.F.; McCune, J.S.; Kvalheim, O.M.; Oberlies, N.H.; Cech, N.B. Comparison of Metabolomics Approaches for Evaluating the Variability of Complex Botanical Preparations: Green Tea (*Camellia Sinensis*) as a Case Study. *J. Nat. Prod.* **2017**, *80*, 1457–1466. [CrossRef] [PubMed]
72. Hernando, M.D.; Suárez-Barcena, J.M.; Bueno, M.J.M.; Garcia-Reyes, J.F.; Fernández-Alba, A.R. Fast Separation Liquid Chromatography–Tandem Mass Spectrometry for the Confirmation and Quantitative Analysis of Avermectin Residues in Food. *J. Chromatogr. A* **2007**, *1155*, 62–73. [CrossRef] [PubMed]
73. Liigand, J.; Kruve, A.; Liigand, P.; Laaniste, A.; Girod, M.; Antoine, R.; Leito, I. Transferability of the Electrospray Ionization Efficiency Scale between Different Instruments. *J. Am. Soc. Mass Spectrom.* **2015**, *26*, 1923–1930. [CrossRef]

MDPI

Review

Non-Targeted Screening Approaches for Profiling of Volatile Organic Compounds Based on Gas Chromatography-Ion Mobility Spectroscopy (GC-IMS) and Machine Learning

Charlotte Capitain and Philipp Weller *

Institute for Instrumental Analytics and Bioanalytics, Mannheim University of Applied Sciences, 68163 Mannheim, Germany; c.capitain@hs-mannheim.de
* Correspondence: p.weller@hs-mannheim.de; Tel.: +49-(0)621-292-6484

Abstract: Due to its high sensitivity and resolving power, gas chromatography-ion mobility spectrometry (GC-IMS) is a powerful technique for the separation and sensitive detection of volatile organic compounds. It is a robust and easy-to-handle technique, which has recently gained attention for non-targeted screening (NTS) approaches. In this article, the general working principles of GC-IMS are presented. Next, the workflow for NTS using GC-IMS is described, including data acquisition, data processing and model building, model interpretation and complementary data analysis. A detailed overview of recent studies for NTS using GC-IMS is included, including several examples which have demonstrated GC-IMS to be an effective technique for various classification and quantification tasks. Lastly, a comparison of targeted and non-targeted strategies using GC-IMS are provided, highlighting the potential of GC-IMS in combination with NTS.

Keywords: gas chromatography ion mobility spectroscopy (GC-IMS); volatile organic compounds (VOCs); non-targeted screening (NTS) using machine learning

Citation: Capitain, C.; Weller, P. Non-Targeted Screening Approaches for Profiling of Volatile Organic Compounds Based on Gas Chromatography-Ion Mobility Spectroscopy (GC-IMS) and Machine Learning. *Molecules* **2021**, *26*, 5457. https://doi.org/10.3390/molecules26185457

Academic Editor: Thomas Letzel

Received: 24 July 2021
Accepted: 1 September 2021
Published: 8 September 2021

1. Introduction

Quality control and early detection of hazard chemicals, allergens, or biological contaminants are critical to ensure product safety. Environmental pollutants, pesticides, or toxins, among others, can compromise food safety and pose a public health risk [1]. Furthermore, food adulteration and food fraud, accelerated by globalization, continue to cause economic losses and customer dissatisfaction and emphasize the need for robust, inexpensive, and fast analytical methods [2]. While new scientific findings continuously identify potential hazardous or allergenic compounds [3], commonly employed methods, which focus on the detection and identification of a particular compound or class of compounds, lack the ability to identify new or unknown compounds. Due to the inherent diversity of biogenic samples, as observed in food analysis, and the chemical complexity of the sample matrices, analysis often requires advanced sample preparation strategies [4]. For systematic monitoring of product quality, it is therefore desirable to develop analytical methods capable of discovering unknown or non-targeted compounds from the complex sample matrices. This approach, also referred to as NTS, requires comprehensive extraction and analysis of compounds of interest. Analysis of the volatile organic compounds (VOCs) of samples, also known as VOC profiling, allows for the detection of compounds in complex sample matrices without the need for detailed a priori knowledge of the molecular composition. Due to its high sensitivity and resolving power on the one hand and its simplicity and robustness on the other, ion mobility spectrometry (IMS) has gained popularity for the analysis of VOCs [5]. Moreover, gas chromatography coupled to ion mobility spectroscopy (GC-IMS) has been shown to be an easy-to-handle and yet highly effective tool for VOC profiling [6]. As a result, non-targeted VOC profiling based on GC-IMS in combination with machine learning has emerged as a promising method for sample monitoring.

Since the 1970s, when IMS was first known as 'plasma chromatography', IMS has developed into a highly sensitive technique for the analysis of VOCs at ultratrace concentration levels, which accounts for additional information regarding the ion's mobility [7–9]. Due to the robust and easy-to-handle instrumentation, a wide range of application fields have been found for IMS today, such as food flavor analysis [5], process monitoring [10,11], and quality control [12], as well as detection and quantification of warfare agents [13] and explosives [14,15].

With IMS, analytes are first ionized in the ionization region of the instrument. The most common ionization method is the atmospheric pressure chemical ionization [16] by beta emitters, which frequently use nickel-63 (Ni-63) [11,15,17,18] or the less hazardous beta-emitting tritium (H-3) [19] or alpha-emitting americium-241 (Am-241) [20,21]. Other ionization methods are atmospheric pressure photo ionization (APPI) [22], which uses ultraviolet light (UV) [23,24] or corona discharge (CD) atmospheric pressure chemical ionization [17,25–27], where a high electric field between a needle and a metal plate or discharge electrode is used. Yet another method is the laser desorption/ionization technique (LDI), which employs a laser pulse as ion source [28].

According to the European Union directive, the exemption limit for the total activity for tritium was set to 1 GBq [29]. Therefore, the usage of a low-radiation tritium ion source with an activity of 300 MBq is not subject to authorization, hence leading to a broad adoption of tritium ion sources in a number of commercially available systems on the market [30–34]. Beta particles, which are emitted by the tritium source, initiate a gas-phase reaction cascade of the drift gas (nitrogen or air), resulting in predominant proton-water clusters $H^+[H_2O]_n$, which are commonly referred to as 'reactant ions' [35]. The number of water molecules (n) depends on the gas temperature and the moisture content of the gas atmosphere [8]. Depending on the proton affinity, molecules entering the ionization region react with the reactant ions to protonated monomers $MH^+[H_2O]_{n-x}$, while decreasing the intensity of the reactant ion peak (RIP). At higher analyte concentrations, proton-bound dimers $M_2H^+[H_2O]_{m-x}$ are formed by the attachment of additional analyte molecules. When the concentration further increases, the formation of higher molecular cluster ions, such as trimers or tetramers, is possible; however, due to their low stability and short lifetime, higher molecular cluster ions are rarely observed [36]. In general, nonlinear behaviors are observed for the ratio of the RIP and the distribution between the protonated monomer and the proton-bound dimer [36,37]. The principles of a drift-time IMS including a H-3 ionization source are shown in Figure 1.

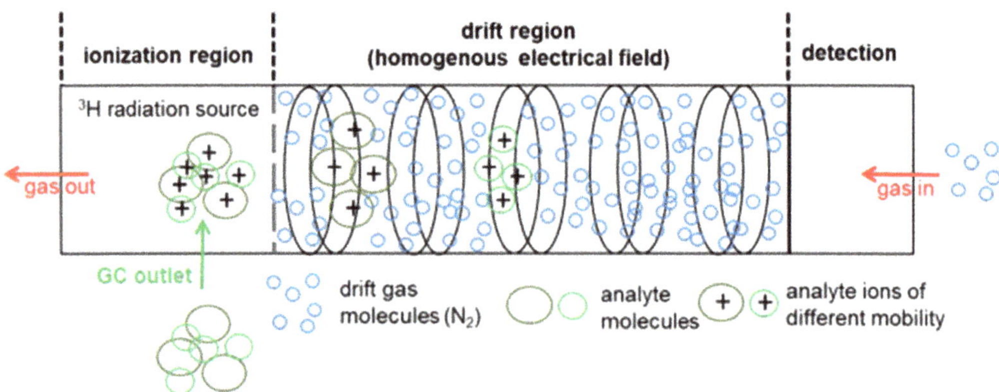

Figure 1. Setup of a drift-time IMS with a tritium (H-3) ionization source, adopted from [38] with permission (ID5138730886281).

Subsequent to ionization, the analyte ions enter the drift region, where they are accelerated towards the detector, typically a Faraday plate, and are separated by their drift time (or mobility) in an electrical field at ambient pressure. The ions are slowed down by the collision with counterflowing drift gas molecules in the collision cross-section (CCS). Due to an equilibrium between the acceleration by an electric field and deceleration by the collision with the drift gas molecules, the ions move with a constant velocity to the detector. Depending on the characteristic mass, charge, and structure, the ions are separated in the drift tube and reach the detector at different drift times [39]. For identification of the analyte, the inverse of the measured drift time is normalized to the drift length and the electric field resulting in the spectrum of ion mobility. The reduced ion mobility K_0 (see Equation (1)), which is independent of ambient conditions and experimental setup, is obtained after further normalization to pressure and temperature.

$$K_0 = \frac{L}{E \cdot t_D} \cdot \frac{p}{p_0} \cdot \frac{T_0}{T} \tag{1}$$

With

K_0 = reduced ion mobility in $cm^2V^{-1}s^{-1}$
L = drift length in cm
E = electric field strength in Vcm^{-1}
t_D = drift time in s
p = pressure of the drift gas in hPa
p_0 = ambient pressure: p_0 = 1013.2 hPa
T = temperature of the drift gas in K
T_0-ambient temperature: T_0 = 273.2 K

Instead of measuring temperature and pressure, the normalization is often carried out using the known mobility of the ions produced in the pure drift gas or by adding a reference analyte [40]. The signal intensity is proportional to the concentration and enables the quantification in ppb_v (for some compounds even ppt_v) levels within a few milliseconds.

The state-of-the-art IMS technologies can be classified into time-dispersive, space-dispersive, and trapping technologies [41]. Time-dispersive IMS separates ions as a function of their mobility in a neutral gas, whereas space-dispersive IMS separates ions by the ratio of low-field to high-field mobilities [42]. Examples of time-dispersive IMS are drift tube ion mobility spectrometry (DTIMS) and travelling tube ion mobility spectrometry (TWIMS). High-field asymmetric waveform ion mobility spectrometry (FAIMS) and differential ion mobility spectrometry (DIMS or DMS) are examples for space-dispersive techniques [43]. The third class is represented by trapped ion mobility spectrometry (TIMS), which contains a trapping technology able to confine and release ions.

IMS alone has been applied for quantification [11,15,17,27] and classification [5] tasks in controlled environments. However, due to the inherent diversity of biogenic samples, the applications of IMS with direct sample introduction are often not sufficient, requiring prior purification or separation. The commonly used purification methods for VOC profiling in combination with IMS are solvent extraction [20,26,44] and solid-phase microextraction (SPME) [13]. SPME devices are constructed of a silica fiber coated with a thin layer of a suitable polymeric sorbent or immobilized liquid, used for the direct extraction of analytes from gaseous and liquid media [45]. While SPME coupled to IMS has been successfully used for quantification tasks, such as the detection and quantification of precursors and degradation products of chemical warfare agents [13], SPME is commonly extended by column separation techniques [18,46,47].

To avoid clustering in the ionization or drift region, IMS devices are commonly coupled to column separation techniques, such as liquid chromatography (LC) or gas chromatography (GC). Column separation coupled to drift-time IMS separates analytes into two orthogonal features, first the retention time through chromatography and second, the drift time or mobility through IMS, resulting in a two-dimensional (2D) highly resolved GC-

IMS spectrum [6,38]. In LC analysis, any soluble compound can be separated, but sample preparation is a critical step for the data quality [48]. A comprehensive extraction method which enables the extraction of a wide range of compounds with minimized potentially interfering coextractives is needed for NTS approaches, since unspecific compounds are being targeted [49]. LC-MS in combination with NTS has been applied for the detection of food contaminants and environmental hazards [50,51].

In GC analysis, the volatility of a sample is a prerequisite. Headspace (HS)-based techniques allow for the analysis of untreated samples, avoiding the time-consuming sample pretreatment steps [52]. The analysis of non-volatile samples may be achieved through the derivatization with a functional group onto the molecule of interest. Although the modification of the functional group enables the analysis of compounds that otherwise could not be easily monitored by GC, NTS approaches usually do not incorporate derivatization, in particular due to the high level of variance.

The advantages of GC-IMS in comparison to established techniques, such as mass spectrometry, are its simple and inexpensive design primarily due to being operated at atmospheric pressure and hence not requiring vacuum pumps [8]. Furthermore, the use of radioactive ionization sources allows for portability, miniaturization, and mechanical robustness and therefore is suitable for field and benchtop applications [52]. Due to efficient ionization, in combination with its fast and sensitive detection, IMS is a universal technique for the analysis of organic and inorganic molecules, atoms, or particles [38]. One potential challenge of IMS analysis is that spectra may contain interference due to widespread ionization, which results in low selectivity. The addition of suitable dopant substances, however, has been shown to overcome these limitations [53,54]. A nonlinear concentration range was previously described for IMS, requiring the careful monitoring of sample concentration to avoid sample saturation. Furthermore, the separation, which is based on CCS, often provides limited information regarding specific qualities concerning size and shape of analytes. However, the drawback of interference caused by spectral complexity and nonlinearities can be overcome by using computer-based analysis tools [55].

The complexity of biological samples results from the presence of a variety of compounds, which provide in their entirety a characteristic HS-GC-IMS spectrum, often referred to as the VOC profile or 'fingerprint' [56,57]. HS-GC-IMS has been demonstrated to be an effective technique for the evaluation of VOC profiles of biological samples due to its simple system setup, robustness, and price [44,58–61]. The chemical fingerprinting of food and beverages in combination with chemometric analysis is widely used for food authentication and ultimately to identify food adulteration and fraud [62]. Furthermore, the VOC profile is influenced by production processes as well as storage conditions. Consequently, process control and quality assurance, such as the control of food freshness or food safety, are topics of interest for NTS using HS-GC-IMS [63,64] techniques.

2. Motivation for Non-Targeted Screening Using HS-GC-IMS

Labelling fraud, e.g., of organic certifications or geographic origin, is the most common type of fraud in agricultural and food markets [65,66]. According to the European Commission, honey and olive oil are particularly affected by mislabeled botanical origin, as well as dilution with inferior or less expensive products [2,67]. Moreover, food adulteration and food fraud have led to cases of economic loss and may pose health risks [2]. The detection of food fraud or adulteration often involves the identification of compounds of unknown molecular composition. Since no identified chemical markers or sets of markers are commonly accessible for a target-based analysis, an analytical approach covering a multitude of parameters in parallel paired with strong discrimination power is required. The currently used methods to determine quality and authenticity, such as sensory analysis and physicochemical analysis [68], are time- and resource-consuming, while lacking sensitivity as well as prediction accuracy, not at least due to univariate analysis. To overcome the limitations of traditional, wet-chemistry-based assays, targeted and non-targeted approaches using chromatographic methods [69,70], often in combination with mass spectrometry [71–73], as well as infrared (IR)-based spectroscopy [74,75], and proton nuclear magnetic resonance (^{1}H NMR) spectroscopy [76,77] have been discussed for various applications. However, to obtain the required reproducibility needed for chemometric analysis, time-consuming sample preparation, including precise adjustments of pH, water content or particle size, have been reported in combination with the mentioned methods. Furthermore, the high costs of ownership and maintenance, as well as the requirement for expert knowledge, may limit applications. Finally, high-end instrumentation also requires suitable laboratory infrastructures, which are usually not available at the point of care. Thus, robust, inexpensive, and fast analytical methods, such as HS-GC-MS, are needed, which require little or no sample preparation but deliver high selectivity.

Application examples for HS-GC-IMS with NTS:

A plethora of studies have shown the potential of HS-GC-IMS in combination with NTS for monitoring food quality or confirmation of geographical or botanical origin, despite the complexity of the samples. For example, HS-GC-IMS with NTS has been widely applied for the classification of olive oil between high-priced type 1 extra-virgin olive oil (EVOO), medium-priced type 2 virgin olive oil (VOO or OO), and non-edible type 3 olive oil, also known as pomace olive oil (POO) or lampante (virgin) olive oil (L(V)OO) [32,33,78,79]. Furthermore, HS-GC-IMS with NTS was successfully used for reliable classification of geographical origins for both olive oil (EVOO) [34,80,81] and wine [30]. Moreover, HS-GC-IMS with NTS was applied for the classification of honey according to botanical origin [52,81,82], as well as for the detection and quantification of honey adulterated with sugar cane or corn syrups [83,84]. Recently, HS-GC-IMS with NTS has been applied to assess the freshness of food [85] and for the detection of mold formation on milled rice [86], peanut kernels [87], and wheat kernels [88]. Further examples of recent studies using HS-GC-IMS with NTS are provided in Table 1.

Table 1. Recent studies of HS-GC-IMS with NTS.

Reference (Main Author/Yeat)	Aim of Reference	Matrice (Number of Samples)	IMS Type (Ionization Source)	Separation (Length × Inner Diameter (ID), Film Thickness (ft))	Unsupervised and Supervised Methods	Complementary Analysis	Method and Number of Compounds Identified	Data-Split; (Cross-) Validation
Arroyo-Manzanares/2017 [89]	classification between pig's food sources	dry-cured Iberian ham (24)	FlavourSpec by GAS (tritium, 6.5 keV)	FS-SE-54-CB stationary phase (30 m × 0.32 mm ID, 0.25 µm ft; 94% methyl, 5% phenyl, 1% vinyl silicone)	PCA PCA-LDA (90%), *k*NN (*k* = 3, 90%)		library search (LS) 5	80/20
Arroyo-Manzanares/2019 [84]	detection of adulteration (sugar cane or corn syrup)	honey (198)	FlavourSpec by GAS (tritium, 6.5 keV)	HP-5MS UI (nonpolar) (30 m × 0.25 mm ID, 0.25 µm ft)	OPLS-DA (97.4% prediction of class, 93.75% prediction of adulteration degree)		0	80/20
Brendel/2021 [90]	detection of botanical origin	citrus juice (47)	OEM IMS by GAS (tritium, 300 MBq)	ZB-5 ms column (30 m × 0.25 mm, 0.25 µm ft; 5% phenyl-methylpolysiloxane)	PCA PCA-LDA (91.5%)	HS-GC-MS/IMS	reference substances (RS) 9	leave-one-out (LOO) 4- and 6-fold CV
Cavanna/2018 [85]	detection of food freshness	egg (132)	FlavourSpec	FS-SE-54-CB-1 stationary phase (15 m × 0.53 mm ID, 1µm ft; 94% methyl, 5% phenyl, 1% vinyl silicone)	PCA OPLS-DA (97%)		RS, SPME-GC-MS 5	leave-one-out CV
Chen/2020 [30]	detection of geographical origin	Chinese yellow wine (122)	FlavourSpec by GAS (tritium, 6.5 keV)	nonpolar column (30 m, 95% methyl, 5% phenyl)	PCA QDA (95.35%)		LS 12	70/30
del Contreras/2019 [33]	quality assess-ment/classification	olive oil (701)	FlavourSpec by GAS (tritium, 6.5 keV)	SE-54-CB (30 m × 0.32 mm, 0.25 µm ft; 94% methyl, 5% phenyl, 1% vinyl silicone)	PCA non-targeted (PCA-LDA, *k*NN *k* = 3: 79.4%), targeted OPLS-DA: 74.3%)		LS, RS 20	80/20; 7-fold CV
Garrido-Delgado/2011 [80]	quality assess-ment/classification	olive oil (49)	portable UV–IMS instrument (UV ionization source: 10.6 eV) and FlavourSpec by GAS (tritium, 6.5 keV)	MCC (nonpolar) OV-5 (20 cm)	PCA UV-IMS: *k*NN (*k* = 3, 86.1%); GC-IMS: *k*NN (*k* = 3, 100%)		0	71/29 and 64/26; leave-one-out CV
Garrido-Delgado/2012 [78]	quality assess-ment/classification	olive oil (98)	FlavourSpec by GAS (tritium, 6.5 keV)	MCC (nonpolar) OV-5 (20 cm, 1000 parallel glass capillaries)	PCA Targeted: *k*NN (*k* = 3, 79%), non-targeted: *k*NN (*k* = 3, 87%)	Organoleptic analysis	LS, RS 10	bootstrap validation (B = 100)
Garrido-Delgado/2015 [32]	quality assess-ment/classification	olive oil (55)	FlavourSpec by GAS (tritium, 6.5 keV)	two different types of columns: MCC (20 cm × 3 mm ID, 900 parallel glass capillaries: 40µm ID, 0.2µm ft) and CC (30 m × 0.25 mm, 0.5µm ft)	PCA MCC-IMS: *k*NN (*k* = 3, 79%); CC-IMS: *k*NN (*k* = 3, 83%)		LS, RS 26	bootstrap validation
Gerhardt/2017 [34]	detection of geographical origin	olive oil (40)	FlavourSpec by GAS (tritium, 6.5 keV)	NB-225 (25 m × 0.32 mm × 0.25 µm ft; 25% phenyl, 25% cyanopropyl methyl siloxane)	PCA PCA-LDA (98%) and *k*NN (*k* = 5, 92%)		RS 4	10-fold CV
Gerhardt/2018 [52]	detection of botanical origin	honey (74)	advanced IMS by GAS (tritium, 300 MBq)	DB-225 (25 m × 0.32 mm, 0.25 µm ft; 25% phenyl, 25% cyanopropyl methyl siloxane)	PCA PCA–LDA (98.6%), *k*NN (*k* = 5, 86.1%), PLS-DA (PLS = 5, 97.0%)	^{1}H NMR	RS 18	87/13; 10-fold CV

Table 1. *Cont.*

Reference (Main Author/Yeat)	Aim of Reference	Matrice (Number of Samples)	IMS Type (Ionization Source)	Separation (Length × Inner Diameter (ID), Film Thickness (ft))	Unsupervised and Supervised Methods	Complementary Analysis	Method and Number of Compounds Identified	Data-Split; (Cross-) Validation
Gerhardt/2019 [79]	quality assess-ment/classification	olive oil (94)	FlavourSpec by GAS (tritium, 6.5 keV)	DB-225 (25 m × 0.32 mm × 0.25 μm ft; 25% phenyl, 25% cyanopropyl methyl siloxane)	PCA, HCA PCA-LDA (10 PCs, 83.3%), *k*NN (73.8%), SVM (88.1%)	Organoleptic analysis	RS 25	10-fold CV
Gu/2020 [88]	detection of quality changes during storage	wheat kernels infected with mold (90)	FlavourSpec by GAS (tritium, 6.5 keV)	MXT-WAX (nonpolar) (30 m × 0.53 mm ID, 0.1 μm ft)	PCA, HCA GA-SVM, Artificial samples: (Classification: 100%), (Quantification > 93.9%); Natural samples: (Classification: 60–86.7%, Quantification: 72.6–88.9%)		LS 13	67/33
Gu/2020 [86]	detection of quality changes during storage	milled rice infected with Aspergillus species (108)	FlavourSpec by GAS (tritium, 6.5 keV)	SE-54-CB	PCA *k*NN ($k = 3$, 94.44%), PLSR (90.9%)	electric nose (E-nose)	LS, partially RS 25	67/33; 10-fold CV
Gu/2021[87]	detection of quality changes during storage	peanut kernel infected with aflatoxigenic fungi (180)	FlavourSpec	Rtx-WAX (30 m × 0.32 mm ID, 0.25 μm ft, RT-12,424)	PCA OPLS-DA (93.3%), low-level data fusion (90%), mid-level data fusion (96.7%)	fluorescence analysis	LS 3	67/33; 7-fold CV, 200× permutation test
Li/2019 [5]	classification between aroma	green tea (23)	real-time IMS: self-made positive photoionization (PP-IMS) with KR lamp	-	PCA PLS-DA (95.6%)	sensory evaluation	0	10-fold CV, 1000× permutation test
Schwolow/2019 [81]	detection of botanical and geographical origin	honey and olive oil (64 honey, 54 EVOOs)	advanced IMS by GAS (tritium, 300 MBq)	DB-225 (25 m × 0.32 mm, 0.25 μm ft)	PCA Honey: PCA-LDA (33%); Olive oil: PCA-LDA (78%), PCA-LDA + data fusion (100%)	FT-MIR	previous studies (18 honey, 7 olive oil)	87/13; 10-fold CV
Shuai/2014 [44]	detection of adulteration (with vegetable oils)	flaxseed oil (78)	IMS-KS-100 by Wuhan Syscan Technology (pulse glow discharge)	n-hexane extraction	recursive support vector machine (R-SVM) (93.1%)		0	10-fold CV
Vega-Marquez/2019 [91]	quality assess-ment/classification	olive oil (701)	FlavourSpec by GAS (tritium, 6.5 keV)	SE-54-CB (30 m × 0.32 mm, 0.25 μm ft; 94% methyl, 5% phenyl, 1% vinyl silicone)	Deep learning (88.8%), SVM (83.1%), *k*NN (84.5%), Tree (78.3%), Regressor (85.5%), XGBoost (85.7%)		LS, RS 20	80/20
Wang/2019 [82]	detection of botanical origin	honey (40)	FlavourSpec by GAS (tritium, 6.5 keV)	FS–SE–54–CB-0.5 (15 m × 0.53 mm ID)	PCA OPLS-DA (95%) (VIP > 1.5)		HS-SPME-GC-MS8	200× permutation test
Wang/2019 [83]	detection of botanical origin	honey (120)	FlavourSpec by GAS (tritium, 6.5 keV)	FS–SE–54–CB-0.5 (15 m × 0.53 mm ID)	PCA PLS-DA (84%)		LS 25	CV-ANOVA
Yuan/2020 [92]	metabolomic studies	rat feces (30)	FlavourSpec	FS-SE-54-CB-1 (15 m × 0.53 mm ID)	PCA OPLS-DA (86.6)		LS 11	200× permutation test
Zhu/2020 [31]	quality assess-ment/classification	wine (143)	FlavourSpec	MXT-WAX (polar) (30 m × 0.53 mm ID, 0.5 μm ft, polyethylene glycol)	PCA PCA-LDA (65.7%), PLS-DA (58.7%), *k*NN ($k = 5$, 60.8%), SVM (51.8%), XGBoot (81.8%), ANN (89.5%)	sensory evaluation	RS >30	85/15; 10-fold CV

3. NTS-Workflow

NTS aims to identify the compounds of unknown molecular composition. The workflow for NTS generally consists of sample preparation, instrumental analysis, and post-acquisition data processing [93]. Since little or no a priori knowledge of the chemical structures and behavior of compounds is required, NTS approaches benefit from gentile sample preparation, robust instrumental analysis, and standardized data processing. The workflow for NTS using GC-IMS is shown in Figure 2. The first step, data acquisition, involves sample preparation and subsequent extraction and separation of VOCs. The collected data are then preprocessed and analyzed in the data-processing step. Since no pre-existing knowledge is used, the entire spectral fingerprint obtained by HS-GC-IMS analysis is subject to data analysis and classification or quantification models being built using machine learning tools. In the third step, model interpretation, key compounds are identified, which are extracted through back projection of loadings. Complementary analyses, such as sensory analysis of GC-MS measurements, provide further insight into sample composition and can be used to improve model accuracy. The model coherency is evaluated and finally applied for benchtop profiling.

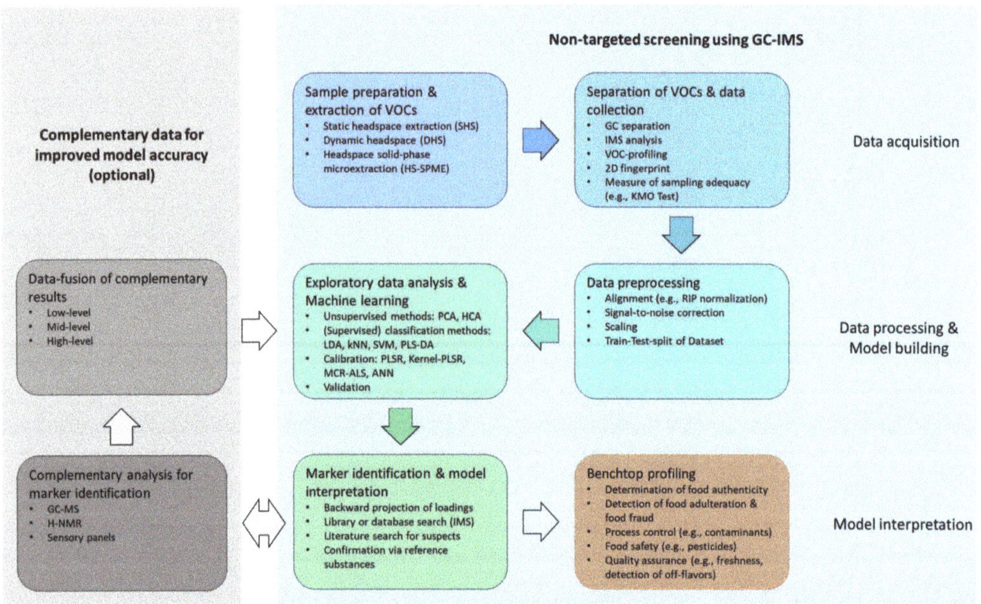

Figure 2. Workflow for non-targeted screening using HS-GC-IMS.

3.1. Data Acquisition

In untargeted approaches, sample preparation strategies need to be suitable for a variety of matrices and enable the extraction of a wide range of compounds [49]. In order to minimize the extraction-related formation of artifacts, static headspace extraction (SHS) is commonly used for the analysis of volatile compounds in samples of different origin [94]. The limited sensitivity and the bias related to the extraction of low-volatile compounds are generally considered as the main limitations of SHS methods, which can be overcome through dynamic headspace extraction (DHS) [95]. If the preconcentration of analytes is needed for analysis, high-concentration-capacity HS techniques, such as SPME can be used, where the selective isolation of compounds of interest from samples with minimal matrix contamination is crucial [95–97]. However, SPME is less commonly applied to IMS with NTS, possibly due to drawbacks which have been associated with commercially available extraction phases, such as the low recovery of polar compounds, insufficient matrix-compatibility, and the reduced life-time and robustness of the coating against matrix macromolecules [98].

VOC profiling based on IMS in combination with NTS has been shown to be an effective tool for various classification tasks, such as the differentiation between various green teas with chestnut-like flavor [5]. Using PLS-DA, the authors were able to obtain an accuracy of 95.6% for the classification of tender, pure, and roasted green tea aromas. However, due to the complexity of many biogenic samples, IMS with direct sample introduction is often not sufficient for NTS approaches, requiring extraction and/or column separation techniques. Shuai and coworkers analyzed adulterated flaxseed oils after n-hexane extraction using a handheld IMS and subsequent data analysis. A recursive support vector machine (R-SVM) led to a model with an accuracy of 93.1% for the identification of adulterated flaxseed oil [44].

To address the complexity of biogenic samples, pre-IMS separation techniques are commonly applied. To avoid the clustering and formation of heteromers in the ionization or drift region, IMS devices are commonly coupled to GC. The spectra obtained by HS-GC-IMS are 2D, with the GC retention time as the first dimension and the IMS drift time as the second dimension. The complexity of biological samples results from a plethora of compounds, which provide in their entirety a characteristic GC-IMS spectrum that is often referred to as VOC profile or 'fingerprint' [56,57]. For GC separation in combination with IMS and NTS, both the capillary column (CC) and multicapillary column (MCC) have been coupled to IMS. Among others, nonpolar CCs, such as fused silica SE-54-CB (94% methyl, 5% phenyl, 1% vinyl silicone) [32,33,86,89,92], as well as polar CCs, such as DB-225 (25% phenyl, 25% cyanopropyl methyl siloxane) [79,81], were used for HS-GC-IMS with NTS. For MCCs, which are composed of a large number of parallel glass capillaries (~1000), nonpolar OV-5 (5% diphenyl, 95% dimethylpolysiloxane) was used [78,99]. Compared to the ordinary CC, MCCs allow the separation to be carried out at an elevated speed and can be operated at increased carrier gas flows, which may pose an advantage for IMS analysis [100]. However, Garrido-Delgado and coworkers obtained the better predictive accuracy for the classification (*k*NN classifier with $k = 3$) between different qualities of olive oil (EVOO, OO, and LVOO), when CC-IMS (83%) versus MCC-IMS (79%) was employed for the separation of the VOCs [32].

The data acquisition is finalized with the evaluation of the suitability of the collected data for factorial analysis. This sampling adequacy is commonly determined using the Kaiser–Meyer–Olkin (KMO) test, which is a measure of the proportion of variance among variables [101].

3.2. Data Processing and Model Building

The ion velocity and thus the signal position are highly dependent on temperature and ambient pressure; therefore, the drift-time alignment is a crucial preprocessing step [36,102]. For baseline correction, drift times are often normalized to the RIP position, which is, however, not sufficient for the determination of absolute mobilities [103]. Alternatively, a reference peak/substance can be used for alignment, which is recommended when GC-IMS measurements are performed with long separation columns. In a second preprocessing step, multiple measurements are averaged, and the background noise is subtracted. Scaling methods, such as unit variance (UnVa) scaling, Pareto (Par) scaling, and mean centering (Ctr) scaling, can be alternatively used for data set normalization [5]. In UnVa scaling, the variance of each variable is unified using the standard deviation, while in Par scaling, the square root of the standard deviation is used for normalization, whereas Ctr scaling emphasizes the analysis of data fluctuations and the large-fold change in the data [82]. For further noise reduction, smoothing algorithms, such as Savitzky-Golay or Gaussian smoothing, are applied to the spectra [52]. Interfering peaks and nonrelevant areas can be removed by the careful selection of relevant GC retention time and drift-time ranges. Prior to data analysis, the 2D spectra (GC retention time $\times$ drift time) are unfolded into arrays, which are then concatenated to the final data set (samples × measurements). For supervised data analyses, the data set is finally split into training and test sets, where the training set is used for model building and the test set for model validation.

Nonlinear behavior has been described for the ratio of the RIP and the distribution between the protonated monomer and the proton-bound dimer [36,37,104]. By increasing the molecular concentration of the analyte, the monomer peak decreases, while the dimer peak increases. These typical nonlinear monomer-dimer distributions in IMS are often accompanied by tailing effects of the monomer or dimer peak [105], requiring careful consideration for quantification. In univariate regression (UR), the peak area or intensity of a single peak is correlated against concentration [46]. Due to the nonlinear behavior between the monomer and dimer peak, linearity can only be approximated for narrow concentration ranges when using the regression analysis based on single-peak analysis [36]. Occasionally, the sum of the volume of monomer and dimer is correlated against the concentration [106]; however, in complex mixtures, quantification is further complicated by overlapping peaks, competitive ionization situations between coeluting analytes and the occurrence of multicomponent cluster ions (heterodimeric ions), which are composed of coeluting analytes [37].

3.2.1. Exploratory Data Analysis and Machine Learning Techniques

Due to the complexity of food matrices, abstract terms such as 'food quality' and 'authenticity' are the sum of multiple characteristics, which implies that the correlation to a single analyte or analytical technique is often not sufficient or even impossible; hence, multivariate variate data analysis (MVA) techniques are required [81]. MVA approaches can be divided into exploratory, classification, and quantitative regression methods. Exploratory methods, such as PCA or HCA, are unsupervised and typically used for pattern recognition, whereas classification methods such as PCA-LDA, *k*NN, or PLS-DA are supervised methods. In this context, PLS-DA is a special case of classification, as it basically uses a regression approach with class boundaries instead of single values, as in quantitative regressions. For sample quantification, the latter methods, such as PLS regression, Kernel-PLSR, MCR-ALS, or ANN are commonly applied.

Principal component analysis (PCA) is a powerful technique for unsupervised discovery of patterns in data, which is further used for dimension reduction [107]. The information extracted from a data matrix is explained by principal components (PCs), which are orthogonal (mathematically independent) to each other. Since PCA models are predicted without labels or validation by test data, they are generally considered unsupervised. Unsupervised statistical methods are exploratory methods that can be used to study data structures and search for clusters of samples [108]. Hierarchical cluster analysis (HCA) of PCA models in a tree-like diagram (dendrogram) is, e.g., used for the visualization of multivariate association and sample similarities [109]. An extension of PCA for processing three-dimensional (3D) data is provided by multiway principal component analysis (MPCA) [110], which has been applied for the feature extraction of GC-IMS matrices, without prior transformation of the 2D data [99]. Further alternatives to PCA and PLS are models based on PARAFAC [111] or Tucker3 [112], which may also be used with 3D data.

Compared to unsupervised techniques, which provide predictions without labels or target variables, supervised techniques aim to build models able to predict target variables. In supervised learning, several data points or samples are described using predictor variables or features and target variables. For classification tasks, the scores obtained by the unsupervised exploratory analysis are combined with subsequent supervised pattern recognition techniques to distinguish samples according to defined categories. Among PCA-based qualitative methods are linear discriminant analysis (LDA) and k-nearest neighbors (*k*NN). Whereas PCA-LDA maximizes the interclass variance, *k*NN assigns the category most common among the *k*-nearest neighbors. The downside of using PCA-based methods is that the correlation between dependent and independent variables are not considered during PCA analysis, which can result in the loss of information included in higher PCs [107]. An alternative is provided by partial least squares (PLS), where the scores are calculated by considering the relationship between the independent and dependent variables.

Other supervised methods used in NTS with HS-GC-IMS are gradient boosting (e.g., XGBoost) [31], decision tree classification (Tree) [91], logistic regression (Regressor) [91], orthogonal partial least-squares discriminant analysis (OPLS-DA), quadratic discriminant analysis (QDA) [30], or soft independent modeling of class analogy (SIMCA) [82]. Furthermore, nonlinear classifications are often performed using support vector machines (SVMs). By using the kernel trick, which transforms the input data into high-dimensional feature spaces, SVMs can perform nonlinear classifications, in addition to performing linear classification [113]. This is of particular importance when no linear hyperplanes are separating the respective classes.

3.2.2. Model Performance and Validation

The performance of a model is usually measured as 'accuracy', which is the fraction of correctly classified samples. The classification accuracy determines the fraction of correctly classified samples for a given sample set. The classification accuracy, however, is susceptible to overfitting and thus should only be used as reference. To prevent overfitting, the data set is split into training and test data. The ratio between training and test data, which is commonly referred to as 'train–test split', is usually between 2:1 [88] and 4:1 [33], and sometimes as low as 6:1 [31]. The test set is used to determine the prediction accuracy, which is usually lower than the classification accuracy but more meaningful [88]. For small and inhomogeneous data sets, a single split of the data into training and test set may give misleading results [114]. An alternative model validation is therefore provided by resample methods, such as cross validation (CV), bootstrapping, or permutation testing, where multiple random subsets are generated. For the CV of samples, a subset of the data is held out for use as a validation set, and a model is fit to the remaining data (training set) and used to predict the validation set. The process of generating a subset of data, model fitting, and evaluation is performed repeatedly, and an overall prediction accuracy is determined by averaging the quality of the predictions across the validation sets. Leave-one-out CV leaves out a single observation at a time, while k-fold CV splits the data into k subsets, which are one by one held out as the validation set [107]. For NTS using HS-GC-IMS, 10-fold CV is commonly performed [31,52], alongside with leave-one-out CV [80,85]. Bootstrapping is a resampling method that can be used as an alternative to CV to estimate the prediction performance of a model with a low number of training samples. Due to drawing with replacement, a bootstrapped data set contains the same number of cases as the original data set, and it can contain multiple instances of the same original cases [114]. Another resampling method is permutation testing, where labels are switched on data points when performing the test statistics. Both bootstrapping [32,78] and permutation tests [82,92] have been applied with HS-GC-IMS and NTS.

The success rate of different chemometric models for classification depends on many factors, such as botanical origin of the samples, number of samples or the selection of PCs, PLS, and k values. Gerhardt and coworkers compared different chemometric methods for the classification of the botanical origins of honey (acacia, canola, and honeydew) using a resolution-optimized HS-GC-IMS. They found a 98.6% accuracy with PCA-LDA, 86.1% with *k*NN ($k = 5$), and 97.0% with PLS-DA after employing 10-fold CV [52]. Quality assessment of olive oils based on temperature-ramped HS-GC-IMS and sensory evaluation conducted by Gerhardt and coworkers reported a 83.3% accuracy with PCA-LDA, 73.8% with *k*NN ($k = 5$) and 88.1% with SVM models, after employing 10-fold CV [79].

Artificial neural networks (ANNs) are a powerful modeling approach, which is vaguely inspired by the biological neural networks in the brain [115]. Due to their hidden layers, ANNs have the ability to capture complex interactions present in biological samples. Furthermore, ANNs can be applied for pattern recognition, classification tasks, and quantification problems, as well as data preprocessing. ANN has recently the gained attention for applications in food science and technology, while often being limited by the requirement of having sufficiently large data sets [116,117]. Zhu and coworkers have shown the superior prediction performance of ANN (89.5%) over PCA-LDA (65.7%), PLS-DA (58.7%), *k*NN ($k = 5$, 60.8%) and SVM (51.8%), and XGBoost (81.8%), for the classification of Sauvignon Blanc via SHS-GC-IMS [31]. Vega-Márquez and coworkers evaluated a deep learning network and five different benchmark methods for the classification of olive oil samples into EVOO, OO, and LVOO, based on HS-GC-IMS spectra obtained for 701 olive oil sample from two different harvests [91]. Among the five benchtop models used for comparison, XGBoost offers the best accuracy of 85.7%, compared to SVM (83.1%), *k*NN (84.5%), Tree (78.3%), and Regressor (85.5%). However, even better results were achieved with the deep learning approach, obtaining an accuracy of 88.8%, which underlines the potential of ANNs.

3.2.3. Quantification Tasks

For quantification tasks, partial least squares regression (PLSR) has become the standard method used in chemometrics, including the fields of sensorial analysis in food chemistry [107,110]. PLSR is used to describe the relationship between two data matrices, X (experimental data) and Y (actual concentrations), which are decomposed into $X = TP^T + E$ and $Y = UQ^T + F$, by finding the maximum covariance and linear relationship between the score matrices T and U. P and Q represent the loading matrices and E and F the matrices of residuals. After multiplying X with a nonlinear function, linear PLS can be applied as described above. After the successful implementation of multivariate models for the interpretation of the training data, the model performance is commonly tested using test data. The model quality can be determined by different figures of merit, such as the determination coefficient (R^2) and the relative percentage error of prediction (RE), as well as root mean square error (RMSE), systematic error (Bias), or standard error of prediction (SEP).

One limitation to PLSR is the typically lower performance on nonlinear and heteroscedastic data, as is partially the case for IMS data. Several studies have analyzed nonlinear IMS data, as shown by the quantification of histamine in tuna stomach [7] or allergenic fragrance compounds, such as citral, in complex cosmetic products [38]. Nonlinear relationships between the matrices T and U can be described by kernel PLSR (also known as nonlinear PLSR), where the data are transformed into higher-dimensional spaces using the kernel trick [118].

An alternative to PLSR or kernel PLSR is provided by multivariate curve resolution alternating least squares (MCR-ALS). MCR decomposes the initial data matrix D with n sample spectra and m data points in the spectra to $D = CS^T + E$. The matrices C ($n \times l$) and S ($l \times m$) represent the concentration and spectra profiles of D for l components, while the matrix E ($n \times m$) contains the residuals not explained by C and S [119]. MCR-ALS may deconvolute overlaying spectra from coelution and reconstitute the pure spectra for quantitation. A comparison of UR, PLSR, MCR-ALS, and Kernel-PLSR for the quantification of fragrances using GC-IMS analysis was provided by Brendel and coworkers. For both, a mixture of geraniol and citral as well as a mixture of the citral and cinnamal, Kernel-PLSR demonstrated the superior ability for the quantification of the nonlinear relationships of GC-IMS data since Kernel-PLSR was able to significantly reduce the RE of prediction and increase R^2 of calibration.

3.3. *Model Interpretation*

Once a robust and reliable model is built, whose predictive abilities are sufficient for the chosen task, the model should be checked for plausibility. Therefore, the influencing variables, also referred to as markers, are extracted, identified, and subsequently used for interpreting and explaining the model. By using PCA for marker extraction, the loadings obtained by multivariate analysis are projected backwards into the original data space and are subsequently evaluated either manually or automatically. This allows one to identify the signals with the main influence for separation in the respective principal components with relation to the original data (Figure 3). Another example is the use of PLS-DA, where volcano plots and variable importance in projection (VIP) analysis have been performed to determine the final markers for each discrimination task [56]. Compounds with a VIP score > 1 are generally considered as suitable markers [120].

Web-based platforms, such as MetaboAnalyst (http://www.metaboanalyst.ca, accessed on 1 July 2021) [121] and XCMS Online (https://xcmsonline.scripps.edu/, accessed on 1 July 2021) [122], are designed to handle comprehensive untargeted metabolomic data. Compared to GC-MS data, which include m/z information, GC-IMS data are intrinsically limited to (normalized) drift times and retention times. While MetaboAnalyst has also been used for the processing of GC-IMS data, including the preprocessing (normalization and scaling) of data [82], VIP analysis [83], and entire chemometric analysis [123], the lack of m/z information limits the use of these metabolomic data processing tools for compound identification. However, the combination of retention indices and normalized drift times is considered as a reliable alternative; as in particular, drift times are highly reproducible. By comparing markers obtained from the GC-IMS data to databases, which are often provided by IMS manufacturers, the substances of interest can be identified [124,125]. Furthermore, a search of the literature and subsequent confirmation by reference substances can be used for substance identification [126–128]. Since coelution and matrix effects can influence GC-IMS data, a common technique is to spike a complex sample with the pure substance. To increase the number of substances identified and to further increase the model's accuracy, complementary data, such as GC-MS [123], can be used as described in Section 3.4. This procedure is optional, as reliable models for classification and quantification tasks can be built solely from HS-GC-IMS data [44,80]. Typically, these MS detectors are unit-resolved and, as such, again limited in their selectivity in comparison to high-resolution systems, such as TOF or Orbitrap systems. However, a full spectral interpretation is not always necessary in NTS approaches. Some studies solely detect markers without subsequent identification of substances [43,80,84].

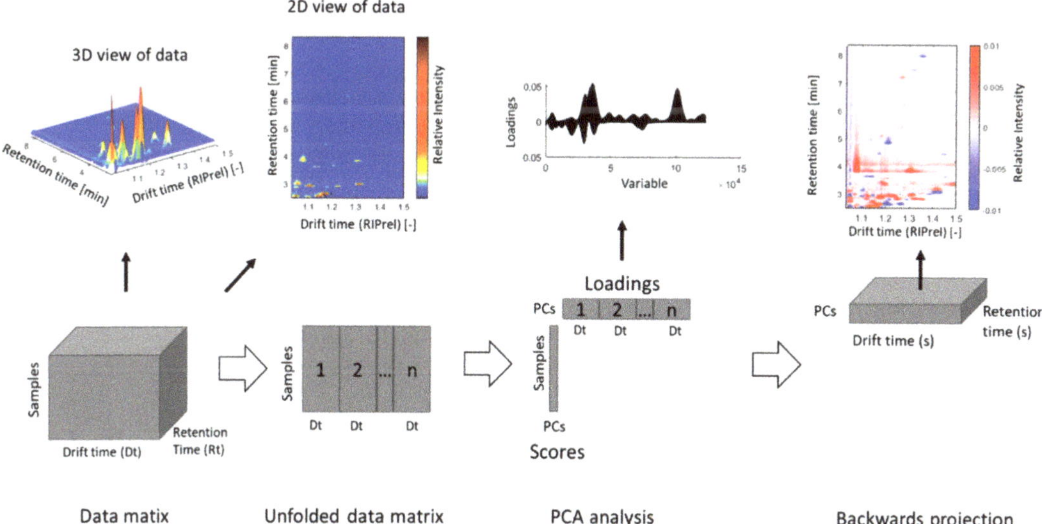

Figure 3. Data analysis and backwards projection of loadings for identification of key substances.

NTS is a powerful approach for complex classification and quantification tasks. However, next to model overfitting, the transferability of a model to new data poses a major challenge. A dramatic example is given by Contreras and coworkers who analyzed 701 olive oil samples from the years 2014–15 and from 2015–16 for the classification into EVOO, OO, and LVOO, using HS-GC-IMS with NTS [33]. A model built with olive oil samples from 2014–15 obtained a prediction percentage of 67.8%. A better prediction success was achieved for a model built with olive oil samples from 2014–15 and 2015–16, obtaining 79.4% accuracy. However, when applying the NTS model built with samples from the year 2014–15 to predict the years 2015–16, a prediction success of only 36.0% was achieved, revealing the low transferability of the model. This is however not an effect solely to be attributed to GC-IMS but rather is a general issue of profiling techniques so far and is mainly driven by typically high analytical variance together with a high biological variance of the samples.

To achieve good predictive abilities and transferability of a model to new data, large sample sets containing independent and diverse samples are necessary; however, studies are often limited by sample availability. To overcome this issue and to reveal influencing factors, a comprehensive evaluation of the model should be performed. The identification of influencing substances, for instance, may reveal the characteristic compounds or classes of substances in which the classes differ. A subsequent comprehensive evaluation can determine which factors may be caused by systematic errors and which factors allow for true differentiation between classes. High predictive accuracies for classification and quantification tasks can be achieved without substance identification, yet a comprehensive evaluation of the model helps to detect strategic errors, such as a narrow sample distribution, and thus should always be part of NTS approaches. Due to its robustness and comparably low prices, HS-GC-IMS is suitable for benchtop profiling. In combination with NTS approaches, HS-GC-IMS data can be used for the implementation of models, which are suitable for classification and quantification task in various fields.

3.4. Complementary Data (Optional)

Due to the complexity of biological matrices, the substance identification with GC-IMS data alone can be challenging, which is why stand-alone IMS are rarely used to investigate the sample composition. Complementary techniques, such as GC-MS [129] or ^{1}H NMR [130], are often used to identify decisive marker substances [131].

Next to marker identification, complementary data can be combined with GC-IMS data to build a multimodal model. The process of integrating multiple data sources, which is commonly known as 'data fusion', has the potential to increase model accuracy and reliability, while reducing interferences and error rates [132]. The process of data fusion is categorized as low-level, mid-level (or intermediate-level), and high-level data fusion, depending on the fusion strategies used [133]. In low-level (or data-level) fusion approaches, data from all sources are preprocessed, concatenated into a common data matrix, and subsequently analyzed using classical multivariate methods, such as PCA or PLS (see Section 3.2). Mid-level data fusion approaches, which are also referred to as feature-level data fusion, are based on the extraction of relevant features from each data source separately [87]. Several latent variables, such as the score values from PCA or PLS, are selected for this feature-based data fusion, concatenated, and subsequently analyzed using classical multivariate methods. The third fusion approach is high-level or decision-level data fusion, where completely independent models are calculated from each data set; this level is the most challenging due to the determination of the ideal parameters for each separate multivariate model. Careful considerations of parameters, together with the implementation of a voting or scoring scheme, which prioritizes results from different data sources, can provide a combined model, which outperforms the individual models [134].

Schwolow and coworkers showed that the data fusion approach significantly increased both the predictive power and the robustness of the resulting classification model for the determination of geographical origin of olive oil [81]. The prediction accuracy obtained for Fourier-transform infrared (FT-IR) data alone was 67% and for GC-IMS 78%, while resulting a perfect score for a low-level data fusion approach using the complementary chemical information provided by GC-IMS and FT-IR analysis. The extra effort needed for data fusion approaches, however, is not always rewarded. Gu and coworkers classified fungal growth on peanut kernels into potentially aflatoxigenic fungi and non-aflatoxin producing fungal species using HS-GS-IMS and fluorescence measurements [87]. The predictive accuracy for HS-GC-IMS measurements alone was 93.3% with an OPLS-DA model, while low-level data fusion reduced the accuracy to 90%. The predictive accuracy of the OPLA-DA after mid-level data fusion using VIP > 1.3 further decreased to 86.7%. Only a mid-level data fusion using 10 PCs increased the model accuracy to 96.7%.

Another approach to increase the discrimination power of a classification model is the parallel analysis of IMS and MS, which has recently gained attention for classification and quantification tasks [90,106,127]. Although IMS and MS cannot be seen as fully complementary methods, in particular due to the identical volatile fraction being monitored by both techniques [90], it was shown that the soft ionization and drift-time-based ion separation on the one hand and a hard ionization and m/z-based separation on the other hand improved substance identification in the case of coelution in hops analysis [127]. While complementary data in general can increase the interpretability and accuracy of a model developed with HS-GC-IMS, they are optional for NTS via GC-IMS.

4. Comparison of NTS and Targeted Strategies

An alternative approach to NTS is targeted screening. While NTS of HS-GC-IMS data uses no pre-existing knowledge and the entire spectral fingerprint is subject to data analysis, for targeted analysis, specific markers are chosen prior to the data analysis. The markers used for targeted screening can be either handpicked or mathematically determined [33,135]. One-way analysis of variance (ANOVA), using for example a Tukey's test, is often applied to identify volatile compounds which exhibit significant differences, commonly quantified at a 5% significance level ($p \leq 0.05$) [59]. Various other methods, such as Gabor filters, local binary pattern, Haar, and histograms of oriented gradients (HOG), have been proposed for feature extraction [136,137]. Chen and coworkers applied MPCA and HOG for feature extraction and data reduction of MCC-IMS data, with subsequent canonical discriminant analysis for the generation of nonlinear boundaries, for the successful quantification of the adulteration degree of canola oil. A predictive accuracy of more than 95.2% was reported for a PLS model, which was obtained using a train–test split of 70–30 [99]. Using targeted approaches and applying PCA-*k*NN, the same authors reported a successful classification of rapeseed oils according to their quality (grade 1–4) and a successful determination of vegetable oil according to its botanical origin (sesame oil, rapeseed oil, and camellia oil). For the classification of the rapeseed oil quality, the colorized differences method was applied to CC-IMS data, resulting in 34 peaks of interest and a predictive accuracy of 100% [138]. Furthermore, Otsu's method and colorized differences method was used for automatic peak detection, resulting in 88 peaks of interest and a predictive accuracy of 98.3% for the classification of vegetable oil using MCC-IMS data [139]. The advantage of preselecting markers with significant differences is the simultaneous reduction of noise in the data, which, however, includes the risk of overlooking valuable information.

NTS approaches (spectral fingerprinting) have also been directly compared to targeted approaches (extraction of specific markers) for the analysis of IMS data. Garrido-Delgado and coworkers compared targeted and NTS approaches for the classification of olive oil into EVOO, OO, and LVOO, using data obtained by MCC coupled to IMS [78]. A PCA-LDA model was used for data reduction and data clustering, followed by *k*NN ($k = 3$) for classification, obtaining a prediction percentage of 79% for the targeted strategy and 85% for NTS strategy. For the classification of olive oil harvested in 2014–15, Contreras and coworkers obtained a prediction percentage of 56.9% for the targeted strategy and 67.8% for the NTS strategy [33]. An improved prediction success was achieved for models built with olive oil samples from 2014–15 and 2015–16, obtaining 74.3% for the targeted strategy and 79.4% for the NTS, hence suggesting superior abilities of the NTS approach versus the selection of specific markers. By contrast, the authors also reported that a targeted model built with samples from the years 2014–15 (prediction success of 51.6%) was superior to the NTS approach (prediction success of 36.0%) when applied to the years 2015–16. Both models built with the data from the years 2014–15 show weak prediction abilities for the prediction of samples from the following year, revealing some fundamental challenges in data science: the predictive ability of a model is highly dependent on the number of samples as well as on the sample diversity. Both approaches include the risk of overfitting to a specific problem [33].

Arroyo-Manzanares and coworkers likewise obtained superior classification accuracy using HS-GC-IMS for a model based on a targeted marker selection (100%) compared to a model based on the whole spectral fingerprint (90%) for the distinction between dry-cured Iberian ham from pigs fattened on acorns and pasture or on feed [89]. However, the model based on marker selection was built using OPLS-DA, while *k*NN ($k = 3$) and PCA-LDA were used for the model based on spectral fingerprints; hence, the differences in the predictability of the models may result from the use of different mathematical tools and do not provide inferences about targeted and non-targeted approaches. Gu and coworkers by contrast obtained better classification results with the NTS versus a targeted approach for distinguishing between fungal infections of wheat kernels, as well as for the quantification of fungal colony counts [88]. In conclusion, NTS and targeted screening approaches are both effective tools for data analysis, with different challenges and application areas.

5. Conclusions

HS-GC-IMS in combination with an NTS approach provides an effective tool for classification and quantification tasks not at least due to low maintenance and easy handling of the instruments [9]. In combination with HS-GC and machine learning, IMS has been demonstrated to be an effective technique for various classification and quantification tasks, such as for the determination of food authenticity as well as for the detection of food adulteration and food fraud. Furthermore, HS-GC-IMS has been applied in the field of process control, for example for the early detection of microbial contaminants or in the field of food safety for the detection of pesticides. HS-GC-IMS is also found in the field of quality assurance for the detection of food freshness or the detection of off-flavors. Recently, HS-GC-IMS in combination with deep learning approaches has shown promising results for further improvement of model accuracies. HS-GC-IMS is suitable for benchtop profiling, due to its robustness and comparably low purchase and running costs. For benchtop analysis, MDV-based methods for HS-GC-IMS devices can be implemented in the laboratory and subsequently applied for food analysis in various fields. A common approach for further market acceptance is the combination of IMS instrumentation with HS-GC-MS, which alone has already been applied to a wide variety of applications.

Author Contributions: Writing—original draft preparation, C.C.; funding acquisition, P.W.; writing—review and editing, P.W. All authors have read and agreed to the published version of the manuscript.

Funding: This research was funded by Zentrales Innovationsprogramm Mittelstand (ZIM) of the Federal Ministry for Economic Affairs and Energy, grant number FZ4014707AW9.

Conflicts of Interest: The authors declare no conflict of interest.

Sample Availability: Samples of the compounds, etc., are available from the authors of the respective publications.

Glossary

Abbreviations	Description
^{1}H NMR	proton nuclear magnetic resonance
2D	two-dimensional
3D	three-dimensional
Am-241	americium-241
ANN	artificial neural networks

ANOVA	one-way analysis of variance
APPI	atmospheric pressure photo ionization
Bias	systematic error
CC	capillary column
CCS	collision cross section
CD	corona discharge
Ctr	mean centering
CV	cross validation
D(I)MS	differential ion mobility spectrometry
DHS	dynamic headspace extraction
DTIMS	drift-tube ion mobility spectrometry
E-nose	electric nose
EVOO	extra-virgin olive oil
FAIMS	high-field asymmetric waveform ion mobility spectrometry
ft	film thickness
FT-IR	Fourier-transform infrared spectroscopy
GC	gas chromatography
GC–IMS	gas chromatography-ion mobility spectrometry
$H^+[H_2O]_n$	proton-water clusters with n number of water molecules
H-3	tritium
HCA	hierarchical cluster analysis
HOG	histograms of oriented gradients
HS	headspace
ID	inner diameter
IMS	ion mobility spectrometry
K_0	reduced ion mobility
KMO	Kaiser–Meyer–Olkin
kNN	k-nearest neighbors
L(V)OO	lampante (virgin) olive oil
LC	liquid chromatography
LDA	linear discriminant analysis
LDI	laser desorption/ionization technique
LOO	leave-one-out
LS	library search
M_2H^+	proton-bound dimers
MCC	multicapillary column
MCR-ALS	multivariate curve resolution alternating least squares
MH^+	protonated monomers
MPCA	multiway principal component analysis
MVA	multivariate variate data analysis
Ni-63	nickel-63
NTS	non-targeted screening
OPLS-DA	orthogonal partial least-squares discriminant analysis
Par	Pareto
PC(A)	principal component (analysis)
PLS(R)	partial least squares (regression)
POO	pomace olive oil
(Q)DA	(quadratic) discriminant analysis
R^2	determination coefficient
RE	relative error of prediction
Regressor	logistic regression
RIP	reactant ion peak
RMSE	root mean square error
RS	reference substances
(R)-SVM	(recursive) support vector machine
SEP	standard error of prediction
SHS	static headspace extraction

SIMCA	soft independent modeling of class analogy
SPME	solid-phase microextraction
TIMS	trapped ion mobility spectrometry
Tree	decision tree classification
TWIMS	travelling tube ion mobility spectrometry
UnVa	unit variance
UR	univariate regression
UV	ultraviolet light
VIP	variable importance in projection
VOC	volatile organic compound
(V)OO	(virgin) olive oil

References

1. Jackson, L.S. Chemical food safety issues in the United States: Past, present, and future. *J. Agric. Food Chem.* **2009**, *57*, 8161–8170. [CrossRef]
2. Moore, J.C.; Spink, J.; Lipp, M. Development and application of a database of food ingredient fraud and economically motivated adulteration from 1980 to 2010. *J. Food Sci.* **2012**, *77*, R118–R126. [CrossRef]
3. Van de Brug, F.J.; Lucas Luijckx, N.B.; Cnossen, H.J.; Houben, G.F. Early signals for emerging food safety risks: From past cases to future identification. *Food Control* **2014**, *39*, 75–86. [CrossRef]
4. Vuckovic, D. Current trends and challenges in sample preparation for global metabolomics using liquid chromatography-mass spectrometry. *Anal. Bioanal. Chem.* **2012**, *403*, 1523–1548. [CrossRef] [PubMed]
5. Li, J.; Yuan, H.; Yao, Y.; Hua, J.; Yang, Y.; Dong, C.; Deng, Y.; Wang, J.; Li, H.; Jiang, Y.; et al. Rapid volatiles fingerprinting by dopant-assisted positive photoionization ion mobility spectrometry for discrimination and characterization of Green Tea aromas. *Talanta* **2019**, *191*, 39–45. [CrossRef] [PubMed]
6. Rodríguez-Maecker, R.; Vyhmeister, E.; Meisen, S.; Rosales Martinez, A.; Kuklya, A.; Telgheder, U. Identification of terpenes and essential oils by means of static headspace gas chromatography-ion mobility spectrometry. *Anal. Bioanal. Chem.* **2017**, *409*, 6595–6603. [CrossRef]
7. Cohen, G.; Rudnik, D.D.; Laloush, M.; Yakir, D.; Karpas, Z. A Novel Method for Determination of Histamine in Tuna Fish by Ion Mobility Spectrometry. *Food Anal. Methods* **2015**, *8*, 2376–2382. [CrossRef]
8. Borsdorf, H.; Eiceman, G.A. Ion Mobility Spectrometry: Principles and Applications. *Appl. Spectrosc. Rev.* **2006**, *41*, 323–375. [CrossRef]
9. Cumeras, R.; Figueras, E.; Davis, C.E.; Baumbach, J.I.; Gràcia, I. Review on ion mobility spectrometry. Part 2: Hyphenated methods and effects of experimental parameters. *Analyst* **2015**, *140*, 1391–1410. [CrossRef]
10. Schmidt, C.; Jaros, D.; Rohm, H. Ion Mobility Spectrometry as a Potential Tool for Flavor Control in Chocolate Manufacture. *Foods* **2019**, *8*, 460. [CrossRef]
11. Karpas, Z.; Guamán, A.V.; Calvo, D.; Pardo, A.; Marco, S. The potential of ion mobility spectrometry (IMS) for detection of 2,4,6-trichloroanisole (2,4,6-TCA) in wine. *Talanta* **2012**, *93*, 200–205. [CrossRef]
12. Tzschoppe, M.; Haase, H.; Höhnisch, M.; Jaros, D.; Rohm, H. Using ion mobility spectrometry for screening the autoxidation of peanuts. *Food Control* **2016**, *64*, 17–21. [CrossRef]
13. Rearden, P.; Harrington, P.B. Rapid screening of precursor and degradation products of chemical warfare agents in soil by solid-phase microextraction ion mobility spectrometry (SPME–IMS). *Anal. Chim. Acta* **2005**, *545*, 13–20. [CrossRef]
14. Cook, G.W.; LaPuma, P.T.; Hook, G.L.; Eckenrode, B.A. Using gas chromatography with ion mobility spectrometry to resolve explosive compounds in the presence of interferents. *J. Forensic Sci.* **2010**, *55*, 1582–1591. [CrossRef] [PubMed]
15. Zalewska, A.; Pawłowski, W.; Tomaszewski, W. Limits of detection of explosives as determined with IMS and field asymmetric IMS vapour detectors. *Forensic Sci. Int.* **2013**, *226*, 168–172. [CrossRef]
16. Eiceman, G.A.; Karpas, Z.; Hill, H.H. *Ion Mobility Spectrometry*, 3rd ed.; CRC Press, Taylor & Francis Group: Boca Raton, FL, USA, 2016.
17. Browne, C.A.; Forbes, T.P.; Sisco, E. Detection and identification of sugar alcohol sweeteners by ion mobility spectrometry. *Anal. Methods* **2016**, *8*, 5611–5618. [CrossRef]
18. Reyes-Garcés, N.; Gómez-Ríos, G.A.; Souza Silva, E.A.; Pawliszyn, J. Coupling needle trap devices with gas chromatography-ion mobility spectrometry detection as a simple approach for on-site quantitative analysis. *J. Chromatogr. A* **2013**, *1300*, 193–198. [CrossRef] [PubMed]
19. Wu, C.; Siems, W.F.; Hill, J.H.H.; Hannan, R.M. Analytical determination of nicotine in tobacco by supercritical fluid chromatography–ion mobility detection. *J. Chromatogr. A* **1998**, *811*, 157–161. [CrossRef]
20. Raatikainen, O.; Reinikainen, V.; Minkkinen, P.; Ritvanen, T.; Muje, P.; Pursiainen, J.; Hiltunen, T.; Hyvönen, P.; Wright, A.v.; Reinikainen, S.-P. Multivariate modelling of fish freshness index based on ion mobility spectrometry measurements. *Anal. Chim. Acta* **2005**, *544*, 128–134. [CrossRef]
21. Pfeifer, K.B.; Rumpf, A.N. Measurement of ion swarm distribution functions in miniature low-temperature co-fired ceramic ion mobility spectrometer drift tubes. *Anal. Chem.* **2005**, *77*, 5215–5220. [CrossRef]

22. Laakia, J.; Adamov, A.; Jussila, M.; Pedersen, C.S.; Sysoev, A.A.; Kotiaho, T. Separation of different ion structures in atmospheric pressure photoionization-ion mobility spectrometry-mass spectrometry (APPI-IMS-MS). *J. Am. Soc. Mass Spectrom.* **2010**, *21*, 1565–1572. [CrossRef]
23. Vautz, W.; Baumbach, J.I.; Jung, J. Beer Fermentation Control Using Ion Mobility Spectrometry—Results of a Pilot Study. *J. Inst. Brew.* **2006**, *112*, 157–164. [CrossRef]
24. Sielemann, S.; Baumbach, J.I.; Schmidt, H.; Pilzecker, P. Quantitative analysis of benzene, toluene, and m-xylene with the use of a UV-ion mobility spectrometer. *Field Anal. Chem. Technol.* **2000**, *4*, 157–169. [CrossRef]
25. Parchami, R.; Kamalabadi, M.; Alizadeh, N. Determination of biogenic amines in canned fish samples using head-space solid phase microextraction based on nanostructured polypyrrole fiber coupled to modified ionization region ion mobility spectrometry. *J. Chromatogr. A* **2017**, *1481*, 37–43. [CrossRef]
26. Sheibani, A.; Tabrizchi, M.; Ghaziaskar, H.S. Determination of aflatoxins B1 and B2 using ion mobility spectrometry. *Talanta* **2008**, *75*, 233–238. [CrossRef] [PubMed]
27. Seyed Khademi, S.M.; Telgheder, U.; Valadbeigi, Y.; Ilbeigi, V.; Tabrizchi, M. Direct detection of glyphosate in drinking water using corona-discharge ion mobility spectrometry: A theoretical and experimental study. *Int. J. Mass Spectrom.* **2019**, *442*, 29–34. [CrossRef]
28. Illenseer, C.; Löhmannsröben, H.-G. Investigation of ion–molecule collisions with laser-based ion mobility spectrometry. *Phys. Chem. Chem. Phys.* **2001**, *3*, 2388–2393. [CrossRef]
29. Council Directive 2013/59/Euratom of 5 December 2013 Laying Down Basic Safety Standards for Protection against the Dangers Arising from Exposure to Ionising Radiation. *Off. J. Eur. Union* **2013**. Available online: https://eur-lex.europa.eu/legal-content/EN/TXT/HTML/?uri=CELEX:32013L0059&rid=6 (accessed on 14 May 2021).
30. Chen, T.; Qi, X.; Chen, M.; Lu, D.; Chen, B. Discrimination of Chinese yellow wine from different origins based on flavor fingerprint. *Acta Chromatogr.* **2020**, *32*, 139–144. [CrossRef]
31. Zhu, W.; Benkwitz, F.; Kilmartin, P. *Assessment of Sauvignon Blanc Aroma and Quality Gradings Based on Static Headspace-Gas Chromatography-Ion. Mobility Spectrometry (SHS-GC-IMS): Merging Analytical Chemistry with Machine Learning*; Cambridge University Press & Assessment: Cambridge, UK, 2020. [CrossRef]
32. Garrido-Delgado, R.; Del Dobao-Prieto, M.M.; Arce, L.; Valcárcel, M. Determination of volatile compounds by GC-IMS to assign the quality of virgin olive oil. *Food Chem.* **2015**, *187*, 572–579. [CrossRef]
33. Del Contreras, M.M.; Jurado-Campos, N.; Arce, L.; Arroyo-Manzanares, N. A robustness study of calibration models for olive oil classification: Targeted and non-targeted fingerprint approaches based on GC-IMS. *Food Chem.* **2019**, *288*, 315–324. [CrossRef]
34. Gerhardt, N.; Birkenmeier, M.; Sanders, D.; Rohn, S.; Weller, P. Resolution-optimized headspace gas chromatography-ion mobility spectrometry (HS-GC-IMS) for non-targeted olive oil profiling. *Anal. Bioanal. Chem.* **2017**, *409*, 3933–3942. [CrossRef]
35. Griffin, G.W.; Dzidic, I.; Carroll, D.I.; Stillwell, R.N.; Horning, E.C. Ion mass assignments based on mobility measuremets. Validity of plasma chromatographic mass mobility correlations. *Anal. Chem.* **1973**, *45*, 1204–1209. [CrossRef]
36. Pomareda, V.; Guamán, A.V.; Mohammadnejad, M.; Calvo, D.; Pardo, A.; Marco, S. Multivariate curve resolution of nonlinear ion mobility spectra followed by multivariate nonlinear calibration for quantitative prediction. *Chemom. Intell. Lab. Syst.* **2012**, *118*, 219–229. [CrossRef]
37. Ewing, R.G.; Eiceman, G.A.; Stone, J. Proton-bound cluster ions in ion mobility spectrometry. *Int. J. Mass Spectrom.* **1999**, *193*, 57–68. [CrossRef]
38. Brendel, R.; Schwolow, S.; Rohn, S.; Weller, P. Comparison of PLSR, MCR-ALS and Kernel-PLSR for the quantification of allergenic fragrance compounds in complex cosmetic products based on nonlinear 2D GC-IMS data. *Chemom. Intell. Lab. Syst.* **2020**, *205*, 104128. [CrossRef]
39. Borsdorf, H.; Rudolph, M. Gas-phase ion mobility studies of constitutional isomeric hydrocarbons using different ionization techniques. *Int. J. Mass Spectrom.* **2001**, *208*, 67–72. [CrossRef]
40. Vautz, W.; Bödeker, B.; Baumbach, J.I.; Bader, S.; Westhoff, M.; Perl, T. An implementable approach to obtain reproducible reduced ion mobility. *Int. J. Ion Mobil. Spectrom.* **2009**, *12*, 47–57. [CrossRef]
41. May, J.C.; McLean, J.A. Ion mobility-mass spectrometry: Time-dispersive instrumentation. *Anal. Chem.* **2015**, *87*, 1422–1436. [CrossRef]
42. Ewing, M.A.; Glover, M.S.; Clemmer, D.E. Hybrid ion mobility and mass spectrometry as a separation tool. *J. Chromatogr. A* **2016**, *1439*, 3–25. [CrossRef]
43. Schneider, B.B.; Nazarov, E.G.; Londry, F.; Vouros, P.; Covey, T.R. Differential mobility spectrometry/mass spectrometry history, theory, design optimization, simulations, and applications. *Mass Spectrom. Rev.* **2016**, *35*, 687–737. [CrossRef]
44. Shuai, Q.; Zhang, L.; Li, P.; Zhang, Q.; Wang, X.; Ding, X.; Zhang, W. Rapid adulteration detection for flaxseed oil using ion mobility spectrometry and chemometric methods. *Anal. Methods* **2014**, *6*, 9575–9580. [CrossRef]
45. Spietelun, A.; Pilarczyk, M.; Kloskowski, A.; Namieśnik, J. Current trends in solid-phase microextraction (SPME) fibre coatings. *Chem. Soc. Rev.* **2010**, *39*, 4524–4537. [CrossRef]
46. Ruzsanyi, V.; Mochalski, P.; Schmid, A.; Wiesenhofer, H.; Klieber, M.; Hinterhuber, H.; Amann, A. Ion mobility spectrometry for detection of skin volatiles. *J. Chromatogr. B Anal. Technol. Biomed. Life Sci.* **2012**, *911*, 84–92. [CrossRef]

47. Thomas, C.F.; Zeh, E.; Dörfel, S.; Zhang, Y.; Hinrichs, J. Studying dynamic aroma release by headspace-solid phase microextraction-gas chromatography-ion mobility spectrometry (HS-SPME-GC-IMS): Method optimization, validation, and application. *Anal. Bioanal. Chem.* **2021**, *413*, 2577–2586. [CrossRef]
48. Gómez-Ramos, M.M.; Ferrer, C.; Malato, O.; Agüera, A.; Fernández-Alba, A.R. Liquid chromatography-high-resolution mass spectrometry for pesticide residue analysis in fruit and vegetables: Screening and quantitative studies. *J. Chromatogr. A* **2013**, *1287*, 24–37. [CrossRef]
49. Baduel, C.; Mueller, J.F.; Tsai, H.; Gomez Ramos, M.J. Development of sample extraction and clean-up strategies for target and non-target analysis of environmental contaminants in biological matrices. *J. Chromatogr. A* **2015**, *1426*, 33–47. [CrossRef]
50. Krauss, M.; Singer, H.; Hollender, J. LC-high resolution MS in environmental analysis: From target screening to the identification of unknowns. *Anal. Bioanal. Chem.* **2010**, *397*, 943–951. [CrossRef]
51. Masiá, A.; Campo, J.; Blasco, C.; Picó, Y. Ultra-high performance liquid chromatography-quadrupole time-of-flight mass spectrometry to identify contaminants in water: An insight on environmental forensics. *J. Chromatogr. A* **2014**, *1345*, 86–97. [CrossRef]
52. Gerhardt, N.; Birkenmeier, M.; Schwolow, S.; Rohn, S.; Weller, P. Volatile-Compound Fingerprinting by Headspace-Gas-Chromatography Ion-Mobility Spectrometry (HS-GC-IMS) as a Benchtop Alternative to 1H NMR Profiling for Assessment of the Authenticity of Honey. *Anal. Chem.* **2018**, *90*, 1777–1785. [CrossRef]
53. Brendel, R.; Rohn, S.; Weller, P. Nitrogen monoxide as dopant for enhanced selectivity of isomeric monoterpenes in drift tube ion mobility spectrometry with 3H ionization. *Anal. Bioanal. Chem.* **2021**, *413*, 3551–3560. [CrossRef]
54. Puton, J.; Nousiainen, M.; Sillanpää, M. Ion mobility spectrometers with doped gases. *Talanta* **2008**, *76*, 978–987. [CrossRef]
55. Gabelica, V.; Marklund, E. Fundamentals of ion mobility spectrometry. *Curr. Opin. Chem. Biol.* **2018**, *42*, 51–59. [CrossRef]
56. Wang, S.; Chen, H.; Sun, B. Recent progress in food flavor analysis using gas chromatography-ion mobility spectrometry (GC-IMS). *Food Chem.* **2020**, *315*, 126158. [CrossRef] [PubMed]
57. Hernández-Mesa, M.; Ropartz, D.; García-Campaña, A.M.; Rogniaux, H.; Dervilly-Pinel, G.; Le Bizec, B. Ion Mobility Spectrometry in Food Analysis: Principles, Current Applications and Future Trends. *Molecules* **2019**, 24. [CrossRef]
58. Márquez-Sillero, I.; Cárdenas, S.; Sielemann, S.; Valcárcel, M. On-line headspace-multicapillary column-ion mobility spectrometry hyphenation as a tool for the determination of off-flavours in foods. *J. Chromatogr. A* **2014**, *1333*, 99–105. [CrossRef] [PubMed]
59. Garrido-Delgado, R.; Dobao-Prieto, M.M.; Arce, L.; Aguilar, J.; Cumplido, J.L.; Valcárcel, M. Ion mobility spectrometry versus classical physico-chemical analysis for assessing the shelf life of extra virgin olive oil according to container type and storage conditions. *J. Agric. Food Chem.* **2015**, *63*, 2179–2188. [CrossRef] [PubMed]
60. Denawaka, C.J.; Fowlis, I.A.; Dean, J.R. Evaluation and application of static headspace-multicapillary column-gas chromatography-ion mobility spectrometry for complex sample analysis. *J. Chromatogr. A* **2014**, *1338*, 136–148. [CrossRef]
61. Krisilova, E.V.; Levina, A.M.; Makarenko, V.A. Determination of the volatile compounds of vegetable oils using an ion-mobility spectrometer. *J. Anal. Chem* **2014**, *69*, 371–376. [CrossRef]
62. Danezis, G.P.; Tsagkaris, A.S.; Camin, F.; Brusic, V.; Georgiou, C.A. Food authentication: Techniques, trends & emerging approaches. *TrAC Trends Anal. Chem.* **2016**, *85*, 123–132. [CrossRef]
63. Karpas, Z. Applications of ion mobility spectrometry (IMS) in the field of foodomics. *Food Res. Int.* **2013**, *54*, 1146–1151. [CrossRef]
64. Hernández-Mesa, M.; Escourrou, A.; Monteau, F.; Le Bizec, B.; Dervilly-Pinel, G. Current applications and perspectives of ion mobility spectrometry to answer chemical food safety issues. *TrAC Trends Anal. Chem.* **2017**, *94*, 39–53. [CrossRef]
65. Charlebois, S.; Schwab, A.; Henn, R.; Huck, C.W. Food fraud: An exploratory study for measuring consumer perception towards mislabeled food products and influence on self-authentication intentions. *Trends Food Sci. Technol.* **2016**, *50*, 211–218. [CrossRef]
66. Ehmke, M.D.; Bonanno, A.; Boys, K.; Smith, T.G. Food fraud: Economic insights into the dark side of incentives. *Aust. J. Agric. Resour. Econ.* **2019**, *63*, 685–700. [CrossRef]
67. Aries, E.; Burton, J.; Carrasco, L.; de Rudder, O.; Maquet, A. Scientific Support to the Implementation of a Coordinated Control Plan with a View to Establishing the Prevalence of Fraudulent Practices in the Marketing of Honey: N° SANTE/2015/E3/JRC/SI2.706828. JRC Technical Report 2016, JRC104749. 2016, p. 38. Available online: https://ec.europa.eu/food/sites/food/files/safety/docs/oc_control-progs_honey_jrc-tech-report_2016.pdf (accessed on 12 May 2021).
68. Louveaux, J.; Maurizio, A.; Vorwohl, G. Methods of Melissopalynology. *BEE World* **1978**, *59*, 139–157. [CrossRef]
69. Cavazza, A.; Corradini, C.; Musci, M.; Salvadeo, P. High-performance liquid chromatographic phenolic compound fingerprint for authenticity assessment of honey. *J. Sci. Food Agric.* **2013**, *93*, 1169–1175. [CrossRef]
70. Zhou, J.; Yao, L.; Li, Y.; Chen, L.; Wu, L.; Zhao, J. Floral classification of honey using liquid chromatography-diode array detection-tandem mass spectrometry and chemometric analysis. *Food Chem.* **2014**, *145*, 941–949. [CrossRef] [PubMed]
71. Karabagias, I.K.; Badeka, A.; Kontakos, S.; Karabournioti, S.; Kontominas, M.G. Characterisation and classification of Greek pine honeys according to their geographical origin based on volatiles, physicochemical parameters and chemometrics. *Food Chem.* **2014**, *146*, 548–557. [CrossRef]
72. Robotti, E.; Campo, F.; Riviello, M.; Bobba, M.; Manfredi, M.; Mazzucco, E.; Gosetti, F.; Calabrese, G.; Sangiorgi, E.; Marengo, E. Optimization of the Extraction of the Volatile Fraction from Honey Samples by SPME-GC-MS, Experimental Design, and Multivariate Target Functions. *J. Chem.* **2017**, *2017*, 6437857. [CrossRef]

73. Jandrić, Z.; Haughey, S.A.; Frew, R.D.; McComb, K.; Galvin-King, P.; Elliott, C.T.; Cannavan, A. Discrimination of honey of different floral origins by a combination of various chemical parameters. *Food Chem.* **2015**, *189*, 52–59. [CrossRef] [PubMed]
74. Woodcock, T.; Fagan, C.C.; O'Donnell, C.P.; Downey, G. Application of Near and Mid-Infrared Spectroscopy to Determine Cheese Quality and Authenticity. *Food Bioprocess Technol.* **2008**, *1*, 117–129. [CrossRef]
75. Gok, S.; Severcan, M.; Goormaghtigh, E.; Kandemir, I.; Severcan, F. Differentiation of Anatolian honey samples from different botanical origins by ATR-FTIR spectroscopy using multivariate analysis. *Food Chem.* **2015**, *170*, 234–240. [CrossRef]
76. Boffo, E.F.; Tavares, L.A.; Tobias, A.C.; Ferreira, M.M.; Ferreira, A.G. Identification of components of Brazilian honey by 1H NMR and classification of its botanical origin by chemometric methods. *LWT* **2012**, *49*, 55–63. [CrossRef]
77. Schievano, E.; Finotello, C.; Uddin, J.; Mammi, S.; Piana, L. Objective Definition of Monofloral and Polyfloral Honeys Based on NMR Metabolomic Profiling. *J. Agric. Food Chem.* **2016**, *64*, 3645–3652. [CrossRef]
78. Garrido-Delgado, R.; Arce, L.; Valcárcel, M. Multi-capillary column-ion mobility spectrometry: A potential screening system to differentiate virgin olive oils. *Anal. Bioanal. Chem.* **2012**, *402*, 489–498. [CrossRef]
79. Gerhardt, N.; Schwolow, S.; Rohn, S.; Pérez-Cacho, P.R.; Galán-Soldevilla, H.; Arce, L.; Weller, P. Quality assessment of olive oils based on temperature-ramped HS-GC-IMS and sensory evaluation: Comparison of different processing approaches by LDA, kNN, and SVM. *Food Chem.* **2019**, *278*, 720–728. [CrossRef]
80. Garrido-Delgado, R.; Mercader-Trejo, F.; Sielemann, S.; de Bruyn, W.; Arce, L.; Valcárcel, M. Direct classification of olive oils by using two types of ion mobility spectrometers. *Anal. Chim. Acta* **2011**, *696*, 108–115. [CrossRef]
81. Schwolow, S.; Gerhardt, N.; Rohn, S.; Weller, P. Data fusion of GC-IMS data and FT-MIR spectra for the authentication of olive oils and honeys-is it worth to go the extra mile? *Anal. Bioanal. Chem.* **2019**, *411*, 6005–6019. [CrossRef] [PubMed]
82. Wang, X.; Rogers, K.M.; Li, Y.; Yang, S.; Chen, L.; Zhou, J. Untargeted and Targeted Discrimination of Honey Collected by Apis cerana and Apis mellifera Based on Volatiles Using HS-GC-IMS and HS-SPME-GC-MS. *J. Agric. Food Chem.* **2019**, *67*, 12144–12152. [CrossRef] [PubMed]
83. Wang, X.; Yang, S.; He, J.; Chen, L.; Zhang, J.; Jin, Y.; Zhou, J.; Zhang, Y. A green triple-locked strategy based on volatile-compound imaging, chemometrics, and markers to discriminate winter honey and sapium honey using headspace gas chromatography-ion mobility spectrometry. *Food Res. Int.* **2019**, *119*, 960–967. [CrossRef]
84. Arroyo-Manzanares, N.; García-Nicolás, M.; Castell, A.; Campillo, N.; Viñas, P.; López-García, I.; Hernández-Córdoba, M. Untargeted headspace gas chromatography–Ion mobility spectrometry analysis for detection of adulterated honey. *Talanta* **2019**, *205*, 120123. [CrossRef]
85. Cavanna, D.; Zanardi, S.; Dall'Asta, C.; Suman, M. Ion mobility spectrometry coupled to gas chromatography: A rapid tool to assess eggs freshness. *Food Chem.* **2019**, *271*, 691–696. [CrossRef] [PubMed]
86. Gu, S.; Chen, W.; Wang, Z.; Wang, J.; Huo, Y. Rapid detection of Aspergillus spp. infection levels on milled rice by headspace-gas chromatography ion-mobility spectrometry (HS-GC-IMS) and E-nose. *LWT* **2020**, *132*, 109758. [CrossRef]
87. Gu, S.; Chen, W.; Wang, Z.; Wang, J. Rapid determination of potential aflatoxigenic fungi contamination on peanut kernels during storage by data fusion of HS-GC-IMS and fluorescence spectroscopy. *Postharvest Biol. Technol.* **2021**, *171*, 111361. [CrossRef]
88. Gu, S.; Wang, Z.; Chen, W.; Wang, J. Targeted versus Nontargeted Green Strategies Based on Headspace-Gas Chromatography-Ion Mobility Spectrometry Combined with Chemometrics for Rapid Detection of Fungal Contamination on Wheat Kernels. *J. Agric. Food Chem.* **2020**, *68*, 12719–12728. [CrossRef] [PubMed]
89. Arroyo-Manzanares, N.; Martín-Gómez, A.; Jurado-Campos, N.; Garrido-Delgado, R.; Arce, C.; Arce, L. Target vs spectral fingerprint data analysis of Iberian ham samples for avoiding labelling fraud using headspace–gas chromatography–ion mobility spectrometry. *Food Chem.* **2018**, *246*, 65–73. [CrossRef] [PubMed]
90. Brendel, R.; Schwolow, S.; Rohn, S.; Weller, P. Volatilomic Profiling of Citrus Juices by Dual-Detection HS-GC-MS-IMS and Machine Learning-An Alternative Authentication Approach. *J. Agric. Food Chem.* **2021**, *69*, 1727–1738. [CrossRef] [PubMed]
91. Vega-Márquez, B.; Nepomuceno-Chamorro, I.; Jurado-Campos, N.; Rubio-Escudero, C. Deep Learning Techniques to Improve the Performance of Olive Oil Classification. *Front. Chem.* **2019**, *7*, 929. [CrossRef]
92. Yuan, Z.-Y.; Li, J.; Zhou, X.-J.; Wu, M.-H.; Li, L.; Pei, G.; Chen, N.-H.; Liu, K.-L.; Xie, M.-Z.; Huang, H.-Y. HS-GC-IMS-Based metabonomics study of Baihe Jizihuang Tang in a rat model of chronic unpredictable mild stress. *J. Chromatogr. B Anal. Technol. Biomed. Life Sci.* **2020**, *1148*, 122143. [CrossRef]
93. Antignac, J.-P.; Courant, F.; Pinel, G.; Bichon, E.; Monteau, F.; Elliott, C.; Le Bizec, B. Mass spectrometry-based metabolomics applied to the chemical safety of food. *TrAC Trends Anal. Chem.* **2011**, *30*, 292–301. [CrossRef]
94. Snow, N.H.; Slack, G.C. Head-space analysis in modern gas chromatography. *TrAC Trends Anal. Chem.* **2002**, *21*, 608–617. [CrossRef]
95. Soria, A.C.; García-Sarrió, M.J.; Sanz, M.L. Volatile sampling by headspace techniques. *TrAC Trends Anal. Chem.* **2015**, *71*, 85–99. [CrossRef]
96. Poole, C.F. (Ed.) *Liquid-Phase Extraction*; Handbooks in Separation Science; Elsevier: Amsterdam, The Netherlands, 2020.
97. Guo, Z.; Zhu, Z.; Huang, S.; Wang, J. Non-targeted screening of pesticides for food analysis using liquid chromatography high-resolution mass spectrometry-a review. *Food Addit. Contam. Part A* **2020**, *37*, 1180–1201. [CrossRef]
98. Souza Silva, E.A.; Risticevic, S.; Pawliszyn, J. Recent trends in SPME concerning sorbent materials, configurations and in vivo applications. *TrAC Trends Anal. Chem.* **2013**, *43*, 24–36. [CrossRef]

99. Chen, T.; Chen, X.; Lu, D.; Chen, B. Detection of Adulteration in Canola Oil by Using GC-IMS and Chemometric Analysis. *Int. J. Anal. Chem.* **2018**, *2018*, 3160265. [CrossRef]
100. Baumbach, J.I.; Eiceman, G.A.; Klockow, D.; Sielemann, S.; Irmer, A.V. Exploration of a Multicapillary Column for Use in Elevated Speed Gas Chromatography. *Int. J. Environ. Anal. Chem.* **1997**, *66*, 225–239. [CrossRef]
101. Cureton, E.E.; D'Agostino, R.B. *Factor Analysis*; Psychology Press: New York, NY, USA, 2013. [CrossRef]
102. Zheng, P.; de Harrington, P.B.; Davis, D.M. Quantitative analysis of volatile organic compounds using ion mobility spectrometry and cascade correlation neural networks. *Chemom. Intell. Lab. Syst.* **1996**, *33*, 121–132. [CrossRef]
103. Kaur-Atwal, G.; O'Connor, G.; Aksenov, A.A.; Bocos-Bintintan, V.; Paul Thomas, C.L.; Creaser, C.S. Chemical standards for ion mobility spectrometry: A review. *Int. J. Ion Mobil. Spectrom.* **2009**, *12*, 1–14. [CrossRef]
104. Liedtke, S.; Seifert, L.; Ahlmann, N.; Hariharan, C.; Franzke, J.; Vautz, W. Coupling laser desorption with gas chromatography and ion mobility spectrometry for improved olive oil characterisation. *Food Chem.* **2018**, *255*, 323–331. [CrossRef] [PubMed]
105. Borsdorf, H.; Neitsch, K. Ion mobility spectra of cyclic and aliphatic hydrocarbons with different substituents. *Int. J. Ion Mobil. Spectrom.* **2009**, *12*, 39–46. [CrossRef]
106. Budzyńska, E.; Sielemann, S.; Puton, J.; Surminski, A.L.R.M. Analysis of e-liquids for electronic cigarettes using GC-IMS/MS with headspace sampling. *Talanta* **2020**, *209*, 120594. [CrossRef]
107. Kessler, W. *Multivariate Datenanalyse: Für die Pharma-, bio- und Prozessanalytik: Ien Lehrbuch*; Wiley-VCH: Weinheim, Germany, 2007.
108. Maimon, O.; Rokach, L. *Data Mining and Knowledge Discovery Handbook*; Springer: New York, NY, USA, 2005.
109. Granato, D.; Santos, J.S.; Escher, G.B.; Ferreira, B.L.; Maggio, R.M. Use of principal component analysis (PCA) and hierarchical cluster analysis (HCA) for multivariate association between bioactive compounds and functional properties in foods: A critical perspective. *Trends Food Sci. Technol.* **2018**, *72*, 83–90. [CrossRef]
110. Wold, S.; Geladi, P.; Esbensen, K.; Öhman, J. Multi-way principal components-and PLS-analysis. *J. Chemom.* **1987**, *1*, 41–56. [CrossRef]
111. Bro, R. PARAFAC. Tutorial and applications. *Chemom. Intell. Lab. Syst.* **1997**, *38*, 149–171. [CrossRef]
112. Kroonenberg, P.M.; Ten Berge, J.M.F. The equivalence of Tucker3 and Parafac models with two components. *Chemom. Intell. Lab. Syst.* **2011**, *106*, 21–26. [CrossRef]
113. Cortes, C.; Vapnik, V. Support-vector networks. *Mach. Learn.* **1995**, *20*, 273–297. [CrossRef]
114. Varmuza, K.; Filzmoser, P. *Introduction to Multivariate Statistical Analysis in Chemometrics*; CRC Press: Boca Raton, FL, USA, 2009.
115. Chen, Y.-Y.; Lin, Y.-H.; Kung, C.-C.; Chung, M.-H.; Yen, I.-H. Design and Implementation of Cloud Analytics-Assisted Smart Power Meters Considering Advanced Artificial Intelligence as Edge Analytics in Demand-Side Management for Smart Homes. *Sensors* **2019**, *19*, 2047. [CrossRef]
116. Torkashvand, A.M.; Ahmadi, A.; Nikravesh, N.L. Prediction of kiwifruit firmness using fruit mineral nutrient concentration by artificial neural network (ANN) and multiple linear regressions (MLR). *J. Integr. Agric.* **2017**, *16*, 1634–1644. [CrossRef]
117. Scheinemann, A.; Sielemann, S.; Walter, J.; Doll, T. Evaluation Strategies for Coupled GC-IMS Measurement including the Systematic Use of Parametrized ANN. *Open J. Appl. Sci.* **2012**, *2*, 257–266. [CrossRef]
118. Rosipal, R. Nonlinear Partial Least Squares An Overview. In *Chemoinformatics and Advanced Machine Learning Perspectives*; Lodhi, H., Yamanishi, Y., Eds.; IGI Global: Hershey, PA, USA, 2011; pp. 169–189. [CrossRef]
119. De Juan, A.; Jaumot, J.; Tauler, R. Multivariate Curve Resolution (MCR). Solving the mixture analysis problem. *Anal. Methods* **2014**, *6*, 4964–4976. [CrossRef]
120. Liu, P.; Duan, J.; Wang, P.; Qian, D.; Guo, J.; Shang, E.; Su, S.; Tang, Y.; Tang, Z. Biomarkers of primary dysmenorrhea and herbal formula intervention: An exploratory metabonomics study of blood plasma and urine. *Mol. Biosyst.* **2013**, *9*, 77–87. [CrossRef] [PubMed]
121. Xia, J.; Sinelnikov, I.V.; Han, B.; Wishart, D.S. MetaboAnalyst 3.0—making metabolomics more meaningful. *Nucleic Acids Res.* **2015**, *43*, W251–W257. [CrossRef]
122. Smith, C.A.; Want, E.J.; O'Maille, G.; Abagyan, R.; Siuzdak, G. XCMS: Processing mass spectrometry data for metabolite profiling using nonlinear peak alignment, matching, and identification. *Anal. Chem.* **2006**, *78*, 779–787. [CrossRef]
123. Zhang, X.; Dai, Z.; Fan, X.; Liu, M.; Ma, J.; Shang, W.; Liu, J.; Strappe, P.; Blanchard, C.; Zhou, Z. A study on volatile metabolites screening by HS-SPME-GC-MS and HS-GC-IMS for discrimination and characterization of white and yellowed rice. *Cereal Chem.* **2020**, *97*, 496–504. [CrossRef]
124. Chen, T.; Liu, C.; Meng, L.; Lu, D.; Chen, B.; Cheng, Q. Early warning of rice mildew based on gas chromatography-ion mobility spectrometry technology and chemometrics. *Food Meas.* **2021**, *15*, 1939–1948. [CrossRef]
125. Lv, W.; Lin, T.; Ren, Z.; Jiang, Y.; Zhang, J.; Bi, F.; Gu, L.; Hou, H.; He, J. Rapid discrimination of Citrus reticulata 'Chachi' by headspace-gas chromatography-ion mobility spectrometry fingerprints combined with principal component analysis. *Food Res. Int.* **2020**, *131*, 108985. [CrossRef] [PubMed]
126. Li, M.; Yang, R.; Zhang, H.; Wang, S.; Chen, D.; Lin, S. Development of a flavor fingerprint by HS-GC-IMS with PCA for volatile compounds of Tricholoma matsutake Singer. *Food Chem.* **2019**, *290*, 32–39. [CrossRef]
127. Brendel, R.; Schwolow, S.; Rohn, S.; Weller, P. Gas-phase volatilomic approaches for quality control of brewing hops based on simultaneous GC-MS-IMS and machine learning. *Anal. Bioanal. Chem.* **2020**, *412*, 7085–7097. [CrossRef] [PubMed]

128. Pu, D.; Zhang, H.; Zhang, Y.; Sun, B.; Ren, F.; Chen, H.; He, J. Characterization of the aroma release and perception of white bread during oral processing by gas chromatography-ion mobility spectrometry and temporal dominance of sensations analysis. *Food Res. Int.* **2019**, *123*, 612–622. [CrossRef]
129. Tang, Z.-S.; Zeng, X.-A.; Brennan, M.A.; Han, Z.; Niu, D.; Huo, Y. Characterization of aroma profile and characteristic aromas during lychee wine fermentation. *J. Food Process. Preserv.* **2019**, *43*, e14003. [CrossRef]
130. Gerhardt, N.; Birkenmeier, M.; Kuballa, T.; Ohmenhaeuser, M.; Rohn, S.; Weller, P. Differentiation of the botanical origin of honeys by fast, non-targeted 1H-NMR profiling and chemometric tools as alternative authenticity screening tool. In Proceedings of the XIII International Conference on the Applications of Magnetic Resonance in Food Science, Karlsruhe, Germany, 7–10 June 2016; IM Publications: Charlton, UK, 2016; p. 33. [CrossRef]
131. Guo, Y.; Chen, D.; Dong, Y.; Ju, H.; Wu, C.; Lin, S. Characteristic volatiles fingerprints and changes of volatile compounds in fresh and dried Tricholoma matsutake Singer by HS-GC-IMS and HS-SPME-GC-MS. *J. Chromatogr. B Anal. Technol. Biomed. Life Sci.* **2018**, *1099*, 46–55. [CrossRef]
132. Klein, L.A. *Sensor and Data Fusion: A Tool for Information Assessment and Decision Making*; SPIE Press: Bellingham, WA, USA, 2010.
133. Borràs, E.; Ferré, J.; Boqué, R.; Mestres, M.; Aceña, L.; Busto, O. Data fusion methodologies for food and beverage authentication and quality assessment—a review. *Anal. Chim. Acta* **2015**, *891*, 1–14. [CrossRef] [PubMed]
134. Kuncheva, L.I. *Combining Pattern Classifiers: Methods and Algorithms*; John Wiley & Sons: New York, NY, USA, 2010.
135. Martín-Gómez, A.; Arroyo-Manzanares, N.; Rodríguez-Estévez, V.; Arce, L. Use of a non-destructive sampling method for characterization of Iberian cured ham breed and feeding regime using GC-IMS. *Meat Sci.* **2019**, *152*, 146–154. [CrossRef] [PubMed]
136. Chatterji, B.N. Feature Extraction Methods for Character Recognition. *IETE Tech. Rev.* **1986**, *3*, 9–22. [CrossRef]
137. Lowe, D.G. Distinctive Image Features from Scale-Invariant Keypoints. *Int. J. Comput. Vis.* **2004**, *60*, 91–110. [CrossRef]
138. Chen, T.; Qi, X.; Chen, M.; Chen, B. Gas Chromatography-Ion Mobility Spectrometry Detection of Odor Fingerprint as Markers of Rapeseed Oil Refined Grade. *J. Anal. Methods Chem.* **2019**, *2019*, 3163204. [CrossRef] [PubMed]
139. Chen, T.; Qi, X.; Lu, D.; Chen, B. Gas chromatography-ion mobility spectrometric classification of vegetable oils based on digital image processing. *Food Meas.* **2019**, *13*, 1973–1979. [CrossRef]

 MDPI

Review

Nontargeted Screening Using Gas Chromatography–Atmospheric Pressure Ionization Mass Spectrometry: Recent Trends and Emerging Potential

Xiaolei Li [1], Frank L. Dorman [2,3], Paul A. Helm [4], Sonya Kleywegt [4], André Simpson [5], Myrna J. Simpson [5] and Karl J. Jobst [1,*]

1 Department of Chemistry, Memorial University of Newfoundland and Labrador, St. John's, NL A1C 5S7, Canada; xiaolei.li@mun.ca
2 Department of Chemistry, The Pennsylvania State University, University Park, State College, PA 16802, USA; fld3@psu.com
3 Department of Chemistry, Dartmouth College, Hannover, NH 03755, USA
4 Ontario Ministry of Environment, Conservation and Parks, Toronto, ON M4V 1M2, Canada; Paul.Helm@Ontario.ca (P.A.H.); Sonya.Kleywegt@ontario.ca (S.K.)
5 Departments of Chemistry and Physical & Environmental Sciences, University of Toronto, Toronto, ON M1C 1A4, Canada; andre.simpson@utoronto.ca (A.S.); myrna.simpson@utoronto.ca (M.J.S.)
* Correspondence: kjobst@mun.ca

Abstract: Gas chromatography–high-resolution mass spectrometry (GC–HRMS) is a powerful nontargeted screening technique that promises to accelerate the identification of environmental pollutants. Currently, most GC–HRMS instruments are equipped with electron ionization (EI), but atmospheric pressure ionization (API) ion sources have attracted renewed interest because: (i) collisional cooling at atmospheric pressure minimizes fragmentation, resulting in an increased yield of molecular ions for elemental composition determination and improved detection limits; (ii) a wide range of sophisticated tandem (ion mobility) mass spectrometers can be easily adapted for operation with GC–API; and (iii) the conditions of an atmospheric pressure ion source can promote structure diagnostic ion–molecule reactions that are otherwise difficult to perform using conventional GC–MS instrumentation. This literature review addresses the merits of GC–API for nontargeted screening while summarizing recent applications using various GC–API techniques. One perceived drawback of GC–API is the paucity of spectral libraries that can be used to guide structure elucidation. Herein, novel data acquisition, deconvolution and spectral prediction tools will be reviewed. With continued development, it is anticipated that API may eventually supplant EI as the de facto GC–MS ion source used to identify unknowns.

Keywords: GC–API; GC–APCI; GC–APLI; GC–APPI; GC–MS; persistent organic pollutants; nontargeted screening; computational mass spectrometry

Citation: Li, X.; Dorman, F.L.; Helm, P.A.; Kleywegt, S.; Simpson, A.; Simpson, M.J.; Jobst, K.J. Nontargeted Screening Using Gas Chromatography–Atmospheric Pressure Ionization Mass Spectrometry: Recent Trends and Emerging Potential. *Molecules* **2021**, *26*, 6911. https://doi.org/10.3390/molecules26226911

Academic Editor: Thomas Letzel

Received: 26 September 2021
Accepted: 12 November 2021
Published: 16 November 2021

1. Introduction

Chemistry is essential to the modern world, producing molecules and materials required in all facets of society. Tens of thousands of chemical substances, representing millions of individual chemical compounds, have now been introduced to the global market and this number is increasing [1]. Since 1982, the number of substances registered in the US Toxic Substances Control Act Chemical Substance Inventory (TSCA) has grown from 62,000 to over 86,557 [2]. According to the United Nations Global Chemicals Outlook [3], the total volume of chemicals is expected to increase at a rate that outpaces population growth over the next decade. Concerns that some of these chemicals can persist in the environment, bioaccumulate and adversely impact human health have led to international efforts to restrict the (un)intentional release of 28 groups of hazardous chemicals, i.e., persistent organic pollutants (POPs) [4]. However, the number of unregulated POPs may be much larger. A recent evaluation of substances compiled in the TSCA and other national chemical inventories has resulted in a list of 3421 chemical substances that may be persistent and bioaccumulative, and have so far evaded detection in the environment. The identification of unknown pollutants, most appropriately using mass spectrometry, is the critical first step to evaluating their potential harm to the environment, establishing policies and guidelines to limit exposure, and preventing global contamination. Such experiments have been coined nontargeted screening (NTS) [5]. Compared to target screening and suspect screening that involve completely or partly known prior information of the compounds in the sample, nontargeted screening is performed without knowledge of exact mass, isotope, adduct or fragmentation behaviour [6].

Gas chromatography– and liquid chromatography–high-resolution mass spectrometry (GC–HRMS and LC–HRMS) are powerful, complementary techniques for NTS of persistent and bioaccumulative organic pollutants. The need for both techniques is underlined by the results of a recent interlaboratory study led by Rostkowski et al. [7,8]. The analysis of an indoor dust sample by over 20 different laboratories using both GC–HRMS and LC–HRMS resulted in the tentative identification of 2350 compounds. Of these compounds, approximately half were identified by GC–HRMS and only 5% of the compounds were detected using both GC and LC. This is because the compounds amenable to GC or LC separation often have different volatility, polarity and ionization behavior. While LC–HRMS has attracted more recent attention, comprehensive NTS of environmental samples cannot be performed without GC–MS.

Most GC–MS instruments employ electron ionization (EI), which produces highly reproducible, structure-diagnostic mass spectra of organic pollutants. A pollutant's identity can be established by searching databases of experimental EI spectra (70 eV) such as the NIST Mass Spectral Library [9]. The use of lower ionization energies in EI typically results in a significant decrease in sensitivity and for this reason is rarely used. However, extensive fragmentation under EI conditions can also produce mass spectra in which the molecular ion is absent, thereby confounding elemental composition determination and structure elucidation. This drawback can potentially be solved by lowering the electron energy in EI or using other vacuum ionization techniques such as chemical ionization (CI) [10], photoionization (PI) [11] and field ionization (FI) [12], which impart significantly less energy to the analyte molecules. The past 30 years have also witnessed the advent of atmospheric pressure ionization techniques and (hybrid) mass analyzers, whose development was primarily driven by the need for LC–MS and direct analysis applications. An unintended consequence is that the concept of performing "GC–MS on an LC–MS instrument" [13] has attracted renewed interest. GC and comprehensive two-dimensional gas chromatography (GC×GC) have recently been hyphenated with a variety of atmospheric pressure ionization sources, including: atmospheric pressure chemical ionization (APCI) [14]; atmospheric pressure photoionization (APPI) [15]; atmospheric pressure laser ionization (APLI) [16]; and electrospray ionization (ESI) [17]. Ionization at elevated pressures offers several advantages. Collisional cooling can often minimize fragmentation and increase the yield of (quasi)molecular ions [18]. The resulting detection limits can be approximately 10~100 times lower than conventional GC–MS experiments [19]. Modifying an LC–MS instrument to perform GC–MS may also reduce costs by minimizing the number of specialized instruments required for analysis. The most attractive benefit, however, may be the fact that GC–API can be adapted to mass analyzers [20] and ion mobility mass spectrometers [21] that result in novel configurations that could potentially better tackle complex (environmental) mixtures. GC–APCI has been successfully applied to complex mixtures, such as phenolic compounds in olive oil [22], metabolites of avocados [23], fatty acids in fish [24], exhaled volatile organic compounds [25], and steroid hormone profiles in human breast adipose tissue [26].

In this review, we summarize the recent developments and applications of GC–API reported during the last five years, using SciFinder with the keywords "GC–APCI", "APGC", "GC–APPI", "GC–APLI" or "GC–ESI". The ion source designs, geometries and ionization mechanisms have been reviewed elsewhere [19,27]. Our contribution will instead focus on the application of various GC–API techniques to identify unknown pollutants, as well as the computational techniques being developed to predict mass spectra and aid in the interpretation of NTS data.

2. Gas Chromatography–Atmospheric Pressure Ionization Techniques

Gas chromatography–atmospheric pressure chemical ionization (GC–APCI) was first developed in the 1970s [14]. It saw only limited usage until McEwan & Mckay [28] and Schiewek et al. [29] adapted the approach to commercially available LC–MS instruments in the early 2000s. The ensuing years witnessed the development of various GC–API ion sources adapted from LC–MS applications. Atmospheric pressure photoionization (APPI), first developed by Robb et al. [30] and Syage et al. [31] as an LC–MS ion source, was later adapted by Revelsky et al. [15] for GC–MS. Schiewek et al. [16] developed an atmospheric pressure laser ionization (APLI) source for GC–MS in 2007 and Brenner et al. [17] were the first to employ an electrospray ionization (ESI) emitter to promote ionization of GC effluent. In 2005 [32], Cody introduced Direct Analysis in Real-Time (DART), a technique that makes use of Penning ionization, wherein a molecule is ionized through collision with an electronically excited (metastable) atom. In 2008, the DART ion source was first hyphenated with GC [33], although a similar technique called metastable atom bombardment (MAB) had been developed as a vacuum ion source almost a decade earlier [34]. Dielectric barrier discharge (DBD) ionization is a recent innovation in GC–MS ion source development [35]. DBD ionization occurs in a low-temperature plasma and this approach has been developed in parallel for LC, GC and direct analysis modes of operation. The ion source designs, geometries, and mechanisms of various GC–API techniques have been reviewed elsewhere [11,18,27]. In the following sections, the techniques are briefly summarized, and their advantages and limitations are discussed in light of the results of recent publications in Table 1.

Table 1. Recent studies that employ GC–API–MS techniques for (non)targeted analysis of environmental pollutants.

	Sample	Detector	(Non)Targeted Chemicals	Method Merits	Ref.
APCI	Food packaging materials	QTOF, HP-5MS	Acrylic adhesives including 2-methyl-1,2-thiazol-3(2*H*)-one, 5-chloro-2-methyl-1,2-thiazol-3(2*H*)-one and 1,2-benzothiazol-3(2*H*)-one		[36]
APCI	Polyurethane foam disks (PUFs), food, and marine samples	Xevo TQ-S QQQ	Hexabromocyclododecane	IDL: 0.10 pg/μL RSD: <7%	[37]
APCI	Surface water, groundwater, wastewater		Pesticides, polycyclic aromatic hydrocarbons (PAHs), polychlorinated biphenyls (PCBs), Polybrominated diphenyl ethers (PBDEs), fragrances, musks, antimicrobials, insect repellents, UV filters, polychloronaphthalenes (PCNs)		[38]
APCI	Indoor air sample	QTOF, HP-1-MS	volatile, intermediate-volatility, and semivolatile organic compound	LOD: 10~100 ppq	[39]
APCI	Chinese mitten crab food webs	Xevo TQ-XS QQQ, DB-5MS	PCBs (mono-to deca-) and polychlorinated dibenzo-p-dioxins/dibenzofurans (PCDD/Fs)	RSD: PCBs: 3.4%~15.5%; PCDD/Fs: 1.7%~7.9% LOD: PCBs: 0.021~0.150 pg/mL; PCDD/Fs: 0.051~0.237 pg/mL	[40]

Table 1. *Cont.*

	Sample	Detector	(Non)Targeted Chemicals	Method Merits	Ref.
APCI	Standard Reference Material (SRM 2585) of household dust	QTOF	191 POPs including PCBs and agricultural drug residues, such as chlordane and degradation products of DDT and Fentichlor, polychlorinated and polybrominated diphenyl ethers (PCDEs and PBDEs), other brominated flame retardants such as tetrabromobisphenol A (TBBPA) and bis(2-ethylhexyl) tetrabromophthalate (BEHTBP), chlorine-containing organophosphate flame retardants tris(1,3-dichloro-2-propyl)phosphate (2 isomers), tris(2-chloroisopropyl)phosphate and tris(2-chloroethyl)phosphate		[41]
APCI	Electronic waste dust	Q-TOF DB5-HT	52 brominated, chlorinated, and organophosphorus compounds identified by suspect screening; 15 unique elemental compositions identified using NTS with 17 chemicals confirmed using standards		[20]
APCI	Low sulfonate lignin	Q-TOF TOF	59 lignin pyrolysis products were positively identified, with 10 chemicals confirmed using standards		[42]
APCI	Urine Blood	QTOF DB- 5MS	Illicit psychostimulant drugs		[43,44]
APGC	Low-temperature coal tar sample and its distillation products	TQ-S DB-35 MS	Phenolic compounds (phenols, indanols, naphthols, and benzenediols)		[45]
APGC	Human serums	Xevo TQ-S	Organochlorine pesticides (OCPs) and PCBs	RSD: <15%	[46]
APCI	Urine samples	QTOF	α-pyrrolidinovalerophenone metabolites		[47]
APGC	Food	FT-ICR, Rtx-1614	Halogenated flame retardants (HFRs)	Recovery: 59~115%; RSD: 5–15%; IQL: 1~5 pg/g; MQL: 0.002~0.04 ng/g	[48]
APCI	Urine	QQQ, HP Ultra 1	Exogenous androgenic anabolic steroids	RSD: 15–25% Most LOD: below 0.5 ng/ mL	[49]
APGC	Seal and egg samples	Xevo TQ-S QQQ, Rtx-1614	PBDEs, their methoxylated derivatives (MeO-PBDEs) and other emerging (brominated flame retardants) BFRs	RSD: <1. IDL: emerging BFRs, BDE 209 and MeO-PBDEs mixtures: 0.075~0.1 pg/μL; Br1–9 PBDEs mixtures: 0.625~6.25 pg/μL	[50]
APGC	Air fine particulate matter (PM 2.5)	Xevo TQ-S QQQ	Nitro-polyaromatic hydrocarbons	IDL: (0.20~2.18 pg/mL MDL: 0.001~0.015 pg/m^3; Recovery: 70%~120%	[51]
APGC	Urine samples	Xevo G2-XS QTOF, DB-17+ custom MXT	1-Hydroxypyrene, 3-hydroxyphenanthrene, 9-hydroxyfluorene	1-Hydroxypyrene LOD: 0.64 ng/L, LOQ 2.16 ng/L; average CV: 11.5%	[52]
APGC	Simulated burn study samples (household and electronics), Particulate matter coating the firefighter's helmets	Xevo TQ-S, Rtx Dioxin-2	Polyhalogenated dibenzo-p-dioxins/dibenzofurans (PXDD/Fs) and polybrominated dibenzo-p-dioxins/dibenzofurans (PBDD/Fs)	total levels of each halogenated homologue group: parts per billion	[53]
APGC	Fish, dust	Xevo TQ-S, Rtx Dioxin-2		Soil: MDL: 0.15~1.4 pg/g, RSD < 11% Fish: 0.21~2.0 pg/g, RSD < 33%	[54]

Table 1. *Cont.*

	Sample	Detector	(Non)Targeted Chemicals	Method Merits	Ref.
APGC	Food and feed	Orbitrap, DB-5MS	Polychlorinated dioxins and polychlorinated biphenyls	S/N: 753 for 40 fg on column Average RSD: 9.8%	[55]
APCI	Dust	Xevo G2-XS qTOF, DB-5 HT	40 PBDEs and 25 emerging HFRs	LOD: HFRs: 0.65 (0.016~9.1) pg/ μL; PBDE: 0.17 (0.0123~2.5) pg μL	[56]
APPI	Drug solutions	HR-LTQ Orbitrap, SLB-5 ms	Triazines and organophosphorus pesticides, PAHs, Drugs (diazepam and methadone)	Pesticide: average 3 pg/mL PAH: 0.1 pg/mL Drugs: average 30 pg/mL	[57]
APPI	Derivazation	oaTOF	Amines, alcohols, carboxylic acids	LOD: pmol~attmol	[58]
APPI	River water, tap water	HRMS (Q-Orbitrap)	fluorotelomer olefins (FTOs), fluorotelomer alcohols (FTOHs), fluoroctanesulfonamides (FOSAs) and sulfonamidoethanols (FOSEs)	LOD: 0.02–15 ng/L; RSD% < 11	[59]
APPI	Fruit and vegetable samples	QTOF	416 pesticides 416 pesticides		[60]
APLI	Human urine	TOF, DB-35	*Trans-anti*-benzo[*a*]pyrene-tetraol (BaP-tetraol) (PAH biomarker)	IOD of 0.5 fg	[61]
APLI	Rocks	HR TOF, RXI-PAH	Triaromatic steroids	LOD: retene: 25 fg on column	[62]
APLI	Coastal and harbor water	HR TOF, RXI-PAH	48 PAHs (alkylated PAHs in suspected target analysis)	Recovery rate: 60.7% to 157.0%, mean 92.1%	[63]
APLI	Reference materials (urban dust, organics in marine sediment, fresh water harbor sediment, and contaminated soil from a former gas plant site) and environment samples (bituminous coal, suspended particulate matter from river and pine needles)	HR TOF, RXI-PAH	59 PAHs	Recovery: 34%~102%, median, 80% mean 78% LODs: 5~50 fg/μL	[64]
ESI	Human urine	LTQ Orbitrap QQQ Ultra-1	Trimethylsilyl (TMS) derivatives of steroids	LOD 0.5~10 ng/mL	[65]
ESI	Soil	QQQ, DB-EUPAH	PAHs	LOD 0.002~10 μg/mL	[66]

LOD: limit of detection. MD(Q)L: method detection (qualification) limit. ID(Q)L: instrument detection (qualification) limit. CV: coefficient of variance.

2.1. Atmospheric Pressure Chemical Ionization (APCI)

Ionization under APCI conditions is a seemingly straightforward process. When GC effluent exits the column, a high flow of make-up gas from the transfer line sweeps the GC effluent towards the corona discharge. A plasma consisting of primary ions (e.g., $N_2^{\bullet+}$ and $N_4^{\bullet+}$ when using N_2 as the make-up gas) and electrons is generated. Analyte molecules (M) may undergo charge exchange with the primary ions if the recombination energy of the primary ions exceeds the ionization energy (IE) of M. In the same vein, the formation of radical anions ($M^{\bullet-}$) will also occur if the electron affinity (EA) of M is sufficiently high. Internal energy in the incipient ions is quickly dissipated by non-reactive collisions with the surrounding make-up gas, thus minimizing fragmentation. This process also depends on the nature of the reagent gas and the presence of other compounds or ions that may be added as dopants.

Nitrogen is the most common reagent gas used in GC–APCI. This is partly because it is conveniently and inexpensively produced (usually by purification of compressed air) at a rate that is necessary for operation of most API ion sources. Although nebulization and desolvation of liquid droplets is not a concern in GC–API, a relatively high volume of nitrogen is still consumed as a skimmer or curtain gas around the orifice of the mass spectrometer to impede neutrals from entering. The ionization energy of N_2 (IE = 15.6 eV) also exceeds that of most organic molecules, making N_2 ideal for the analysis of a wide range of compounds, as is often required for NTS. In some cases, it may be desirable to select a gas with an ionization energy that lies above that of the analyte, and below that of potential interferents. However, in practice, the high consumption of gases by most API sources discourages the use of alternative gases that are more expensive than nitrogen.

Efforts to control the introduction of reagent gases and dopants have mostly been restricted to placing a small vial containing a volatile liquid into the ion source, sometimes with a short piece of capillary tube inserted in the vial's septum to restrict the flow. For example, when H_2O or other protic solvents (S) are introduced to the ion source, they may undergo charge exchange to form ($H_2O^{\bullet+}$ or $S^{\bullet+}$), and subsequently self-protonate to form H_3O^+ or SH^+ according to the reaction: $S^{\bullet+} + S \rightarrow [S - H]^{\bullet} + SH^+$. The protonated solvent molecule SH^+ may then transfer its proton to an analyte molecule to form $[M + H]^+$ if the proton affinity of M exceeds that of S. The formation of $[M + H]^+$ ions by protonation instead of charge exchange may be desirable in cases where the compounds of interest have relatively high proton affinities compared to their potential interferents. Schreckenbach et al. used this approach to confirm the identity of the molecular ion of a previously unknown chlorinated amide [20]. However, the unintentional or uncontrolled introduction of H_2O (e.g., from laboratory air humidity) can be a nuisance, especially if the compound of interest has a low PA and thus can be suppressed under "wet" conditions.

In the negative mode, M can form $M^{\bullet-}$ radical anions through electron capture negative ionization. The presence of low concentrations of oxygen (<1%) may also result in displacement reactions between M and $O_2^{\bullet-}$ to form ions $[M - X + O]^{\bullet-}$ (where X = H, Cl, Br) [26]. The negative mode is important for the identification of halogenated POPs due to their high electron affinities. Reactions with $O_2^{\bullet-}$ have also been shown to be structure-diagnostic, and this will be discussed in Section 3.4. It is also possible to generate Cl^- adducts by placing a vial of chloroform in the ion source, and this reaction appears to be selective towards polyhalogenated alkanes [67].

As the most popular GC–API technique, a number of publications have recently appeared that demonstrate its advantages over EI or CI. Analysis of hydroxypyrene (a PAH metabolite in human urine) using GC–APCI showed lower detection limits and a wider linear range than LC–MS/MS [52]. In the same vein, GC–APCI analysis resulted in >10-fold lower method detection limits for halogenated dioxins and furans in sediments, fish and fire debris as well as 9-nitrophenanthrene and 3-nitrophenanthrene in PM 2.5, as compared to using EI. GC–API techniques may also enable faster analysis. Unlike vacuum ion sources, API is inherently resilient to high flows of both nitrogen and helium carrier gases. For example, Di Lorenzo et al. demonstrated that PBDEs could be separated in less than 7 min and 15 min, respectively, with helium and nitrogen carrier gas [68]. Critical isomers could be separated using nitrogen, which is desirable in the face of looming shortages of helium, a non-renewable resource. A drawback of GC–APCI is increased ion suppression compared to EI. NTS practitioners should be cautious when using APCI to analyze compounds with a wide range of ionization energies. While preserving the sample information is a benefit, an interfering matrix could obfuscate unknown pollutants.

2.2. *Atmospheric Pressure Photoionization (APPI)*

APPI employs UV light from a Xe, Kr, or Ar lamp, to produce 8.4 eV, 10.1 eV and 11.2 eV photons, respectively, for ionization. A Kr lamp is commonly used because its longevity exceeds that of an Ar lamp while also providing energetic photons that are capable of ionizing a wider range of compounds than a Xe lamp. The formation of positive ions $M^{\bullet+}$ will occur if the photon energy exceeds the ionization energy of M, and akin to APCI, negative ions can form by electron attachment or through reactions with $O_2^{\bullet-}$. Introduction of a suitable dopant (e.g., toluene, acetone, anisole, and chlorobenzene, etc.) whose IE is smaller than the photon energy can promote protonation and/or adduct formation with the analyte and significantly enhance ionization efficiency [42,69] and broaden the range of chemicals subject to APPI ionization. For example, neutral perfluoroalkyl and polyfluoroalkyl substances [59] cannot be directly photoionized into $M^{\bullet+}$. Instead, negative ions, including adducts with oxygen, are generated with the assistance of a dopant. In this case, the direct ionization of the dopant (D) results in the formation of $D^{\bullet+}$ radical cations, as well as free electrons that can promote negative ionization.

APPI is a more convenient way of selectively ionizing compounds using different photon energies compared to charge exchange reactions with different gases in APCI. It is desirable in some cases to exclude the ionization of H_2O, which drives the formation of $[M + H]^+$ rather than $M^{\bullet+}$ radical cations by charge exchange. Figure 1a shows the partial mass spectra of PBDE-209 and its $^{13}C_{12}$-labelled analog obtained under both APPI (top) and APCI (bottom) conditions. Since H_2O cannot be ionized by 10 eV photons, no subsequent protonation occurs, and only $M^{\bullet+}$ ions are observed in the APPI spectrum. In contrast, APCI results in a mixture of $M^{\bullet+}$ and $[M + H]^+$ ions, increasing the complexity of the mass spectrum. This contrast is more evident for 1,8-dibromo-2,6-dichloro-9H-carbazole (Figure 1b), which belongs to an emerging class of POPs believed to occur as byproducts of halogenated indigo dyes [68]. $[M + H]^+$ ions dominate the APCI spectrum because the presence of nitrogen increases the PA, but the APPI experiments produce $M^{\bullet+}$ ions only. The ability to control the types of (quasi)molecular ions being generated is an advantage of APPI because: (i) the complexity of the mass spectra is reduced; and (ii) ionization efficiency for a wide range of POPs is more uniform while also excluding potential interfering compounds whose IEs exceed 10 eV.

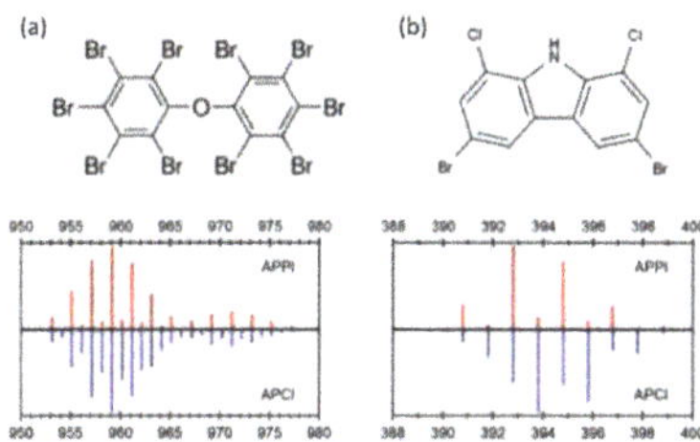

Figure 1. Partial mass spectra obtained using GC–APPI (**top**) and GC–APCI (**bottom**) for (**a**) decabromodiphenylether (BDE-209) and its ^{13}C-labelled counterpart, and (**b**) 1,8-dibromo-2,6-dichloro-9H-carbazole. Reprinted from Di Lorenzo et al. [68]. Copyright 2019, with permission from Elsevier.

Excluding the ionization of compounds whose IEs exceed the photon energy of the lamp can also be a limitation. Obviously, the number of compounds that can be identified in such an experiment will decrease, and the use of dopants can also have the same effect. For example, a comprehensive comparison between APPI with and without dopants was performed for 75 EPA priority environmental pollutants. The study showed that the use of dopants increased the ionization efficiency for many compounds, but ultimately decreased the number of compounds detected [69].

2.3. *Atmospheric Pressure Laser Ionization (APLI)*

APLI is similar to resonance-enhanced multiphoton ionization (REMPI) [27], wherein two absorption steps are involved in ionization. The absorption of the first photon results in the formation of an excited molecule M*, which is subsequently ionized by a second photon to form $M^{\bullet+}$. This process is favorable if (i) the combined energies of the photons are resonant with the energy required for excitation and ionization, and (ii) the excited state M* is sufficiently long-lived to absorb a second photon. These two requirements are usually satisfied by π-electron-rich compounds [70] and most applications of APLI have focused on PAH and other aromatics (see Table 1). The high sensitivity of this technique allows sample dilution of up to 1000-fold, resulting in significantly decreased matrix interference, better separation and improved peak shape [64]. To expand the range of compounds that can be analyzed by APLI, Deibel et al. [58] introduced a series of APLI ionization labels to derivatize amines, alcohols and carboxylic acids. A recent comparison between APPI and APLI revealed that APPI was able to ionize the widest range of analytes (66/77) and halogenated aromatics were much more readily ionized by APPI than by APLI [69].

2.4. Electrospray Ionization (ESI)

ESI involves the creation of ions from charged droplets. In LC–MS, the column effluent is delivered through a charged capillary. Gas phase ions are produced as the solvent molecules are stripped away with the aid of a flow of nitrogen. ESI produces a wide range of ions that originate from the solvent itself, including protonated molecules and dimers, (sodium and potassium) cationized adducts and cluster ions. In the negative ion mode, ESI produces deprotonated molecules, as well as adducts and clusters bridged by common anions present in the solvent or mobile phase, such as Cl^- or formate. The electrosprayed droplet–ions can also serve as vehicles for protons and other reagent ions that may ionize solid, liquid or gaseous samples. For example, in desorption electrospray ionization (DESI), charged droplets are directed towards a solid, extracting molecules from the surface while transferring charge from the ionized solvent to the analyte. At the same time, the droplet bounces from the surface, propelling the analyte ions towards the entrance of the MS for analysis. Similarly, in extractive electrospray ionization (EESI), the charged droplets collide with and ionize a secondary aerosol spray. In principle, the same approach could be used to ionize gaseous molecules exiting a GC column, but to date, GC–ESI has only been attempted by one group [65,66]. The results of Cha et al. [65] showed that GC–ESI could achieve linearity, repeatability, robustness and detection limits comparable to standard GC–MS and LC–MS methods. Even non-polar PAHs could be ionized by GC–ESI, and protonated molecules $[M + H]^+$ dominated their mass spectra. GC–ESI is a relatively unexplored means of introducing reagent ions, such as metal cations and anions [71], that would not be possible using GC–APCI. Whether such ion chemistry will be useful for structure elucidation is a question that deserves more attention.

2.5. Penning Ionization (PI)

Direct Analysis in Real-Time (DART) was first introduced by Cody and Laramee [32,72]. Penning ionization is initiated by glow discharge of a gas (commonly helium), resulting in neutral atoms in a metastable excited state. The internal energy of metastable helium (19.8 eV) exceeds the ionization energies of most common atmospheric gases. While Penning ionization may result in the formation of radical cations $M^{\bullet+}$, water from the ambient environment will also ionize, self-protonate and form cluster ions. Akin to APCI, the formation of protonated molecules $[M + H]^+$ will occur when M has a higher PA than H_2O and its cluster ions. In the vast majority of applications, DART is directly coupled with a mass spectrometer [32]. Penning ionization has also been employed for ionization of GC effluent by Moore et al. [73] and later by Cody et al. [32]. In general, the appearance of mass spectra obtained by GC–DART are similar to those obtained by GC–APCI.

2.6. Dielectric Barrier Discharge Ionization (DBDI)

The ionization process is initiated by a dielectric barrier discharge, which involves the creation of a low temperature plasma (LTP, ~30 °C) that is ignited when a potential is applied between two electrodes separated by a dielectric material. It was first introduced as a GC–MS ion source by Nørgaard and coworkers in 2013 [74]. Similar to DART, LTP also produces $M^{\bullet+}$ and $[M + H]^+$ ions in the positive mode, and $[M - H]^-$ and $M^{\bullet-}$ ions in the negative mode. As a proof of concept, Nørgaard et al. demonstrated that 20 common indoor VOCs, including alkanes, alkenes, alcohols, aromatic compounds, aldehydes, PAHs, phenols, and terpene alcohols, could be detected using this approach [74]. In 2017 [75], Hagenhoff and coworkers developed a similar DBDI source, albeit with electrodes configured in a different geometry. The exquisite sensitivity of the approach enabled the detection of femtogram levels of 28 pesticides and 14 illicit drugs. Their ion source also showed promise when applied to NTS:GC–DBD was used to screen semifluorinated n-alkanes (SFAs) in ski wax samples [76]: SFAs with carbon numbers of 26, 28, 30, and 32 were tentatively identified and 1-(perfluorooctyl)-hexadecane confirmed with an authentic standard.

3. Strategies to Identify Unknowns by GC–API

3.1. Multidimensional Chromatography

Recent interlaboratory studies of NTS methods suggest that a combination of complementary chromatography methods and ionization sources is essential to detect and identify all compounds present in a sample [7,77]. For example, the identification of approximately 1500 organic pollutants in surface and groundwater surrounding a solid-waste treatment plant required the use of multiple techniques, including GC–(EI)TOF, GC–(APCI)QTOF, LC–(ESI)QTOF and LC–(ESI)QqQ [78]. There is also a growing interest in developing multidimensional separation techniques that involve multiple separation stages in a single experiment.

GC×GC can significantly increase the number of identifiable compounds in a sample compared to using single-dimension GC–MS. Ballesteros-Gómez et al. [79] were the first combine GC×GC with APCI and it has since been applied to characterizing plasma [80] and household dust [7]. The contour plot in Figure 2a was obtained from a pooled sample of plasma and it may serve to demonstrate the separation power of GC×GC. It is evident from the plot that many compounds can be separated by the second dimension column that would otherwise coelute in the first dimension. In the interlaboratory study led by Roskowski et al. [7], both GC×GC–APCI and GC×GC–EI were used to tentatively identify >500 compounds, representing a large fraction of the total number of compounds reported by all participants. It is because the improved separation resulted in the collection of higher quality (CID) mass spectra used for identification. Separation is not the only benefit afforded by modulating effluent from the primary column. As shown in Figure 2b, the width of a GC×GC peak is very narrow and more intense compared to one collected using a single-dimension GC. Patterson et al. [81] pioneered the approach called cryogenic zone compression. Their result showed that it could improve the detection limits for trace level contaminants such as PCDDs. When coupled with GC–API and a full-scanning, high-resolution mass spectrometer, the technique could potentially enable the identification of unknown contaminants with volume-limited samples, such as dried blood spots [80].

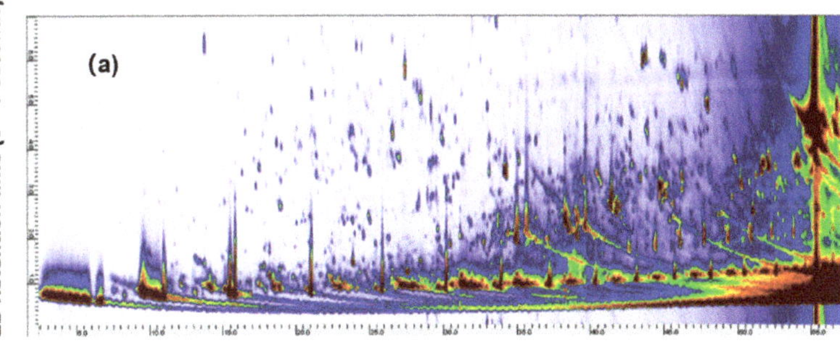

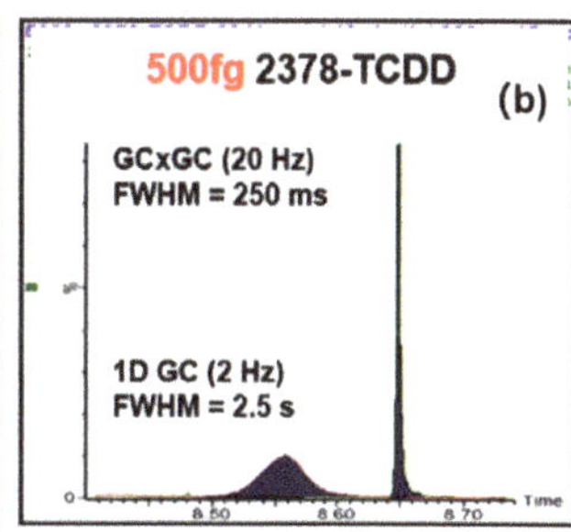

Figure 2. (**a**) Comprehensive two-dimensional gas chromatography (GC×GC) affords unparalleled separation of organic pollutants from in plasma; (**b**) Modulated gas chromatogram obtained from 2,3,7,8-tetrachlorodibenzo-*p*-dioxin, resulting in 10-fold signal-to-noise (S/N) enhancement. Similar S/N enhancement can be achieved using either thermal or flow modulation. Reprinted from Ref. [80]. Copyright 2020, with permission from Elsevier.

Most GC×GC instruments employ cryogenic modulation, whereby the primary column effluent is trapped by a flow of cooled nitrogen gas, and then reinjected into the secondary column by a pulse of heated nitrogen. GC×GC may alternatively be accomplished using a flow modulator, which converts primary peaks to secondary peaks using pulses of gas flow delivered using one or more valves without cryogens. Valve-based modulators do not technically zone compress (i.e., focus) the GC effluent, but the width of the secondary peak may still be controlled by the flow, which can exceed the primary flow by 10 times or more. Such a configuration is not compatible with conventional EI and CI ion sources. In contrast, flows in excess of 100 mL/min are well suited to the conditions of GC–API. A multimode flow modulator designed by J.V. Seeley [82] was adapted to a GC–APCI instrument, resulting in a significant enhancement in sensitivity [83].

The profile of compounds detected by GC×GC–API, sometimes referred to as a "chemical fingerprint", has proven to be vital for identifying potential sources of pollution. For example, Bowman et al. [84] used GC×GC–APCI to characterize the components of coal tar-based sealcoat products in comparison to those in other sources of polycyclic aromatic compounds (PACs). The results revealed that there was a clear difference in the composition of PACs across different types of sealcoat products and sources of PAHs. Figure 3a,b displays the Kendrick mass defect plots obtained from coal tar and asphalt sealcoat products, respectively. They have shown that the petrogenic asphalt sealcoat is characterized by a greater range of alkylated PAH, compared to the pyrogenic coal tar sealcoat. Individual components were also identified by accurate mass measurements in combination with first dimension retention indices. Hierarchical clustering analysis led to the identification of signature compounds that could distinguish coal tar from other pyrogenic sources of PAH pollution, such as creosote (from a railroad tie), and diesel particulate.

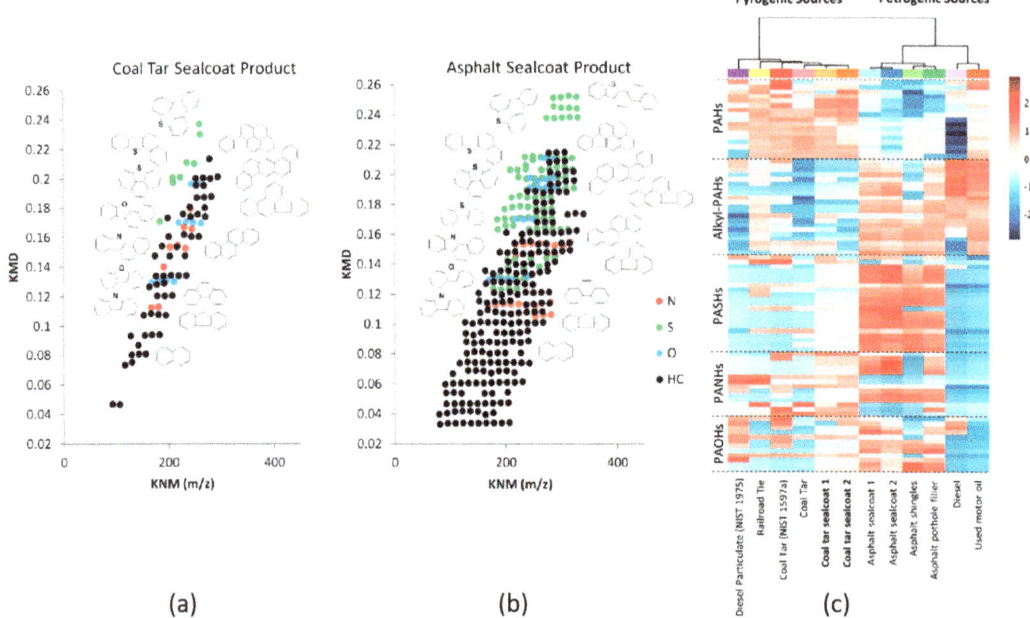

Figure 3. Kendrick mass defect (KMD) plots of (**a**) a coal tar sealcoat product, and (**b**) an asphalt sealcoat product. (**c**) Two-dimensional hierarchical cluster analysis with heat map plot of various environmental PAH sources. The colors red and blue represent the most intense and least intense relative abundance values (log scale). The relative abundances in the heat map plot are presented as an average of three replicates. Reproduced from Bowman et al. [84] with permission. Copyright 2019 American Chemical Society.

3.2. Data-Independent Identification

In NTS, an important benefit afforded by "soft ionization techniques like GC–APCI is the abundant formation of (quasi)molecular ions whose accurate mass can determine an unknown's elemental composition. However, a major weakness is the loss of structure-diagnostic fragmentation normally obtained by EI. Instead, the structural identity of newly discovered contaminants is typically achieved using tandem mass spectrometric techniques, such as collision-induced dissociation (CID). Two strategies have emerged for automated collection of CID mass spectra during GC (×GC) separation, viz. data-dependent acquisition and data-independent acquisition (DIA).

Data-independent acquisition [85] was first introduced for the analysis of peptide mixtures in combination with LC–MS separation. It was only recently applied to GC–APCI for suspect screening and NTS of organic pollutants [20]. Compared to data-dependent acquisition, which selects ions based on predefined criteria such as mass, intensity or isotopic ratios, the DIA approach enables the automated, unbiased selection and acquisition of the precursor and CID mass spectra of all ions detected in the sample [20]. By cycling between low- and high-collision energy, precursor and product ions can be identified by changes in their intensities: precursor ions will decrease in intensity when subjected to high collision energy, whereas the opposite is true for fragment ions. A computer algorithm then deconvolutes the CID mass spectrum of each compound by grouping together precursor and product ions that share the same GC retention time. A drawback of this approach is its inability to deconvolute CID mass spectra of compounds that coelute. For example, Figure 4a shows the CID mass spectrum of the flame retardant PBDE-47 that was collected using DIA during a short 15 min GC separation [20]. Numerous interfering peaks (*m/z* 206, 253 and 340) can be observed that are absent when a longer GC×GC separation was performed, see Figure 4c. However, it is not always possible to separate coeluting compounds, and in this case, the time requirement for the GC×GC separation exceeded that of the single-dimension experiment by four-fold [20]!

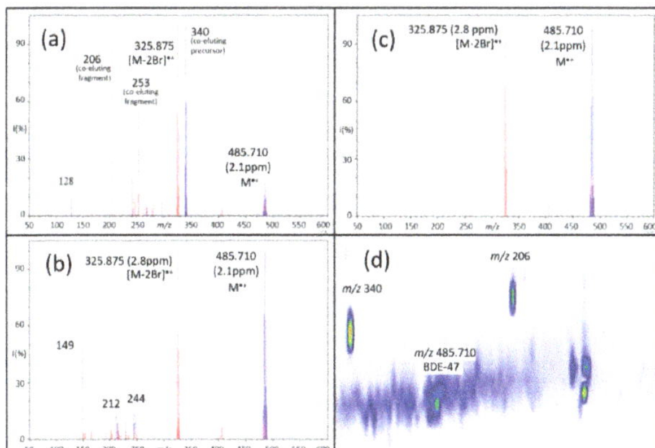

Figure 4. Comparison of (**a**) GC–DIA, (**b**) GC×GC–DIA, and (**c**) deconvolved GC–SQDIA spectra of a tetrabromodiphenyl ether. In all spectra, the low-energy channel is shown in blue and the high-energy channel is shown in red. (**d**) GC×GC–DIA chromatogram showing separation of compounds that coeluted in one-dimensional GC–DIA (as seen in (**a**)). Greater separation in both GC×GC–DIA and GC–SQDIA resulted in spectra with fewer interferences than those obtained using GC–DIA; this separation can be obtained chromatographically, as in GC×GC, or utilizing the quadrupole, as in SQDIA. Adapted with permission from Schreckenbach et al. [20]. Copyright 2021 American Chemical Society.

One way to obtain a better-quality CID mass spectrum is to use the quadrupole analyzer to cycle through narrow isolation windows (typically c. 20~50 amu segments) and sequentially subjecting the selected ions to CID. For example, a mass spectrum with a range of 20~1000 amu can be subdivided into 49 segments or *swaths*, which are isolated and fragmented. This was pioneered by Gillet et al. [86], who coined the approach Sequential Windowed Acquisition of all Theoretical fragment ion–Mass Spectra (SWATH–MS). Scanning quadrupole DIA (SQDIA) is closely related to SWATH–MS, but instead of cycling the quadrupole between *swaths*, it is scanned continuously across the mass range [87]. This significantly reduces coeluting interferences in the resulting CID mass spectra, as illustrated by comparing the DIA and SQDIA mass spectra obtained for PBDE-47 shown in Figure 4a,b, respectively. SQDIA can produce CID mass spectra without sacrificing the time required for a more rigorous chromatographic separation. The gain in selectivity and speed provided by SQDIA, however, also comes with a cost in sensitivity: while the quadrupole is scanning, only a small subset of the ions proportional to the isolation window (IW) are transmitted to the detector and the remaining ions are lost. The theoretical transmission can be calculated using the equation: IW (amu)/mass range (amu) = transmission (%).

GC–SQDIA has been used to screen 2542 suspected organic contaminants listed in the AMAP (Arctic Monitoring and Assessment Programme) 2016 Chemicals of Emerging Arctic Concern in a dust sample collected from an electronics recycling facility. The procedure for structure assignment involved comparing the measured mass of the (quasi)molecular ions with the theoretical masses of the 2542 suspected pollutants. Then, a software tool (MassFragment [88]) was used to predict the possible fragments from each of the structures in the library. The structure that produced the greatest number of fragments matching with those observed in the CID mass spectrum was then assigned. In this case, the software tool was part of a commercial software package (UNIFI [89]), but the same structure assignment procedure could be applied using open source software such as: MetFrag [90]; CSI (Compound Structure Identification): FingerID [91]; CFM (competitive fragmentation modeling)-ID [92]; and QCEIMS [93]. A more detailed discussion on the prediction of mass spectra will be presented in Section 3.6. The results of this suspect screening experiment showed that SQDIA significantly lowers the rate of false structure assignment.

Ion mobility can also be used to disentangle the CID mass spectra of coeluting compounds [94]. One requirement of this approach is that compounds that coelute from the GC column must be separable according to their mobility, which can be characterized by an ion's collisional cross section (CCS). At elevated pressures, non-reactive collisions impede the ions akin to an aircraft in flight such that large, bulky ions tend to have larger CCS values and lower mobilities than small, compact ions. CCS values can also be used as confirmatory evidence of a structure assignment [21]. Lipok et al. were the first to combine GC×GC–APCI with ion mobility–mass spectrometry. They used an in-house database consisting of 800 CCS values to screen for drug-like compounds and pesticides [21]. Recently, Olanrewaju et al. [95] have analyzed PAHs and related petroleum hydrocarbons in crude oil using a trapped ion mobility–mass spectrometer hyphenated with GC. In this study, the identification of isomeric PAHs and related unknown aromatic hydrocarbons was accomplished with the aid of CCS measurements. The results also raise the intriguing possibility of separating small contaminant molecules by ion mobility alone. For example, the difference in collisional cross sections (Δ CCS) of the isomers triphenylene and chrysene is only ~2 Å^2. Using the formula $R = CCS/\Delta CCS$, it is anticipated that an ion mobility resolution of ~73 would be sufficient to resolve isomers that are closely eluted by GC. Ion mobility–mass spectrometers capable of $R > 200$ are now commercially available [96], but it has yet to be shown that this separation is fast enough to be compatible with GC (×GC).

3.3. Evaluating Confidence in Structure Assignments

According to Schymanski et al. [77], structure assignments based on accurate mass and isotopic measurements of their (quasi)molecular ions may be considered tentative at best, corresponding to structure confidence levels 4 and 5 on their 5-level scale. The highest confidence score, level 1, requires confirmatory evidence obtained using authentic standards. While a meaningful study of the occurrence and fate of organic pollutants will require authentic standards [5], it is also recognized that this is not always practical at the earliest stage of a contaminant's discovery. In the absence of authentic standards, acquiring complementary information such as CID mass spectra, retention time(s) and CCS can increase the confidence in a tentative identification.

The time-honored approach to identifying an unknown pollutant involves comparing its EI mass spectrum with those compiled in spectral libraries. There are databases containing hundreds of thousands of EI spectra (e.g., the NIST Mass Spectral Library), but equivalent libraries compiling CID mass spectra are orders of magnitude smaller. For example, Mesihää et al. [44] have developed an in-house GC–APCI–QTOFMS library that includes 29 psychoactive substances. However, creating spectral libraries is time-consuming and costly. To bridge this gap, practitioners of GC–API can take advantage of workflows that were originally conceived for LC–MS. Instead of relying on spectral libraries, one could search structure libraries (e.g., PubChem and ChemSpider) and then compare the experimental CID mass spectra with those predicted by in silico methods [97]. This approach typically involves comparing the measured mass of the (quasi)molecular ions with the theoretical masses of all compounds in the structure library. Then rules-based or combinatorial fragmentation predictors are used to predict the possible fragments from each of the structures in the library whose molecular ions fall within a preselected mass range (usually 1–5 ppm) of the experimental mass. The structure that produces the greatest number of fragments that match with those observed in the CID mass spectrum is then assigned.

Su et al. [98] have suggested a modified version of the confidence scale proposed by Schymanski et al. [77]. Their approach hinges on comparing results obtained by GC–APCI-MS and GC–EI–MS, taking advantage of both structural and spectral libraries [99]. Briefly, the criteria for a level 3 identification are: (i) a compound's EI spectrum must match that of a library spectrum with a match factor >700; and (ii) the mass of the compound's (quasi)molecular ion peak must fall within 5 ppm of the theoretical mass. Confidence in the proposed structure increases to level 2 when complementary evidence, such as retention index or CCS, is used. In the absence of a good quality spectral library match, a structure library search can also be used in combination with careful interpretation of the CID mass spectrum. This requires a firm understanding of the dissociation chemistry of organic ions and their reactivity with gas molecules used in the ion source and collision cell. Ultimately, a tentative identification must be confirmed with an authentic standard.

3.4. Ion-Molecule Reactions for Separation and Structural Elucidation

The conditions of the GC–APCI ion source can promote ion–molecule reactions that are structure-diagnostic and, in some cases, can differentiate between toxic and non-toxic isomers. A prime example is the reaction between dioxygen and (mixed) halogenated dibenzo-p-dioxins in the negative ion mode. Quasimolecular ions $[M - Cl + O]^-$ are generated by reactions between the analyte molecules and $O_2^{\bullet -}$. Mitchum and Korfmacher et al. [100–102] showed that the reaction between 2,3,7,8-tetrachlorodibenzo-*p*-dioxin (2,3,7,8-TCDD) and $O_2^{\bullet -}$ also results in cleavage at the ether bonds of 2,3,7,8-TCDD, as shown in Figure 5a. This specific reaction was shown to distinguish 2,3,7,8-TCDD from many other common interfering species, including its isomers. For example, the negative ion APCI mass spectrum of 2,3,7,8-TCDD displays an intense peak at *m/z* 176, corresponding to the ether cleavage product shown in Figure 5a. In contrast, 1,2,3,4-TCDD cannot produce a peak at *m/z* 176 because all four of its chlorine atoms are present on the same ring. As shown in Figure 5b–e, this difference can be exploited to reduce the burden on GC to separate TCDD isomers.

The ubiquity of brominated flame retardants in everyday household items increases the likelihood that PBDDs will be formed during (accidental) fires [104]. Highly brominated contaminants, including the tetrabromoodibenzop-dioxins (TBDDs), are challenging to analyze by GC–MS because of their thermal lability. To minimize thermal decomposition during chromatographic separation, a relatively short (~15 m) GC column is used along with a relatively thin (0.1 μm) stationary phase. Therefore, separating toxic from non-toxic isomers is a major challenge that cannot be solved using GC alone, as witnessed by the coeluting isomers 2,3,7,8-TBDD (toxic) and 1,2,3,4-TBDD (less toxic) in Figure 5d. Fernando et al. [103] showed that ion–molecule reactions with oxygen could be exploited to separate the coeluting isomers because 2,3,7,8-TBDD reacts with oxygen to produce the ether cleavage product $C_6H_2BrO_2^{\bullet-}$ (*m/z* 265.840), whereas 1,2,3,4-TBDD cannot. When applied to samples collected from a major industrial fire, this ion chemistry could differentiate isomers of PXDDs (where X = Cl, Br) that would not be feasible using EI.

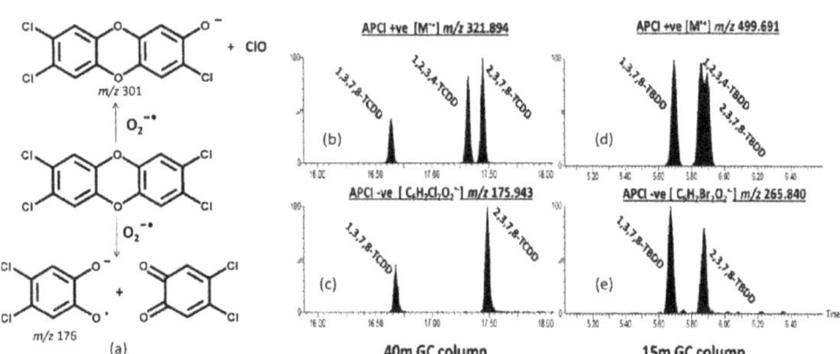

Figure 5. (**a**) Scheme of reaction between 2,3,7,8-TCDD and $O_2^{-\bullet}$; (**b**–**e**) APCI− spectra of 2,3,7,8-TCDD, 2,3,7,8-TBDD, and 2,3 Br-7,8 Cl-DD. Ether cleavage products (ECPs) are observed at *m/z* 176 and 266 for the 2,3,7,8-TCDD and 2,3,7,8-TBDD species. Both of these ECPs are observed for the 2,3 Br-7,8 Cl-DD. Reproduced with permission from Ref. [103]. Copyright 2016 American Chemical Society.

Structure-diagnostic reactions have also been observed in the positive ion mode[68]. For example, Di Lorenzo et al. [68] observed that the PBDE flame retardants can undergo isomer-specific photooxidation in a GC–APPI source, viz. that PBDE-71 produces an $[M - Br + O]^+$ ion in its APPI mass spectrum that is absent in that of PBDE-49. EPA method 1614 requires the separation of isomers PBDE-49 and PBDE-71, but the observation that GC–APPI can differentiate isomeric PBDEs raises the possibility that this requirement may be relaxed. The group of R. G. Cooks [105] has shown that corona discharge under solvent-free conditions can promote oxidation reactions of alkanes that would otherwise require catalysis. Megson et al. [106] have also observed similar reactions under GC–APCI conditions: the ubiquitous plasticizer and flame retardant tricresyl phosphate (TCP) undergoes an uncatalyzed oxidative transformation into the metabolite 2-(ortho-cresyl)-4*H*-1,3,2-benzodioxa-phosphoran-2-one (CBDP), which is responsible for the neurotoxicity of TCP. This ion–molecule reaction is specific to ortho-substituted triaryl phosphates, which are toxic, unlike the non-toxic meta- and para-substituted isomers. The reaction also mirrors the microsome/enzyme-promoted transformation that occurs in vivo. There is currently little fundamental understanding of this reactivity, but the implications for NTS are significant: substituted aryl phosphates are widespread environmental contaminants and an indoor dust sample may contain hundreds of (unknown) homologs. GC–APCI could potentially be used to identify the neurotoxic ortho-substituted isomers in such a mixture selectively.

3.5. Retrospective Analysis and Compound Discovery

A key advantage of all HRMS techniques is that full scan results can be digitally archived and exploited retrospectively. Lai et al. [107] recently developed a prescreening and identification workflow implemented as an R package to support regulatory environmental monitoring. One of the challenges of retrospective analysis is developing a strategy to recognize pollutants from among the many thousands of chemicals detected by HRMS. Zhang et al. [41] recently developed an approach to identify unknown persistent and bioaccumulative organics using mass spectrometry and applied this approach using GC–APCI.

Most POPs contain three or more Cl or Br atoms, making them easy to recognize based on their isotope patterns. Even polyfluorinated compounds can be recognized based on a relatively weak ^{13}C-isotopic peak compared to non-fluorinated compounds, see Figure 6.

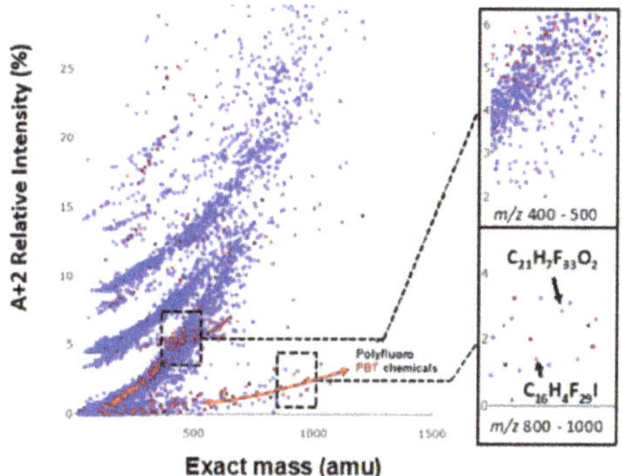

Figure 6. Distribution of 610 prioritized persistent bioaccumulative compounds (in red) and commercial chemicals (in blue) from the North American chemical inventories in the compositional spaces defined by *m/z* of the molecular ion and the ratio of its isotopic peaks (A + 2): A, where "A" is the most intense isotopic peak. Adapted from Zhang et al. [41]. Copyright 2019, with permission from Elsevier.

Data collected using GC–APCI is ideally suited for retrospective analysis because the molecular ion is preserved. Zhang et al. [41] identified 191 isotopic clusters from a housedust standard reference material using their prioritization strategy. The identified chemicals included PCBs, agricultural drug residues, polychlorinated and polybrominated diphenyl ethers and other brominated flame retardants. Previously unknown chlorofluoro flame retardants were also discovered in this study, including thermal decomposition products of 2,3,4,5-Tetrachloro-6-((3-(trideca-fluorohexyl)sulfonyloxy)phenylaminocarbonyl)benzoic acid.

3.6. Computational Tools to Predict Mass Spectra

Libraries of experimental CID mass spectra are much smaller than those compiled for EI. To bridge this gap, novel computational tools have emerged to predict CID spectra. These tools may be broadly classified into three types, reviewed by Scheubert et al. [97], and have been widely used to predict CID mass spectra collected during LC–MS experiments. The most common type utilizes either a rules-based or a combinatorial approach to predict the fragmentation of an ion. Examples include MetFrag [90], MassFragment [108] and Mass Frontier [109]. These methods do not technically predict the spectrum, but rather identify the number of experimentally observed peaks that can be explained by a given structure. This approach is popularly used for suspect screening and structural database searching, but one limitation is the fact that an unknown's structure must be present in the database in order for a search to be successful. Another limitation is that these methods do not predict peak intensities, which could be used to assign a structure more reliably. In contrast, recently developed machine learning methods, such as CFM-ID [92], can predict whole mass spectra, including relative intensities, but their accuracy depends on the size of the training set used. This is problematic for GC–API because there are few experimental CID spectra available. CSI: FingerID [91] combines fragmentation tree computation and machine learning to predict the molecular fingerprint of the unknown compounds. Another spectra prediction tool is based on computational chemistry. Quantum chemical electron ionization mass spectrometry (QCEIMS) [93] employs semiempirical quantum mechanical and/or density functional theory methods, and Born-Oppenheimer molecular dynamics to predict the dissociation behavior of radical cations. While this approach is the most accurate way to predict a mass spectrum, it is also the most resource-demanding, requiring approximately 1000 core hours to compute the spectrum of a small molecule pollutant.

CSI: Finger ID, CFM-ID and QCEIMS have all been used to predict CID spectra and guide the interpretation of GC–MS data. CSI: Finger ID was used by Larson et al. to identify products of lignin pyrolysis [42]. QCEIMS was evaluated by Schreckenbach et al. [110] to predict the mass spectra of selected halogenated and organophosphorus flame retardants. While QCEIMS is designed for EI spectral prediction, it can also be informative when predicting the CID mass spectra of ions $M^{\bullet+}$ generated by charge exchange (GC–APCI) or photoionization (GC–APPI). This is because the unimolecular dissociation behavior of an ion is largely determined by the potential energy surface that does not depend on the ionization technique. The results showed that QCEIMS predicted the mass spectra of 35 organic pollutants as accurately as the less computationally demanding CFM-ID method. QCEIMS is best suited for compounds that are truly unknown and thus not present in any library or training set. For example, QCEIMS accurately predicted the EI mass spectrum of the dioxin-like compound 1,8-dibromo-3,6-dichlorocarbazole (shown in Figure 7), an emerging contaminant of the Laurentian Great Lakes. A recent study [111] reports on the development of QCxMS, which extends QCEIMS for the prediction of both EI and CID spectra prediction. A test with six standards showed the calculated data was in reasonable agreement with the experiment (see Figure 7b). These methods can help guide the analyst in selecting authentic standards.

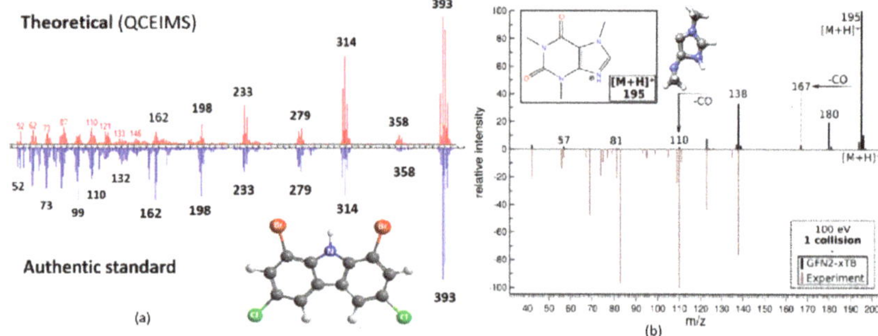

Figure 7. (**a**) EI mass spectrum calculated by QCEIMS and experimental EI mass spectrum of 3,6-dichloro-1,8-dibromo-carbazole. Reproduced from Ref. [110]. Copyright 2021 American Chemical Society. (**b**) calculated spectrum and 40 eV literature spectrum of the most populated caffeine protomer at 600 K, single collision at 100 eV in the laboratory frame. Reprinted from Ref. [111]. Copyright 2021 American Chemical Society.

4. Summary and Outlook

GC–API techniques generally minimize fragmentation and preserve the (quasi)molecular ion, resulting in simplified mass spectra and improved detection limits. It is relatively facile to modify an existing LC–MS instrument and increase its analytical power. The conditions of the ion source are compatible with high flows that enable faster analysis as well as ion–molecule reactions that can aid in structure analysis. Moreover, spectacular advances in mass spectrometry have led to the development of novel mass analyzers and ion mobility–mass spectrometers that have not previously been coupled to GC, enlarging the scope of GC–MS analysis. It is now possible to use techniques and workflows, such as data-independent acquisition, to significantly decrease the false-positive rate for unknown structure assignments.

GC–API is already a complementary to EI and it may eventually supplant it as the de facto GC–MS ion source used to identify unknowns, but to achieve this a number of challenges will need to be surmounted. First, the absence of libraries of CID mass spectra is the most obvious challenge. However, EI mass spectra can still inform the interpretation of GC–API experiments, considering both GC–API and EI can produce the same types of ions. The dissociation chemistry of a radical cation $M^{\bullet+}$ is only partly dependent on how it is formed, and it is likely that the low-energy reactions will be observed in both experiments. There are also emerging computational techniques that promise to predict CID mass spectra reliably and construct in silico spectral libraries. Second, the ion chemistry that occurs under GC–API conditions has not been fully exploited and in some cases, it is not completely understood. Recent studies have shown that uncatalyzed reactions occurring in the ionization source can aid in structure analysis, e. g. oxygen or nitrogen insertion or displacement reactions that can differentiate toxic from non-toxic isomers [105]. To fully exploit these reactions, their mechanisms will need to be revealed by an integrated experimental and computational approach. Finally, GC–API, when coupled with HRMS, ion mobility and other sophisticated analyzers, is capable of detecting tens of thousands of chemical compounds. New strategies are required for retrospective analysis, annotation of mass spectra and to prioritize the identification of environmentally relevant compounds.

Funding: Funding for this work was provided by the Natural Sciences and Engineering Research Council (NSERC). Computing resources were provided by ACENET (www.ace-net.ca) and Compute Canada (www.computecanada.ca).

Institutional Review Board Statement: Not applicable.

Informed Consent Statement: Not applicable.

Conflicts of Interest: The authors declare no conflict of interest.

References

1. Muir, D.C.G.; Howard, P.H. Are There Other Persistent Organic Pollutants? A Challenge for Environmental Chemists. *Environ. Sci. Technol.* **2006**, *40*, 7157–7166. [CrossRef]
2. McGuire, J.M.; Collette, T.W.; Thurston, A.D.; Richardson, S.D.; Payne, W.D. *Multispectral Identification and Confirmation of Organic Compounds in Wastewater Extracts*; Environmental Research Laboratory: Athens, GA, USA, 1990.
3. UN. UN Report: Urgent Action Needed to Tackle Chemical Pollution as Global Production is Set to Double by 2030. Available online: https://www.unep.org/news-and-stories/press-release/un-report-urgent-action-needed-tackle-chemical-pollution-global (accessed on 9 August 2021).
4. Stockholm Convention. What are POPs? Available online: http://www.pops.int/TheConvention/ThePOPs/tabid/673/Default.aspx (accessed on 9 August 2021).
5. Hites, R.A.; Jobst, K.J. Is Nontargeted Screening Reproducible? *Environ. Sci. Technol.* **2018**, *52*, 11975–11976. [CrossRef]
6. Chiaia-Hernandez, A.C.; Schymanski, E.L.; Kumar, P.; Singer, H.P.; Hollender, J. Suspect and nontarget screening approaches to identify organic contaminant records in lake sediments. *Anal. Bioanal. Chem.* **2014**, *406*, 7323–7335. [CrossRef] [PubMed]
7. Rostkowski, P.; Haglund, P.; Aalizadeh, R.; Alygizakis, N.; Thomaidis, N.; Arandes, J.B.; Nizzetto, P.B.; Booij, P.; Budzinski, H.; Brunswick, P.; et al. The Strength in Numbers: Comprehensive Characterization of House Dust Using Complementary Mass Spectrometric Techniques. *Anal. Bioanal. Chem.* **2019**, *411*, 1957–1977. [CrossRef] [PubMed]
8. Andreasen, B.; van Bavel, B.; Fischer, S.; Haglund, P.; Rostkowski, P.; Reid, M.J.; Samanipour, S.; Schlabach, M.; Veenaas, C.; Dam, M. *Maximizing Output from Non-Target Screening*; Nordisk Ministerråd: Copenhagen, Denmark, 2021. [CrossRef]
9. Hertz, H.S.; Hites, R.A.; Biemann, K. Identification of Mass Spectra by Computer-Searching a File of Known Spectra. *Anal. Chem.* **1971**, *43*, 681–691. [CrossRef]
10. Harrison, A.G. *Chemical Ionization Mass Spectrometry*, 2nd ed.; CRC Press: Boca Raton, FL, USA, 1992.
11. Kauppila, T.J.; Syage, J.A.; Benter, T. Recent Developments in Atmospheric Pressure Photoionization-Mass Spectrometry. *Mass Spectrom. Rev.* **2017**, *36*, 423–449. [CrossRef]
12. Gross, J.H. From the Discovery of Field Ionization to Field Desorption and Liquid Injection Field Desorption/Ionization-Mass Spectrometry—A Journey from Principles and Applications to a Glimpse into the Future. *Eur. J. Mass Spectrom.* **2020**, *26*, 241–273. [CrossRef] [PubMed]
13. McEwen, C.N. GC/MS on an LC/MS Instrument Using Atmospheric Pressure Photoionization. *Int. J. Mass Spectrom.* **2007**, *259*, 57–64. [CrossRef]
14. Horning, E.C.; Horning, M.G.; Carroll, D.I.; Dzidic, I.; Stillwell, R.N. New Picogram Detection System Based on a Mass Spectrometer with an External Ionization Source at Atmospheric Pressure. *Anal. Chem.* **1973**, *45*, 936–943. [CrossRef]
15. Revelsky, I.A.; Yashin, Y.S.; Sobolevsky, T.G.; Revelsky, A.I.; Miller, B.; Oriedo, V. Electron Ionization and Atmospheric Pressure Photochemical Ionization in Gas Chromatography—Mass Spectrometry Analysis of Amino Acids. *Eur. J. Mass Spectrom.* **2003**, *507*, 497–507. [CrossRef] [PubMed]
16. Schiewek, R.; Schellentra, M.; Mo, R.; Lorenz, M.; Giese, R.; Brockmann, K.J.; Ga, S. Ultrasensitive Determination of Polycyclic Aromatic Compounds with Atmospheric-Pressure Laser Ionization as an Interface for GC/MS. *Anal. Chem.* **2007**, *79*, 4135–4140. [CrossRef]

17. Brenner, N.; Haapala, M.; Vuorensola, K.; Kostiainen, R. Simple Coupling of Gas Chromatography to Electrospray Ionization Mass Spectrometry. *Anal. Chem.* **2008**, *80*, 8334–8339. [CrossRef] [PubMed]
18. Niu, Y.; Liu, J.; Yang, R.; Zhang, J.; Shao, B. Atmospheric Pressure Chemical Ionization Source as an Advantageous Technique for Gas Chromatography-Tandem Mass Spectrometry. *TrAC—Trends Anal. Chem.* **2020**, *132*, 116053. [CrossRef]
19. Fang, J.; Zhao, H.; Zhang, Y.; Lu, M.; Cai, Z. Atmospheric Pressure Chemical Ionization in Gas Chromatography-Mass Spectrometry for the Analysis of Persistent Organic Pollutants. Trends Environ. *Anal. Chem.* **2020**, *25*, e00076. [CrossRef]
20. Schreckenbach, S.A.; Simmons, D.; Ladak, A.; Mullin, L.; Muir, D.C.G.; Simpson, M.J.; Jobst, K.J. Data-Independent Identification of Suspected Organic Pollutants Using Gas Chromatography-Atmospheric Pressure Chemical Ionization-Mass Spectrometry. *Anal. Chem.* **2021**, *93*, 1498–1506. [CrossRef]
21. Lipok, C.; Hippler, J.; Schmitz, O.J. A Four Dimensional Separation Method Based on Continuous Heart-Cutting Gas Chromatography with Ion Mobility and High Resolution Mass Spectrometry. *J. Chromatogr. A* **2018**, *1536*, 50–57. [CrossRef]
22. García-Villalba, R.; Pacchiarotta, T.; Carrasco-Pancorbo, A.; Segura-Carretero, A.; Fernández-Gutiérrez, A.; Deelder, A.M.; Mayboroda, O.A. Gas Chromatography-Atmospheric Pressure Chemical Ionization-Time of Flight Mass Spectrometry for Profiling of Phenolic Compounds in Extra Virgin Olive Oil. *J. Chromatogr. A* **2011**, *1218*, 959–971. [CrossRef]
23. Hurtado-Fernández, E.; Pacchiarotta, T.; Mayboroda, O.A.; Fernández-Gutiérrez, A.; Carrasco-Pancorbo, A. Metabolomic Analysis of Avocado Fruits by GC-APCI-TOF MS: Effects of Ripening Degrees and Fruit Varieties. *Anal. Bioanal. Chem.* **2015**, *407*, 547–555. [CrossRef]
24. Garlito, B.; Portolés, T.; Niessen, W.M.A.; Navarro, J.C.; Hontoria, F.; Monroig, O.; Varó, I.; Serrano, R. Identification of Very Long-Chain (>C 24) Fatty Acid Methyl Esters Using Gas Chromatography Coupled to Quadrupole/Time-of-Flight Mass Spectrometry with Atmospheric Pressure Chemical Ionization Source. *Anal. Chim. Acta* **2019**, *1051*, 103–109. [CrossRef]
25. Allers, M.; Langejuergen, J.; Gaida, A.; Holz, O.; Schuchardt, S.; Hohlfeld, J.M.; Zimmermann, S. Measurement of Exhaled Volatile Organic Compounds from Patients with Chronic Obstructive Pulmonary Disease (COPD) Using Closed Gas Loop GC-IMS and GC-APCI-MS. *J. Breath Res.* **2016**, *10*, 026004. [CrossRef]
26. Hennig, K.; Antignac, J.P.; Bichon, E.; Morvan, M.-L.; Miran, I.; Delaloge, S.; Feunteun, J.; le Bizec, B. Steroid Hormone Profiling in Human Breast Adipose Tissue Using Semi-Automated Purification and Highly Sensitive Determination of Estrogens by GC-APCI-MS/MS. *Anal. Bioanal. Chem.* **2018**, *410*, 259–275. [CrossRef]
27. Li, D.X.; Gan, L.; Bronja, A.; Schmitz, O.J. Gas Chromatography Coupled to Atmospheric Pressure Ionization Mass Spectrometry (GC-API-MS): Review. *Anal. Chim. Acta* **2015**, *891*, 43–61. [CrossRef] [PubMed]
28. Mcewen, C.N.; Mckay, R.G. A Combination Atmospheric Pressure LC/MS:GC/MS Ion Source: Advantages of Dual AP-LC/MS:GC/MS Instrumentation. *J. Am. Soc. Mass Spectrom.* **2005**, *16*, 1730–1738. [CrossRef] [PubMed]
29. Schiewek, R.; Lorenz, M.; Giese, R.; Brockmann, K.; Benter, T.; Gäb, S.; Schmitz, O.J. Development of a Multipurpose Ion Source for LC-MS and GC-API MS. *Anal. Bioanal. Chem.* **2008**, *392*, 87–96. [CrossRef] [PubMed]
30. Robb, D.B.; Covey, T.R.; Bruins, A.P. Atmospheric Pressure Photoionization: An Ionization Method for Liquid Chromatography-Mass Spectrometry. *Anal. Chem.* **2000**, *72*, 3653–3659. [CrossRef]
31. Syage, J.; Matthew, A.; Evans, D.; Hanold, K.A. Photoionization Mass Spectrometry. *Am. Lab.* **2000**, *32*, 24–29.
32. Cody, R.B.; Laramée, J.A.; Durst, H.D. Versatile New Ion Source for the Analysis of Materials in Open Air under Ambient Conditions. *Anal. Chem.* **2005**, *77*, 2297–2302. [CrossRef] [PubMed]
33. Cody, R.B. GC/MS with a DARTTM Ion Source. In *LC GC North America*; Advanstar Communications: Duluth, MN, USA, 2008; p. 59.
34. Faubert, D.; Paul, G.J.C.; Giroux, J.; Bertrand, M.J. Selective Fragmentation and Ionization of Organic Compounds Using an Energy-Tunable Rare-Gas Metastable Beam Source. *Int. J. Mass Spectrom. Ion Process.* **1993**, *124*, 69–77. [CrossRef]
35. Nudnova, M.M.; Zhu, L.; Zenobi, R. Active Capillary Plasma Source for Ambient Mass Spectrometry. *Rapid Commun. Mass Spectrom.* **2012**, *26*, 1447–1452. [CrossRef]
36. Canellas, E.; Vera, P.; Domeño, C.; Alfaro, P.; Nerín, C. Atmospheric Pressure Gas Chromatography Coupled to Quadrupole-Time of Flight Mass Spectrometry as a Powerful Tool for Identification of Non Intentionally Added Substances in Acrylic Adhesives Used in Food Packaging Materials. *J. Chromatogr. A* **2012**, *1235*, 141–148. [CrossRef]
37. Sales, C.; Portolés, T.; Sancho, J.V.; Abad, E.; Ábalos, M.; Sauló, J.; Fiedler, H.; Gómara, B.; Beltrán, J. Potential of Gas Chromatography-Atmospheric Pressure Chemical Ionization-Tandem Mass Spectrometry for Screening and Quantification of Hexabromocyclododecane. *Anal. Bioanal. Chem.* **2016**, *408*, 449–459. [CrossRef] [PubMed]
38. Hernández, F.; Ibáñez, M.; Portolés, T.; Cervera, M.I.; Sancho, J.V.; López, F.J. Advancing towards Universal Screening for Organic Pollutants in Waters. *J. Hazard. Mater.* **2015**, *282*, 86–95. [CrossRef]
39. Sheu, R.; Marcotte, A.; Khare, P.; Charan, S.; Ditto, J.C.; Gentner, D.R. Advances in Offline Approaches for Chemically Speciated Measurements of Trace Gas-Phase Organic Compounds via Adsorbent Tubes in an Integrated Sampling-to-Analysis System. *J. Chromatogr. A* **2018**, *1575*, 80–90. [CrossRef] [PubMed]
40. Li, X.; Zhen, Y.; Wang, R.; Li, T.; Dong, S.; Zhang, W.; Cheng, J.; Wang, P.; Su, X. Application of Gas Chromatography Coupled to Triple Quadrupole Mass Spectrometry (GC-(APCI)MS/MS) in Determination of PCBs (Mono-to Deca-) and PCDD/Fs in Chinese Mitten Crab Food Webs. *Chemosphere* **2021**, *265*, 129055. [CrossRef] [PubMed]
41. Zhang, X.; di Lorenzo, R.A.; Helm, P.A.; Reiner, E.J.; Howard, P.H.; Muir, D.C.G.; Sled, J.G.; Jobst, K.J. Compositional Space: A Guide for Environmental Chemists on the Identification of Persistent and Bioaccumulative Organics Using Mass Spectrometry. *Environ. Int.* **2019**, *132*, 104808. [CrossRef]

42. Larson, E.A.; Hutchinson, C.P.; Lee, Y.J. Gas Chromatography-Tandem Mass Spectrometry of Lignin Pyrolyzates with Dopant-Assisted Atmospheric Pressure Chemical Ionization and Molecular Structure Search with CSI:FingerID. *J. Am. Soc. Mass Spectrom.* **2018**, *29*, 1908–1918. [CrossRef]
43. Ojanperä, I.; Mesihää, S.; Rasanen, I.; Pelander, A.; Ketola, R.A. Simultaneous Identification and Quantification of New Psychoactive Substances in Blood by GC-APCI-QTOFMS Coupled to Nitrogen Chemiluminescence Detection without Authentic Reference Standards. *Anal. Bioanal. Chem.* **2016**, *408*, 3395–3400. [CrossRef]
44. Mesihää, S.; Ketola, R.A.; Pelander, A.; Rasanen, I.; Ojanperä, I. Development of a GC-APCI-QTOFMS Library for New Psychoactive Substances and Comparison to a Commercial ESI Library. *Anal. Bioanal. Chem.* **2017**, *409*, 2007–2013. [CrossRef]
45. Ma, S.; Ma, C.; Qian, K.; Zhou, Y.; Shi, Q. Characterization of Phenolic Compounds in Coal Tar by Gas Chromatography/Negative-Ion Atmospheric Pressure Chemical Ionization Mass Spectrometry. *Rapid Commun. Mass Spectrom.* **2016**, *30*, 1806–1810. [CrossRef] [PubMed]
46. Geng, D.; Jogsten, I.E.; Dunstan, J.; Hagberg, J.; Wang, T.; Ruzzin, J.; Rabasa-Lhoret, R.; van Bavel, B. Gas Chromatography/Atmospheric Pressure Chemical Ionization/Mass Spectrometry for the Analysis of Organochlorine Pesticides and Polychlorinated Biphenyls in Human Serum. *J. Chromatogr. A* **2016**, *1453*, 88–98. [CrossRef] [PubMed]
47. Mesihaa, S.; Rasanen, I.; Ojanpera, I. Quantitative Estimation of α-PVP Metabolites in Urine by GC-APCI-QTOFMS with Nitrogen Chemiluminescence Detection Based on Parent Drug Calibration. *Forensic Sci. Int.* **2018**, *286*, 12–17. [CrossRef] [PubMed]
48. Zacs, D.; Perkons, I.; Bartkevics, V. Evaluation of Analytical Performance of Gas Chromatography Coupled with Atmospheric Pressure Chemical Ionization Fourier Transform Ion Cyclotron Resonance Mass Spectrometry (GC-APCI-FT-ICR-MS) in the Target and Nontargeted Analysis of Brominated and Chlo. *Chemosphere* **2019**, *225*, 368–377. [CrossRef] [PubMed]
49. Raro, M.; Portolés, T.; Pitarch, E.; Sancho, J.V.; Hernández, F.; Garrostas, L.; Marcos, J.; Ventura, R.; Segura, J.; Pozo, O.J. Potential of Atmospheric Pressure Chemical Ionization Source in Gas Chromatography Tandem Mass Spectrometry for the Screening of Urinary Exogenous Androgenic Anabolic Steroids. *Anal. Chim. Acta* **2016**, *906*, 128–138. [CrossRef]
50. Geng, D.; Kukucka, P.; Jogsten, I.E. Analysis of Brominated Flame Retardants and Their Derivatives by Atmospheric Pressure Chemical Ionization Using Gas Chromatography Coupled to Tandem Quadrupole Mass Spectrometry. *Talanta* **2017**, *162*, 618–624. [CrossRef]
51. Zhang, Y.; Chen, Y.; Li, R.; Chen, W.; Song, Y.; Hu, D.; Cai, Z. Determination of PM2.5-Bound Polyaromatic Hydrocarbons and Their Hydroxylated Derivatives by Atmospheric Pressure Gas Chromatography-Tandem Mass Spectrometry. *Talanta* **2019**, *195*, 757–763. [CrossRef]
52. Gill, B.; Mell, A.; Shanmuganathan, M.; Jobst, K.; Zhang, X.; Kinniburgh, D.; Cherry, N.; Britz-McKibbin, P. Urinary Hydroxypyrene Determination for Biomonitoring of Firefighters Deployed at the Fort McMurray Wildfire: An Inter-Laboratory Method Comparison. *Anal. Bioanal. Chem.* **2019**, *411*, 1397–1407. [CrossRef] [PubMed]
53. Organtini, K.L.; Myers, A.L.; Jobst, K.J.; Reiner, E.J.; Ross, B.; Ladak, A.; Mullin, L.; Stevens, D.; Dorman, F.L. Quantitative Analysis of Mixed Halogen Dioxins and Furans in Fire Debris Utilizing Atmospheric Pressure Ionization Gas Chromatography-Triple Quadrupole Mass Spectrometry. *Anal. Chem.* **2015**, *87*, 10368–10377. [CrossRef]
54. Organtini, K.L.; Haimovici, L.; Jobst, K.J.; Reiner, E.J.; Ladak, A.; Stevens, D.; Cochran, J.W.; Dorman, F.L. Comparison of Atmospheric Pressure Ionization Gas Chromatography-Triple Quadrupole Mass Spectrometry to Traditional High-Resolution Mass Spectrometry for the Identification and Quantification of Halogenated Dioxins and Furans. *Anal. Chem.* **2015**, *87*, 7902–7908. [CrossRef]
55. Portoles, T.; Sales, C.; Abalos, M.; Saulo, J.; Abad, E. Evaluation of the Capabilities of Atmospheric Pressure Chemical Ionization Source Coupled to Tandem Mass Spectrometry for the Determination of Dioxin-like Polychlorobiphenyls in Complex-Matrix Food Samples. *Anal. Chim. Acta* **2016**, *937*, 96–105. [CrossRef]
56. Megson, D.; Robson, M.; Jobst, K.J.; Helm, P.A.; Reiner, E.J. Determination of Halogenated Flame Retardants Using Gas Chromatography with Atmospheric Pressure Chemical Ionization (APCI) and a High-Resolution Quadrupole Time-of-Flight Mass Spectrometer (HRqTOFMS). *Anal. Chem.* **2016**, *88*, 11406–11411. [CrossRef]
57. Mirabelli, M.F.; Zenobi, R. Solid-Phase Microextraction Coupled to Capillary Atmospheric Pressure Photoionization-Mass Spectrometry for Direct Analysis of Polar and Nonpolar Compounds. *Anal. Chem.* **2018**, *90*, 5015–5022. [CrossRef] [PubMed]
58. Deibel, E.; Klink, D.; Schmitz, O.J. New Derivatization Strategies for the Ultrasensitive Analysis of Non-Aromatic Analytes with APLI-TOF-MS. *Anal. Bioanal. Chem.* **2015**, *407*, 7425–7434. [CrossRef] [PubMed]
59. Ayala-Cabrera, J.F.; Contreras-Llin, A.; Moyano, E.; Santos, F.J. A Novel Methodology for the Determination of Neutral Perfluoroalkyl and Polyfluoroalkyl Substances in Water by Gas Chromatography-Atmospheric Pressure Photoionization-High Resolution Mass Spectrometry. *Anal. Chim. Acta* **2020**, *1100*, 97–106. [CrossRef]
60. Cervera, M.I.; Portolés, T.; López, F.J.; Beltrán, J.; Hernández, F. Screening and Quantification of Pesticide Residues in Fruits and Vegetables Making Use of Gas Chromatography-Quadrupole Time-of-Flight Mass Spectrometry with Atmospheric Pressure Chemical Ionization. *Anal. Bioanal. Chem.* **2014**, *406*, 6843–6855. [CrossRef]
61. Richter-Brockmann, S.; Dettbarn, G.; Jessel, S.; John, A.; Seidel, A.; Achten, C. GC-APLI-MS as a Powerful Tool for the Analysis of BaP-Tetraol in Human Urine. *J. Chromatogr. B* **2018**, *1100*, 1–5. [CrossRef]
62. Leider, A.; Richter-Brockmann, S.; Nettersheim, B.J.; Achten, C.; Hallmann, C. Low-Femtogram Sensitivity Analysis of Polyaromatic Hydrocarbons Using GC-APLI-TOF Mass Spectrometry: Extending the Target Window for Aromatic Steroids in Early Proterozoic Rocks. *Org. Geochem.* **2019**, *129*, 77–87. [CrossRef]

63. Dohmann, J.F.; Thiäner, J.B.; Achten, C. Ultrasensitive Detection of Polycyclic Aromatic Hydrocarbons in Coastal and Harbor Water Using GC-APLI-MS. *Mar. Pollut. Bull.* **2019**, *149*, 110547. [CrossRef]
64. Große Brinkhaus, S.; Thiäner, J.B.; Achten, C.; Grosse Brinkhaus, S.; Thiaener, J.B.; Achten, C. Ultra-High Sensitive PAH Analysis of Certified Reference Materials and Environmental Samples by GC-APLI-MS. *Anal. Bioanal. Chem.* **2017**, *409*, 2801–2812. [CrossRef]
65. Cha, E.; Sook, E.; Cha, S.; Lee, J. Analytica Chimica Acta Coupling of Gas Chromatography and Electrospray Ionization High Resolution Mass Spectrometry for the Analysis of Anabolic Steroids as Trimethylsilyl Derivatives in Human Urine. *Anal. Chim. Acta* **2017**, *964*, 123–133. [CrossRef] [PubMed]
66. Cha, E.; Jeong, E.S.; Han, S.B.; Cha, S.; Son, J.; Kim, S.; Oh, H.B.; Lee, J. Ionization of Gas-Phase Polycyclic Aromatic Hydrocarbons in Electrospray Ionization Coupled with Gas Chromatography. *Anal. Chem.* **2018**, *90*, 4203–4211. [CrossRef]
67. Ayala-Cabrera, J.F.; Galceran, M.T.; Moyano, E.; Santos, F.J. Chloride-Attachment Atmospheric Pressure Photoionisation for the Determination of Short-Chain Chlorinated Paraffins by Gas Chromatography-High-Resolution Mass Spectrometry. *Anal. Chim. Acta* **2021**, *1172*, 338673. [CrossRef]
68. Di Lorenzo, R.A.; Lobodin, V.V.; Cochran, J.; Kolic, T.; Besevic, S.; Sled, J.G.; Reiner, E.J.; Jobst, K.J. Fast Gas Chromatography-Atmospheric Pressure (Photo)Ionization Mass Spectrometry of Polybrominated Diphenylether Flame Retardants. *Anal. Chim. Acta* **2019**, *1056*, 70–78. [CrossRef]
69. Kauppila, T.J.; Kersten, H.; Benter, T. Ionization of EPA Contaminants in Direct and Dopant- and Atmospheric Pressure Laser Ionization. *J. Am. Soc. Mass Spectrom.* **2015**, *26*, 1036–1045. [CrossRef] [PubMed]
70. Stader, C.; Beer, F.T.; Achten, C. Environmental PAH Analysis by Gas Chromatography-Atmospheric Pressure Laser Ionization-Time-of-Flight-Mass Spectrometry (GC-APLI-MS). *Anal. Bioanal. Chem.* **2013**, *405*, 7041–7052. [CrossRef]
71. Mollah, S.; Pris, A.D.; Johnson, S.K.; Gwizdala, A.B., III; Houk, R.S. Identification of Metal Cations, Metal Complexes, and Anions by Electrospray Mass Spectrometry Inthe Negative Ion Mode. *Anal. Chem.* **2000**, *72*, 985–991. [CrossRef]
72. Cody, R.B. Observation of Molecular Ions and Analysis of Nonpolar Compounds with the Direct Analysis in Real Time Ion Source. *Anal. Chem.* **2009**, *81*, 1101–1107. [CrossRef]
73. Moore, S. Use of the Metastable Atom Bombardment (MAB) Ion Source for the Elimination of PCDE Interference in PCDD/PCDF Analysis. *Chemosphere* **2002**, *49*, 121–125. [CrossRef]
74. Nørgaard, A.W.; Kofoed-Sørensen, V.; Svensmark, B.; Wolkoff, P.; Clausen, P.A. Gas Chromatography Interfaced with Atmospheric Pressure Ionization-Quadrupole Time-of-Flight-Mass Spectrometry by Low-Temperature Plasma Ionization. *Anal. Chem.* **2013**, *85*, 28–32. [CrossRef] [PubMed]
75. Mirabelli, M.F.; Wolf, J.C.; Zenobi, R. Atmospheric Pressure Soft Ionization for Gas Chromatography with Dielectric Barrier Discharge Ionization-Mass Spectrometry (GC-DBDI-MS). *Analyst* **2017**, *142*, 1909–1915. [CrossRef] [PubMed]
76. Hagenhoff, S.; Korf, A.; Markgraf, U.; Brandt, S.; Schütz, A.; Franzke, J.; Hayen, H. Screening of Semifluorinated N-Alkanes by Gas Chromatography Coupled to Dielectric Barrier Discharge Ionization Mass Spectrometry. *Rapid Commun. Mass Spectrom.* **2018**, *32*, 1092–1098. [CrossRef]
77. Schymanski, E.L.; Singer, H.P.; Slobodnik, J.; Ipolyi, I.M.; Oswald, P.; Krauss, M.; Schulze, T.; Haglund, P.; Letzel, T.; Grosse, S.; et al. Non-Target Screening with High-Resolution Mass Spectrometry: Critical Review Using a Collaborative Trial on Water Analysis. *Anal. Bioanal. Chem.* **2015**, *407*, 6237–6255. [CrossRef]
78. Pitarch, E.; Cervera, M.I.; Portolés, T.; Ibáñez, M.; Barreda, M.; Renau-Pruñonosa, A.; Morell, I.; López, F.; Albarrán, F.; Hernández, F. Comprehensive Monitoring of Organic Micro-Pollutants in Surface and Groundwater in the Surrounding of a Solid-Waste Treatment Plant of Castellón, Spain. *Sci. Total Environ.* **2016**, *548*, 211–220. [CrossRef]
79. Ballesteros-Gómez, A.; de Boer, J.; Leonards, P.E.G. Novel Analytical Methods for Flame Retardants and Plasticizers Based on Gas Chromatography, Comprehensive Two-Dimensional Gas Chromatography, and Direct Probe Coupled to Atmospheric Pressure Chemical Ionization-High Resolution Time-of-Flight-Mass Spectrometry. *Anal. Chem.* **2013**, *85*, 9572–9580. [CrossRef]
80. Jobst, K.J.; Arora, A.; Pollitt, K.G.; Sled, J.G. Dried Blood Spots for the Identification of Bioaccumulating Organic Compounds: Current Challenges and Future Perspectives. *Curr. Opin. Environ. Sci. Health* **2020**, *15*, 66–73. [CrossRef]
81. Patterson, D.G.; Welch, S.M.; Turner, W.E.; Sjödin, A.; Focant, J.-F. Cryogenic Zone Compression for the Measurement of Dioxins in Human Serum by Isotope Dilution at the Attogram Level Using Modulated Gas Chromatography Coupled to High Resolution Magnetic Sector Mass Spectrometry. *J. Chromatogr. A* **2011**, *1218*, 3274–3281. [CrossRef]
82. Seeley, J.V.; Schimmel, N.E.; Seeley, S.K. The Multi-Mode Modulator: A Versatile Fluidic Device for Two-Dimensional Gas Chromatography. *J. Chromatogr. A* **2018**, *1536*, 6–15. [CrossRef] [PubMed]
83. Jobst, K.J.; Seeley, J.V.; Reiner, E.J.; Mullin, L.; Ladak, A. Enhancing the Sensitivity of Atmospheric Pressure Ionization Mass Spectrometry Using Flow Modulated Gas Chromatography. *Curr. Trends Mass Spectrom.* **2018**, *16*, 15–19.
84. Bowman, D.T.; Jobst, K.J.; Helm, P.A.; Kleywegt, S.; Diamond, M.L. Characterization of Polycyclic Aromatic Compounds in Commercial Pavement Sealcoat Products for Enhanced Source Apportionment. *Environ. Sci. Technol.* **2019**, *53*, 3157–3165. [CrossRef] [PubMed]
85. Venable, J.D.; Dong, M.; Wohlschlegel, J.; Dillin, A.; Yates, J.R., III. Automated Approach for Quantitative Analysis of Complex Peptide Mixtures from Tandem Mass Spectra. *Nat. Methods* **2004**, *1*, 39–45. [CrossRef]

86. Gillet, L.C.; Navarro, P.; Tate, S.; Ro, H.; Selevsek, N.; Reiter, L.; Bonner, R.; Aebersold, R. Targeted Data Extraction of the MS/MS Spectra Generated by Data-Independent Acquisition: A New Concept for Consistent and Accurate Proteome Analysis. *Mol. Cell. Proteom.* **2012**, *11*, 1–17. [CrossRef] [PubMed]
87. Moseley, M.A.; Hughes, C.J.; Juvvadi, P.R.; Soderblom, E.J.; Lennon, S.; Perkins, S.R.; Thompson, J.W.; Steinbach, W.J.; Geromanos, S.J.; Wildgoose, J.; et al. Scanning Quadrupole Data-Independent Acquisition, Part A: Qualitative and Quantitative Characterization. *J. Proteome Res.* **2018**, *17*, 770–779. [CrossRef]
88. Waters. MassFragment for Structural Elucidation in Metabolite ID Using Exact Mass MS. Available online: https://www.waters.com/waters/library.htm?locale=en_US&lid=10064396 (accessed on 9 August 2021).
89. Waters. UNIFI Scientific Information System. Available online: https://www.waters.com/waters/en_US/UNIFI-Scientific-Information-System/nav.htm?cid=134801359&locale=en_US (accessed on 9 August 2021).
90. Wolf, S.; Schmidt, S.; Müller-Hannemann, M.; Neumann, S. In Silico Fragmentation for Computer Assisted Identification of Metabolite Mass Spectra. *BMC Bioinform.* **2010**, *11*, 148. [CrossRef]
91. Dührkop, K.; Shen, H.; Meusel, M.; Rousu, J.; Böcker, S.; Dührkop, K.; Shen, H.; Meusel, M.; Rousu, J.; Böcker, S. Searching Molecular Structure Databases with Tandem Mass Spectra Using CSI: FingerID. *Proc. Natl. Acad. Sci. USA* **2015**, *112*, 12580–12585. [CrossRef]
92. Allen, F.; Pon, A.; Wilson, M.; Greiner, R.; Wishart, D. CFM-ID: A Web Server for Annotation, Spectrum Prediction and Metabolite Identification from Tandem Mass Spectra. *Nucleic Acids Res.* **2014**, *42*, W94–W99. [CrossRef]
93. Grimme, S. *Towards First Principles Calculation of Electron Impact Mass Spectra of Molecules. Angewandte Chemie*; Verlag Chemie: Weinheim/Bergstrasse, Germany; New York, NY, USA, 2013; pp. 6306–6312. [CrossRef]
94. Kanu, A.B.; Dwivedi, P.; Tam, M.; Matz, L.; Hill, H.H. Ion Mobility-Mass Spectrometry. *J. Mass Spectrom.* **2008**, *43*, 1–22. [CrossRef] [PubMed]
95. Olanrewaju, C.A.; Ramirez, C.E.; Fernandez-Lima, F. Comprehensive Screening of Polycyclic Aromatic Hydrocarbons and Similar Compounds Using GC-APLI-TIMS-TOFMS/GC-EI-MS. *Anal. Chem.* **2021**, *93*, 6080–6087. [CrossRef]
96. Giles, K.; Ujma, J.; Wildgoose, J.; Pringle, S.; Richardson, K.; Langridge, D.; Green, M. A Cyclic Ion Mobility-Mass Spectrometry System. *Anal. Chem.* **2019**, *91*, 8564–8573. [CrossRef] [PubMed]
97. Scheubert, K.; Hufsky, F.; Böcker, S. Computational Mass Spectrometry for Small Molecules. *J. Cheminform.* **2013**, *5*, 12. [CrossRef] [PubMed]
98. Su, Q.Z.; Vera, P.; van de Wiele, C.; Nerín, C.; Lin, Q.B.; Zhong, H.N. Non-Target Screening of (Semi-)Volatiles in Food-Grade Polymers by Comparison of Atmospheric Pressure Gas Chromatography Quadrupole Time-of-Flight and Electron Ionization Mass Spectrometry. *Talanta* **2019**, *202*, 285–296. [CrossRef] [PubMed]
99. Cherta, L.; Portolés, T.; Pitarch, E.; Beltran, J.; López, F.J.; Calatayud, C.; Company, B.; Hernández, F. Analytical Strategy Based on the Combination of Gas Chromatography Coupled to Time-of-Flight and Hybrid Quadrupole Time-of-Flight Mass Analyzers for Non-Target Analysis in Food Packaging. *Food Chem.* **2015**, *188*, 301–308. [CrossRef]
100. MItchum, R.K.; Korfmacher, W.A.; Molar, G.F.; Stalling, D.L. Capillary Gas Chromatography/Atmospheric Pressure Negative Chemical Ionization Mass Spectrometry of the 22 Isomeric T Etrachlorodlbenzo-p-Dioxins. *Anal. Chem.* **1982**, *54*, 719–722. [CrossRef]
101. Korfmacher, W.A.; Rowland, K.R.; Mitchum, R.K.; Daly, J.J.; McDaniel, R.C.; Plummer, M.V. Analysis of Snake Tissue and Snake Eggs for 2,3,7,8-Tetrachlorodibenzo-p-Dioxin via Fused Silica GC Combined with Atmospheric Pressure Ionization MS. *Chemosphere* **1984**, *13*, 1229–1233. [CrossRef]
102. Korfmacher, W.A.; Hansen, E.B.; Rowland, K.L. Tissue Distribution of 2,3,7,8-TCDD in Bullfrogs Obtained from a 2,3,7,8-TCDD-Contaminated Area. *Chemosphere* **1986**, *15*, 121–126. [CrossRef]
103. Fernando, S.; Green, M.K.; Organtini, K.; Dorman, F.; Jones, R.; Reiner, E.J.; Jobst, K.J. Differentiation of (Mixed) Halogenated Dibenzo-p-Dioxins by Negative Ion Atmospheric Pressure Chemical Ionization. *Anal. Chem.* **2016**, *88*, 5205–5211. [CrossRef] [PubMed]
104. Ishaq, R.; Näf, C.; Zebühr, Y.; Broman, D.; Järnberg, U. PCBs, PCNs, PCDD/Fs, PAHs and Cl-PAHs in Air and Water Particulate Samples—Patterns and Variations. *Chemosphere* **2003**, *50*, 1131–1150. [CrossRef]
105. Ayrton, S.T.; Jones, R.; Douce, D.S.; Morris, M.R.; Cooks, R.G. Uncatalyzed, Regioselective Oxidation of Saturated Hydrocarbons in an Ambient Corona Discharge. *Angew. Chem.* **2018**, *130*, 777–781. [CrossRef]
106. Megson, D.; Hajimirzaee, S.; Doyle, A.; Cannon, F.; Balouet, J.-C. Investigating the Potential for Transisomerisation of Trycresyl Phosphate with a Palladium Catalyst and Its Implications for Aircraft Cabin Air Quality. *Chemosphere* **2019**, *215*, 532–534. [CrossRef]
107. Lai, A.; Singh, R.R.; Kovalova, L.; Jaeggi, O.; Kondić, T.; Schymanski, E.L. Retrospective Non-Target Analysis to Support Regulatory Water Monitoring: From Masses of Interest to Recommendations via in Silico Workflows. *Environ. Sci. Eur.* **2021**, *33*, 1–21. [CrossRef]
108. Waters. MassFragment. Available online: https://www.waters.com/waters/en_US/MassFragment-/nav.htm?cid=1000943&locale=en_US (accessed on 9 August 2021).
109. Thermo Fisher. Mass Frontier Spectral Interpretation Software. Available online: https://www.thermofisher.com/ca/en/home/industrial/mass-spectrometry/liquid-chromatography-mass-spectrometry-lc-ms/lc-ms-software/multi-omics-data-analysis/mass-frontier-spectral-interpretation-software.html (accessed on 9 August 2021).

110. Schreckenbach, S.A.; Anderson, J.S.M.; Koopman, J.; Grimme, S.; Simpson, M.J.; Jobst, K.J. Predicting the Mass Spectra of Environmental Pollutants Using Computational Chemistry: A Case Study and Critical Evaluation. *J. Am. Soc. Mass Spectrom.* **2021**, *32*, 1508–1518. [CrossRef]
111. Koopman, J.; Grimme, S. From QCEIMS to QCxMS: A Tool to Routinely Calculate CID Mass Spectra Using Molecular Dynamics. *J. Am. Soc. Mass Spectrom.* **2021**, *32*, 1735–1751. [CrossRef] [PubMed]

 molecules

Review

Towards Non-Targeted Screening of Lipid Biomarkers for Improved Equine Anti-Doping

Kathy Tou [1,*], Adam Cawley [2], Christopher Bowen [3], David P. Bishop [4] and Shanlin Fu [1]

1 Centre for Forensic Science, University of Technology Sydney, Sydney, NSW 2007, Australia
2 Australian Racing Forensic Laboratory, Racing NSW, Sydney, NSW 2000, Australia
3 Mass Spectrometry Business Unit, Shimadzu Scientific Instruments (Australasia), Sydney, NSW 2116, Australia
4 Hyphenated Mass Spectrometry Laboratory (HyMAS), University of Technology, Sydney, NSW 2007, Australia
* Correspondence: kathy.tou@student.uts.edu.au

Abstract: The current approach to equine anti-doping is focused on the targeted detection of prohibited substances. However, as new substances are rapidly being developed, the need for complimentary methods for monitoring is crucial to ensure the integrity of the racing industry is upheld. Lipidomics is a growing field involved in the characterisation of lipids, their function and metabolism in a biological system. Different lipids have various biological effects throughout the equine system including platelet aggregation and inflammation. A certain class of lipids that are being reviewed are the eicosanoids (inflammatory markers). The use of eicosanoids as a complementary method for monitoring has become increasingly popular with various studies completed to highlight their potential. Studies including various corticosteroids, non-steroidal anti-inflammatories and cannabidiol have been reviewed to highlight the progress lipidomics has had in contributing to the equine anti-doping industry. This review has explored the techniques used to prepare and analyse samples for lipidomic investigations in addition to the statistical analysis and potential for lipidomics to be used for a longitudinal assessment in the equine anti-doping industry.

Keywords: lipidomics; review; analytical; corticosteroids; NSAIDs

Citation: Tou, K.; Cawley, A.; Bowen, C.; Bishop, D.P.; Fu, S. Towards Non-Targeted Screening of Lipid Biomarkers for Improved Equine Anti-Doping. *Molecules* **2023**, *28*, 312. https://doi.org/10.3390/molecules28010312

Academic Editor: Thomas Letzel

Received: 8 December 2022
Revised: 22 December 2022
Accepted: 24 December 2022
Published: 30 December 2022

1. Biomarkers for Equine Anti-Doping

The current approach to equine anti-doping is focused on the targeted detection of prohibited substances [1]. However, as new substances are rapidly being developed, the need for complimentary methods of monitoring is important to ensure the integrity of the racing industry is upheld [1]. The use of biomarkers for the detection of doping abuse is a significant advancement for sports anti-doping. Teale et al. [2] define biomarkers as an "individual biological parameter or substance (metabolite, protein or transcript); the concentration of which is indicative of the use or abuse of a drug or therapy". With the discovery of novel biomarkers for detecting doping abuse, the potential exists for a larger number of drugs to be indirectly detected and over longer periods of time. However, with indirect detection, there is the possibility of the method not being specific and the increased likelihood of inconsistent results [3]. This review will focus on the use of lipidomics for indirect detection of substances in equine racing, contributing to an intelligence-based anti-doping strategy [4].

2. Current Challenges for Targeted Screening

Doping practices are increasingly sophisticated for all sports with testing laboratories consistently improving their workflows to identify the ever-growing list of prohibited substances. Current detection methods are, however, by nature limited in scope to a defined number of substances and the applicability of the analytical technique used [5].

Maintaining a contemporary scope of testing makes direct detection continually difficult due to availability of reference materials. An "omics" approach may provide an alternative to direct detection of doping [5,6]. The use of metabolomics has been utilised in many different laboratories to measure metabolites at low levels relative to time-related biological responses of a drug administration [3,5,6]. This provides a framework for non-targeted detection, particularly for drugs that have a short half-life but long lasting effect on any individual system [5].

3. Lipids

Lipidomics is a growing field involved in the characterisation of lipids, their function and metabolism in a biological system [7,8]. Lipids are non-polar molecules with a diverse chemistry and functionality [9,10]. In conjunction with carbohydrates, lipids are the main energy source for equine striated muscles [11]. There are a number of different classes of lipids including monounsaturated fatty acids (MUFAs) and polyunsaturated fatty acids (PUFAs) [12,13]. MUFAs are lipids that have a single double bond present in the compound and usually only exist in seeds or marine organisms, however, are naturally rare [12]. PUFAs comparatively contain more than one double bond, are more commonly found [12] and have various biological effects including platelet aggregation and inflammation [14]. In animals, common PUFAs include arachidonic acid (AA), eicosapentaenoic acid (EPA) and docosahexaenoic acid (DHA) [12]. The most relevant and important oxygenated products for the racing industry are lipids known as the eicosanoids [12]. Eicosanoids are a large subclass predominately defined by the 20 carbon chain containing over 100 lipid mediators including prostaglandins, thromboxanes, leukotrienes, hydroxy fatty acids and lipoxins [12,15] with the majority derived from AA, an omega-6 fatty acid [16]. Eicosanoids are believed to act as inflammatory mediators since they have the ability to mimic inflammatory symptoms and decrease in the presence of anti-inflammatory drugs [8,14]. Disruption of eicosanoids can cause a range of inflammatory pathological conditions including asthma, chronic obstructive pulmonary disease, fevers, pain, a range of cardiovascular diseases and cancers [15]. Eicosanoids are synthesised at the site of injury in order to control and regulate the inflammatory response [7].

AA is released from membrane phospholipids through the activation of phospholipase A_2 enzyme (PLA_2) [14]. It can be further converted into other eicosanoids in the cascade (Figure 1). In a non-targeted sense, it should be possible to utilise the AA cascade to determine which lipids are being affected following drug administration. This cascade includes prostaglandin D_2 (PGD_2) [17], prostaglandin E_2 (PGE_2) [18], prostaglandin $F_{2\alpha}$ ($PGF_{2\alpha}$) [7], thromboxane B_2 (TXB_2) [19], 11-dehydro thromboxane B_2 (11-Dehydro TXB_2) [20], 6-keto prostaglandin $F_{1\alpha}$ (6-Keto $PGF_{1\alpha}$) [21], 15(S)-hydroxyeicosatetraenoic acid (15-HETE) [7], 5(S)-hydroxyeicosatetraenoic acid (5-HETE) [22], leukotriene B_4 (LTB_4) [23–25], leukotriene D_4 (LTD_4) [23,25] and leukotriene E_4 (LTE_4) [23,25]. The analogues of AA are also of interest including arachidonoyl ethanolamide (AEA) [26,27] and oleoyl ethanolamide (OEA) [26].

AEA is an endogenous cannabinoid ligand that has binding activity resulting in pharmacological effects of tetrahydrocannabinol (THC) such as euphoria and calmness [26,27]. In a variety of cells, cannabinoid agonists have caused an increase in the amount of AA [26]. This has been hypothesised to result from the combination of PLA_2 and acyltransferase inhibition [26]. AA can also be converted into N-(4-hydroxyphenyl) arachidonylamide (AM404) which displays analgesic properties and the ability to lower body temperature [28,29]. AM404 is produced when acetaminophen is metabolised in the body to produce *p*-aminophenol, which is then conjugated with AA [29]. AM404 has been reported to inhibit the cyclooxygenase (COX) pathways leading to the decreased formation of PGE_2, demonstrating effectiveness as a COX-2 enzyme inhibitor to reduce the production of prostaglandins by consumption of AA [29]. The COX and lipoxygenase (LOX) pathways are of particular interest for equine anti-doping due to their augmentation following anti-inflammatory treatments.

Figure 1. Arachidonic Acid Cascade, adapted from various sources [7,8,14,17–29]. Abbreviations of eicosanoids: arachidonoyl ethanolamide (AEA), oleoyl ethanolamide (OEA), prostaglandin (PG), thromboxane (Tx), hydroperoxyeicosatetraenoic acids (HPETES), hydroxyeicosatetraenoic acid (HETE) and leukotriene (LT).

Prostaglandins are monocarboxylic acids with two side chains at carbon 7 and 8 attached to a central, five-membered ring [14]. Prostaglandins have oxygen-containing substituents in various positions in the molecule with the naming of the prostaglandins ranging from PGA to PGI depending on the basis of the substituents in the ring, and further sub-grouped into three series depending on the degree of unsaturation [14]. Prostaglandins are one of the key compounds in the generation of the inflammatory response due to an increase in concentration in inflamed tissue [30]. Prostaglandins are formed from AA being converted by the COX enzyme to prostaglandin G_2 and H_2 [17]. Prostaglandin endoperoxide-D-isomerase can convert PGH_2 into a mixture of PGD_2, PGE_2 and $PGF_{2\alpha}$ [17]. PGH_2 can also produce prostacyclin (PGI_2) that may further metabolise to a more stable compound, 6-keto $F_{1\alpha}$ [19]. PGEs and PGIs have been identified to mimic the signs of inflammation caused by vasodilation and swelling due to an increase in vascular permeability [14].

Prostaglandins can also convert into thromboxane A_2 (TXA_2), an unstable intermediate in the production of TXB_2, which further metabolises to form 11-dehydro TXB_2 [20]. Thromboxanes are the major products of prostaglandin endoperoxides in platelets, lungs and the spleen [14]. The production of new platelets could potentially have a high capacity to further increase the synthesis of TXA_2, leading to increased amounts of TXB_2 and 11-dehydro TXB_2 [19]. The use of non-steroidal anti-inflammatory drugs (NSAIDs) results in PGI_2 and thromboxane synthesis being inhibited [19].

Using the LOX pathways, AA converts into esterified hydroperoxyeicosatetraenoic acids (HPETEs) [31]. HPETEs are further reduced to their corresponding hydroxyeicosatetraenoic acids (HETEs) [7]. For example, using the 15-LOX enzyme, AA will metabolise into 15-HPETE, and then further metabolises to 15-HETE. Similarly, the 5-LOX enzyme converts AA to 5-HPETE and further to 5-HETE. The leukotrienes are an oxygenated metabolite of polyunsaturated fatty acids, but the initial formation is catalysed by lipoxygenases [23]. Leukotrienes are produced from AA using the 5-lipoxygenase (5-LOX) enzyme to produce Leukotriene A_4 (LTA_4), an unstable epoxide. LTA_4 is hydrolysed to LTB_4 or conjugated with glutathione to yield Leukotriene C_4 (LTC_4) and its metabolites LTD_4 and LTE_4. Leukotrienes are known for their strong vascular effect with the most effective being LTB_4, in comparison to LTC_4 and LTD_4 [14,15]. In the presence of more leukocytes, the leukotrienes also have a role in the inflammatory process by increasing blood pressure [23].

4. Analytical Techniques for Lipidomics

4.1. Sample Preparation

With a wide range of analytical techniques applicable for doping analysis, optimised sample preparation methods have been developed [32]. There are typically two types of sample preparation strategies: protein precipitation (PP) which is monophasic; and lipid extractions which are biphasic [33]. For the latter, solid phase extraction (SPE) and liquid-liquid extraction (LLE) are commonly used [32,33]. LLE efficiency is improved if a mixture of solvents is used in comparison to only a single solvent [34]. SPE is a useful sample preparation method as it allows for the preconcentration of low sample volumes, which provides sensitive detection [35].

Chambers et al. [32] conducted a study comparing the various sample preparation methods including PP, SPE and LLE to optimise sample preparation for the analysis of eicosanoids. While PP was rapid and simple, it did not often provide a clean extract for analysis, however, the type of organic solvent used in the PP did influence the cleanliness of the extract. The MeOH extract contained 40% more potentially interfering phospholipids in comparison to the ACN extract. This result is consistent with the work of Bruce et al. [36] and Stojiljkovic et al. [5] showing that ACN or acetone was the best solvent for PP-based methods. ACN is more efficient in protein removal due to the higher dielectric constant and lower viscosity that it possesses [5]. Ammonium sulphate can also be used, however it has to be combined with SPE or a secondary LLE to ensure a cleaner extract before liquid-chromatography mass spectrometry (LC-MS) analysis [5]. SPE stationary phases are selected along similar principles to chromatographic column phases, with the size, polarity and charge of analyte considered. The pure cation exchange stationary phase has produced cleaner extracts, however, polymetric mixed-mode strong cation exchange can enable hydrophobic phospholipids containing two alkyl chains and lysophospholipids to be removed more efficiently. In comparison, LLE has been widely investigated and provides a cleaner method of sample preparation in comparison to PP [32]. The cleanliness of the extracts were comparable with cation-exchange mixed-mode SPE using three methods: a 3:1 ratio of methyl tert-butyl ether (MTBE) to human plasma, a 3:1 ratio of basified MTBE (5% NH_4OH in MTBE) to human plasma and basified MTBE in a two-step extraction [32].

4.2. Analysis

A broad range of analytical techniques can be used to separate, detect and quantify eicosanoids including high performance liquid chromatography with ultraviolet detection (HPLC-UV) [37,38], enzyme-linked immunosorbent assay (ELISA) [39], nuclear magnetic resonance (NMR) [5], gas chromatography-mass spectrometry (GC-MS) [40] and LC-MS [15,39]. The main disadvantage of using HPLC-UV is the limited sensitivity and specificity of UV detection in complex biological matrices such as urine or plasma [15,37,38].

There is also the possibility of the lack of active chromophores in lipids that absorb UV light at the appropriate wavelengths [38]. ELISA is often limited to one analyte per assay and due to eicosanoids having a large number of isomers, there is the possibility of high cross-reactivity affecting quantification [39]. The use of NMR has its advantages as a non-destructive analysis involving minimal sample preparation, however the main disadvantage is the low sensitivity in comparison to GC-MS or LC-MS and the relatively large sample requirement [5]. GC-MS boasts high sensitivity and resolution of eicosanoids that have numerous isomers, however, complex sample preparation and derivatisation is usually required [40]. One of the most commonly used LC-MS instruments used is triple quadrupole mass spectrometry due to the high sensitivity and specificity obtained using multiple reaction monitoring acquisition for extremely low level lipids in the equine system [7]. With this high sensitivity, quanititation of eicosanoids has been possible which is further discussed in chapter 6 of this review.

The use of high-resolution mass spectrometry (HRMS) in metabolomics has enabled the indirect detection of substances that have a short half-life but long-lasting effects on biological systems [5]. Following this approach, studies have utilised LC-HRMS for lipidomics to investigate the potential to distinguish isomeric lipids [41]. Lipidomic studies are increasingly using LC-HRMS due to high sensitivity, mass accuracy, resolution and acquisition rates providing the ability to detect subtle differences in complex biological matrices such as plasma and urine [42–45]. The two types of HRMS instruments include the Quadrople Time-of-Flight mass spectrometry (QToF-MS) and the Orbitraps. These two HRMS instruments have the ability to achieve a mass accuracy of below 5 ppm for higher accuracy in the identification of compounds, with some instruments being adveritsed to have the ability to achieve less than 1 ppm difference [46,47]. Commerically available QToF instruments have resolving power between 35,000 and 70,000 full width at half maximum (FWHM) whilst comparatively, the Orbitraps have over 1,000,000 FWHM [47–49]. Whilst the Orbitraps do outperform the QToFs in mass resolving power, the disadvantage would be the longer accumulation times, which result in the Orbitraps being less suitable for accurate quanitification of compounds with narrow LC peaks [46]. Therefore, the identification of unknown lipids using LC-HRMS is slightly difficult due to the isomeric and isobaric nature of lipids, but the accuracy of these instruments has the potential to narrow the search to a particular class of lipids rather than the individual compound. The use of LC-HRMS allows for a complementary targeted/non-targeted approach to detect new and emerging drugs that could potentially be used as doping agents, and the metabolic signature that can result from such [5].

4.3. Data Acquisition Modes

Non-targeted lipidomics workflows allow for the generation of new hypotheses for supporting evidence in biological interpretations or for complementary 'omics' data. Targeted lipidomics is often used for the validation of biomarkers following the untargeted discovery phase or whilst trying to measure certain lipids that are associated with disease or are present in low abundance [50].

Two common modes of data acquisition for LC-HRMS analyses supporting lipidomic workflows are data dependent acquisition (DDA) and data independent acquisition (DIA) [47]. DDA utilises precursor ions that have exceeded an abundance threshold, predefined isotopic pattern and the presence of diagnostic product ions [47,48]. These ions, once detected, facilitate the acquisition of MS2 data, however, this is limited to ions of relatively high intensity [51]. There is also the possibility of precursor ions and relevant analytes being missed [52]. DDA is well suited to targeted screening, however reproducibility and variation in spectral databases often results in unknown metabolites being difficult to interpret [47].

Comparatively, DIA or sequential window acquisition of all theoretical fragmentation ion (SWATH) allows for simultaneous screening of all precursor ions and their fragmentation pattern within a specified mass to charge (*m/z*) window regardless of intensity [47–49,51,53,54]. The use of DIA for lipidomics was demonstrated after its potential was seen in proteomics [50]. DIA has been shown to be an effective means to produce a larger number of quantifiable results in short time periods with fewer errors relating to reproducibility across replicate analyses [55]. The freedom of ion choices reduces the need for data reacquisition due to unexpected ion formation or unrecognisable compounds [52]. Therefore, with the ability to screen multiple precursors, it is often used for untargeted screening. The major disadvantage of DIA is the potential for the incorrect precursor ion being identified as there is no criteria for ions to undergo dissociation [43]. It is also possible to obtain a mixture of (i.e., chimeric) spectra due to the wide scanning range of the *m/z* windows [43]. Therefore, in lipidomics, with a large number of isomers and levels of unsaturation, the ability to separate between lipid classes is more difficult [52]. Class separation for lipids is generally limited to DDA workflows where the ability to isolate unit mass would be more appropriate [52]. At the time of writing this review, the authors have identified a gap in the research where there is a lack of research focusing on the use of DIA on eicosanoids, specifically. There is research completed on lipids as an entirety focusing on whole classes but lacking on eicosanoids due to the vast number of lipid isomers and levels of unsaturation [52,56]. Another limitation of DIA is the possibility of isomeric or isobaric compounds being indistinguishable due to co-elution with an increased possibility of incorrect compound assignment as these compounds can act as interferences [43]. There is also the lack of spectral databases available for non-targeted screening as most databases utilise DDA [48].

5. Statistical Analysis

A comprehensive lipidomics study of molecules in any biological system can result in a large amount of data generated. There are two common approaches to process lipidomic data: chemometric and quantitative [57]. Chemometric analysis is performed on spectral patterns or signal intensity data to identify any lipids through non-targeted screening. Chemometric analysis can be automated, however, there is the uncertainty of strict sample uniformity and repetition. Quantitative analysis in comparison is more applicable to biological studies as it allows for the identification of all lipids proceeding to the analysis of the lipid data.

The information collated from quantitative studies are either univariate or multivariate depending on the study being conducted. Saccenti et al. discuss the advantages and disadvantages of univariate and multivariate analyses of metabolomics data [58]. Univariate analysis is performed when only one variable is analysed, and it includes methods to test different sets of samples such as ANOVA or *t*-tests. There are, however, disadvantages with univariate analysis with higher possibilities of equivocal findings due to the requirement for multiple testing corrections. Univariate methods are usually the preferred choice of statistical analysis for biologists due to the ease of interpretation the analysis provides. In comparison, multivariate analysis consists of data that have two or more variables. There are many tools available for multivariate analysis including correlation analysis and simple linear regression which can analyse a data set that contains several hundred or more variables simultaneously, thereby eliminating the need for multiple univariate methods. Multivariate statistical analysis provides advantages that univariate may not, including the ability of independent variables to complement each other for the prediction of a dependent variable. Lipidomic studies are often multivariate in nature due to a large number of compounds investigated, however, it is also common practice for both univariate and multivariate approaches to be used in a complementary manner [58].

Data are often analysed using both unsupervised and supervised strategies [42]. Unsupervised studies allow the discovery of groups or trends in the data with the most common tool being principal component analysis (PCA). Supervised methods are then often used to discover new biomarkers with partial least squares (PLS) being one of the most common. PLS is a multidimensional method that utilises a data matrix containing independent variables and relates it to dependent variables [42]. As biomarker research usually involves multiple approaches, it is recommended to combine both unsupervised and supervised methods [58]. Various statistical software programs have been utilised to implement supervised and unsupervised methods. One such software program is the Mass Spectrometry-Data Independent Analysis software version 4 (MS-DIAL 4), where a non-targeted lipidomics platform has been utilised to provide a comprehensive lipid database [59]. This contains retention times, mass-to-charge ratios, isotopic ions, adduct information and other mass spectral information for the possible identification of unknowns and lipid pathways [59]. However, there are many limitations that arise with trying to determine the putative structure of an unknown given the limited information provided from statistical analysis [1,60,61]. Confirmations of putative biomarkers is difficult as comparisons need to be made with authentic reference standards if possible, if not, there is the possibility of custom synthesis, but these are highly cost inefficient and may incur timely delays for a highly purified sample [60]. The use of these statistical analyses will help provide the information required to further improve current routine testing using non-targeted workflows to monitor for new biomarkers indicative of doping.

6. Studies Monitoring Eicosanoids for Equine Anti-Doping Screening

The administration of corticosteroids and NSAIDs to alleviate pain and inflammation is known to affect the amount of AA, with the COX and LOX pathways subsequently reducing the concentration of eicosanoids present [62–67]. The use of approved corticosteroids and NSAIDs, whilst being legitimate therapeutics for racehorses out-of-competition, are controlled for race day competition. Therefore, the monitoring of terminal compounds in the AA cascade may prove beneficial for the detection of these prohibited substances. Glucocorticoids are potent anti-inflammatory agents that increase the tolerance for pain which may allow horses to compete under conditions which could compromise the health of the individual horse and safety of riders. Early studies on eicosanoids were performed using ELISA and high-performance liquid chromatography, however, these techniques are known to lack specificity [7]. Currently, advanced technology of hyphenated mass spectrometry such as triple quadrupole mass spectrometry and QToF instruments are being used. These instruments have enabled an enhanced scope of eicosanoid monitoring [7] for targeted screening, however, the use of non-targeted methods have not been extensively explored. Therefore, there is potential in future studies for untargeted methods to provide evidence of an exogenous administration.

6.1. Lipid Screening in Equine Samples

Jackson et al. [18] was one of the first studies to explore the use of lipids (specifically PGE_2) in lipopolysaccharide (LPS)-stimulated blood for a comprehensive NSAID screening tool using ELISA. However, there was a moderate degree of variability between the different horses tested due to the LPS-stimulation. Even so, PGE_2 production was less than 50% between 8- and 12-h post-administration. Therefore, it was proposed that an LPS-stimulated plasma concentration of PGE_2 of less than 500 pg/mL could potentially indicate the administration of an NSAID given the terminal position of PGE_2 in the cascade.

Nolazco et al. [11] studied the detection of lipids in racehorses before and after supramaximal exercise utilising an LC-QToF-MS. A non-targeted approach discovered 933 lipids present in the plasma of which 130 were known based on library matches. One-tenth (i.e., 13 out of 130) were deemed statistically different compared to baseline concentrations. From these, three unsaturated fatty acids and six phospholipids displayed an increase in signal intensity, whilst four saturated fatty acids (including n-eicosanoid acid) and five triacylglycerols had a decrease in signal intensity. Various hypotheses were proposed to explain the results which included that during exercise, lipolysis causes the ratio of unsaturated to saturated fatty acids for triacylglycerols contained in adipose tissue to be higher than non-esterified fatty acids. Another possibility is that during exercise, the use of saturated fatty acids as an energy source is preferred over unsaturated fatty acids, causing an increase in concentration. Whilst this study provided information about the lipidome changes that occur during exercise in racehorses, limitations include the use of a treadmill rather than a racetrack and the small sample size due to the availability of horses. This study demonstrated that non-targeted lipid profiling aimed at detecting anti-inflammatory drug use needs to account for the effects of exercise.

6.2. Lipid Inflammatory Markers and Their Effect from a Corticosteroid Administration

Corticosteroids including dexamethasone, triamcinolone acetonide (TACA) and flumetasone are commonly used throughout the racing industry to alleviate symptoms of inflammation and prevent further tissue damage caused by inflammatory markers [64].

Mangal et al. [7] and Knych et al. [64] explored the use of dexamethasone on the effects of inflammatory mediators. Mangal explored 25 eicosanoids in equine plasma using stable isotope dilution reversed-phased LC-MS, whilst Knych utilised triple quadrupole mass spectrometry to monitor six eicosanoids. Both Mangal and Knych employed calcium ionophore (CI) A23187 or LPS-simulation of whole blood, whilst Mangal also used AA added exogenously to determine whether extra enzymatic activity would increase the production of eicosanoids. The use of CI allows for any free arachidonic acid to be released from surrounding cellular membranes to produce eicosanoids through either COX or LOX enzyme activity [62,63]. The use of LPS-stimulation allows for the modelling of COX enzymes for the production of prostaglandins and thromboxanes from arachidonic acid [30,62,63]. Mangal concluded from the 25 monitored eicosanoids, 9 resulted in reduced levels [7]. Comparatively, Knych concluded that whilst dexamethasone does not have a direct effect on the COX-1 enzyme, the effect could be seen on the COX-2 enzyme due to LPS stimulation [64]. Knych also observed that the LOX pathways were affected, specifically 5-LOX, due to the nuclear-factor (NF) kappa beta enzyme which is a known inducer of the 5-LOX enzyme [64]. From CI-stimulation, significant down-regulation was observed for four eicosanoids, while LPS-stimulation showed five eicosanoids to be significantly down-regulated [64].

TACA is a commonly used, long-lasting and potent glucocorticoid that is known to bind to glucocorticoid responsive genes. This can result in increased concentrations of anti-inflammatory mediators whilst decreasing the inflammatory markers, as demonstrated by another study conducted by Mangal et al. using a triple quadrupole mass spectrometer [62]. CI and LPS stimulation were again used to explore the inflammatory mediators in equine plasma. For CI-stimulation, a significant increase for two inflammatory mediators was observed but only one proceeded to show a significant decrease. Comparatively, for LPS-stimulation, four eicosanoids showed significant reduction in concentration, then a significant increase in concentration prior to a return to basal levels. This study concluded that major products expressed through the COX-2 enzymatic activity are down-regulated under the influence of an external glucocorticoid suggesting inhibition of the COX-2 enzyme [62]. It is worth noting that there is speculation that PGE_2 has the potential to have a binary role in inflammation given that it is produced using the COX-1 enzyme as an inflammatory mediator, however delayed production of PGE_2 by the m-PGES-1 enzyme, coupled to COX-2 gives PGE_2 its anti-inflammatory properties [62,68].

Flumetasone is a potent corticosteroid often used in the treatment of performance related injuries associated with strenuous exercise [69]. Knych et al. explored the effects of Flumetasone on the inflammatory markers under the stimulation of CI and LPS using a triple quadrupole mass spectrometer. The authors concluded that Flumetasone had a direct inhibitory effect on the COX and LOX enzymes, however, further studies will be required for the specific effect on the 15-LOX enzyme [69]. This was demonstrated with significant change for two eicosanoids in the 5-LOX enzymatic pathway following CI-stimulation [69]. Comparatively, utilising LPS-stimulation, six eicosanoids displayed a significant reduction indicating a direct effect on the COX enzyme [69].

6.3. Lipid Inflammatory Markers and Their Effect from an NSAID Administration

NSAIDs are known to predominately inhibit the COX-2 enzyme rather than COX-1 [65]. Cuniberti et al. utilised several NSAIDs, namely eltenac, Naproxen, tepoxalin, SC-560 and NS 398, to study their effects on both COX-1 and COX-2 enzymes using a chromatographic assay [65]. This study concluded that certain NSAIDs (e.g., SC-560, NS 398 and eltenac) can act as either COX-1 or COX-2 enzyme inhibitors whilst other NSAIDs (e.g., tepoxalin and Naproxen) can act as dual inhibitors [65]. This was seen with an in vitro model demonstrating 100% inhibition of three eicosanoids by all drugs whilst ex vivo models also showed signs of inhibition but not to the same extent.

Phenylbutazone (PBZ) is one of the most commonly used NSAIDs in equine medicine for the treatment of training and performance related injuries [63]. The use of PBZ is highly regulated due to the potential to mask injuries during pre-race (i.e., fitness-to-race) examinations [63]. Knych et al. explored how PBZ affected the biomarkers of inflammation following intravenous (IV) and oral administration whilst also utilising stimulation with CI and LPS [63]. The technique used for this study was a triple quadrupole mass spectrometer. The duration of monitored effects of PBZ exceeded the direct detection of PBZ in plasma [63]. Utilising CI-stimulation, there was a significant change for two eicosanoids whilst LPS-simulation showed four eicosanoids to significantly change indicating PBZ is an effective inhibitor of the COX-1 enzyme [63].

6.4. Lipid Inflammatory Markers and Their Effect from a Cannabidiol Administration

Cannabidiol (CBD), a known antidote for inflammation in humans, has recently been of concern with growing interest from horse trainers considering the use of hemp-based supplements [70]. The effects of CBD on the equine system have not been extensively studied, however, it is believed that there are multiple mechanisms that CBD follows which potentially impact the endocannabinoid pathway including inflammatory, analgesia and stress responses [70]. There is a hypothesis that CBD inhibits COX and LOX enzymes which decreases the amount of pro-inflammatory eicosanoids that are produced during the inflammatory response [70]. This hypothesis follows a similar mechanism for NSAIDs which have been demonstrated to affect the inflammatory pathways. Ryan et al. recently evaluated the anti-inflammatory effects of CBD with results indicating that both COX-1 and COX-2 enzymes were being stimulated [70] utilising an Agilent 1260 chromatography system coupled to a triple quadrupole mass spectrometer. Whole blood samples were stimulated using either CI or LPS to monitor the inflammatory mediators. During LPS-stimulation, four eicosanoids were observed to have significant change indicating the effects of both COX-1 and COX-2 enzymes, however, there is also evidence of CBD being dose dependent with a higher dose resulting in stronger enzyme inhibition [70]. During CI-stimulation, three eicosanoids following the LOX pathways were affected [70]. The major limitation with this study is that the endogenous levels of these inflammatory mediators are not known since the study used an ex vivo model of inflammation.

7. Potential for Lipidomics to Contribute Longitudinal Assessments for Equine Anti-Doping

The primary focus of anti-doping analyses is the detection and quantification of exogenous drugs or substances in biological matrices such as urine and plasma [4]. However, issues such as the clear distinction between endogenous and exogenous origins of naturally occurring substances and varying individual levels of biomarkers have been evident for decades. Therefore, the need for a longitudinal perspective is necessary given that it can also be an indicator of pharmaceutical manipulation. The equine biological passport (EBP) can provide a basis for complementing targeted analysis (detecting prohibited substances) with the non-targeted monitoring of biomarkers, which may be present after the original substance has been eliminated, metabolised or excreted [71]. The establishment of individual reference limits has been useful in monitoring levels of endogenous biomarkers to determine their respective basal levels, which may differ from the general population. This form of monitoring allows for an indirect approach to monitor biomarkers in order to potentially increase the detection window of administered drugs. The EBP has dual purposes: allowing for the possible detection of novel doping agents together with directing the resources of racing authorities to participants that may be engaging in prohibited practices [71]. It also allows for the monitoring of substances considered relevant to integrity but which do not have an agreed international threshold that exists to control specific misuse [71]. The EBP can provide a deterrent to the use of prohibited substances, since it has the potential to measure the biological effect of an administered substance for longer periods than the presence of the substance itself [71]. As outlined in this review, a lipidomic component may provide a useful contribution to an EBP. For future work, it would be ideal to include lipids into the EBP based off of lipids that were identified from complementary targeted and untargeted research. To the authors' knowledge at the time of writing this review, this potential has not been presented before highlighting the novelty of the passport and the use of lipids throughout the passport.

8. Conclusions

In conclusion, the improved use of biomarkers for a complementary targeted/non-targeted approach has been explored to potentially extend the time of detection for an administered substance. The current issues involving the use of targeted metabolomics and lipidomics such as short detection windows and matrix effects require investigation of non-targeted screening strategies. Lipidomics is an expanding field within the broader area of metabolomics focusing on the lipidome in any system with common polyunsaturated fatty acids being a starting point in the potential for a non-targeted approach. Whilst current research still utilises targeted methods to monitor biomarkers as seen in the various equine studies (including corticosteroids, NSAIDs and cannabidiol), there is considerable potential for non-targeted studies in the future. This can be attributed to improved sample preparation in conjunction to the increased use of high-resolution accurate mass spectrometry for non-targeted methods. The use of DIA and DDA to monitor biomarkers not already currently monitored, in addition to more sophisticated statistical analyses, will likely realise the potential for non-targeted lipidomic studies in the equine racing industry.

Author Contributions: Writing-Original draft preparation: K.T.; writing -review and editing: A.C., D.P.B. and S.F.; supervision: S.F., A.C., D.P.B. and C.B. All authors have read and agreed to the published version of the manuscript.

Funding: This research is funded by the Australian Government Research Training Program Scholarship.

Institutional Review Board Statement: Not Applicable.

Informed Consent Statement: Not Applicable.

Data Availability Statement: Not Applicable.

Acknowledgments: The authors thank the Australian Government Research Training Program Scholarship provided for Kathy Tou.

Conflicts of Interest: The authors declare no conflict of interest.

References

1. Fragkaki, A.G.; Kioukia-Fougia, N.; Kiousi, P.; Kioussi, M.; Tsivou, M. Challenges in detecting substances for equine anti-doping. *Drug Test. Anal.* **2017**, *9*, 1291–1303. [CrossRef] [PubMed]
2. Teale, P.; Barton, C.; Driver, P.M.; Kay, R.G. Biomarkers: Unrealized potential in sports doping analysis. *Bioanalysis* **2009**, *1*, 1103–1118. [CrossRef] [PubMed]
3. Narduzzi, L.; Dervilly, G.; Audran, M.; Le Bizec, B.; Buisson, C. A role for metabolomics in the antidoping toolbox? *Drug Test. Anal.* **2020**, *12*, 677–690. [CrossRef] [PubMed]
4. Marclay, F.; Mangin, P.; Margot, P.; Saugy, M. Perspectives for Forensic Intelligence in anti-doping: Thinking outside of the box. *Forensic Sci. Int.* **2013**, *229*, 133–144. [CrossRef] [PubMed]
5. Stojiljkovic, N.; Paris, A.; Garcia, P.; Popot, M.-A.; Bonnaire, Y.; Tabet, J.-C.; Junot, C. Evaluation of horse urine sample preparation methods for metabolomics using LC coupled to HRMS. *Bioanalysis* **2014**, *6*, 785–803. [CrossRef]
6. Reichel, C. OMICS-strategies and methods in the fight against doping. *Forensic Sci. Int.* **2011**, *213*, 20. [CrossRef]

7. Mangal, D.; Uboh, C.E.; Soma, L.R. Analysis of bioactive eicosanoids in equine plasma by stable isotope dilution reversed-phase liquid chromatography/multiple reaction monitoring mass spectrometry. *Rapid Commun. Mass Spectrom.* **2011**, *25*, 585–598. [CrossRef]
8. Dass, C. Characterization of Lipids. In *Fundamentals of Contemporary Mass Spectrometry*; John Wiley & Sons: New York, NY, USA, 2007; pp. 423–451. [CrossRef]
9. López-Bascón, M.A.; Calderón-Santiago, M.; Díaz-Lozano, A.; Camargo, A.; López-Miranda, J.; Priego-Capote, F. Development of a qualitative/quantitative strategy for comprehensive determination of polar lipids by LC–MS/MS in human plasma. *Anal. Bioanal. Chem.* **2020**, *412*, 489–498. [CrossRef]
10. Koelmel, J.P.; Li, X.; Stow, S.M.; Sartain, M.J.; Murali, A.; Kemperman, R.; Tsugawa, H.; Takahashi, M.; Vasiliou, V.; Bowden, J.A.; et al. Lipid Annotator: Towards Accurate Annotation in Non-Targeted Liquid Chromatography High-Resolution Tandem Mass Spectrometry (LC-HRMS/MS) Lipidomics Using A Rapid and User-Friendly Software. *Metabolites* **2020**, *10*, 101. [CrossRef]
11. Nolazco Sassot, L.; Villarino, N.F.; Dasgupta, N.; Morrison, J.J.; Bayly, W.M.; Gang, D.; Sanz, M.G. The lipidome of Thoroughbred racehorses before and after supramaximal exercise. *Equine Vet. J.* **2019**, *51*, 696–700. [CrossRef]
12. Harwood, J.L.; Frayn, K.N.; Murphy, D.J.; Michell, R.H.; Gurr, M.I. *Lipids: Biochemistry, Biotechnology and Health*; John Wiley & Sons, Incorporated: Hoboken, UK, 2016.
13. Fahy, E.; Cotter, D.; Sud, M.; Subramaniam, S. Lipid classification, structures and tools. *Biochim. Et Biophys. Acta. Mol. Cell Biol. Lipids* **2011**, *1811*, 637–647. [CrossRef] [PubMed]
14. Granström, E. The arachidonic acid cascade. *Inflammation* **1984**, *8*, S15–S25. [CrossRef] [PubMed]
15. Thakare, R.; Chhonker, Y.S.; Gautam, N.; Nelson, A.; Casaburi, R.; Criner, G.; Dransfield, M.T.; Make, B.; Schmid, K.K.; Rennard, S.I.; et al. Simultaneous LC–MS/MS analysis of eicosanoids and related metabolites in human serum, sputum and BALF. *Biomed. Chromatogr.* **2018**, *32*, e4102. [CrossRef] [PubMed]
16. Toewe, A.; Balas, L.; Durand, T.; Geisslinger, G.; Ferreirós, N. Simultaneous determination of PUFA-derived pro-resolving metabolites and pathway markers using chiral chromatography and tandem mass spectrometry. *Anal. Chim. Acta* **2018**, *1031*, 185–194. [CrossRef] [PubMed]
17. Giles, H.; Leff, P. The biology and pharmacology of PGD2. *Prostaglandins* **1988**, *35*, 277–300. [CrossRef]
18. Jackson, C.A.; Colahan, P.T.; Rice, B. Use of a Commercially Available Prostaglandin E2 Enzyme-Linked Immunosorbent Assay for Non-Steroidal Anti-Inflammatory Drug Screening. In Proceedings of the 16th International Conference of Racing Analysts and Veterinarians, Tokyo, Japan, 21–27 October 2006; Volume 16, pp. 477–482.
19. Lees, P.; Ewins, C.P.; Taylor, J.B.O.; Sedgwick, A.D. Serum thromboxane in the horse and its inhibition by aspirin, phenylbutazone and flunixin. *Br. Vet. J.* **1987**, *143*, 462–476. [CrossRef]
20. Lopez, L.R.; Guyer, K.E.; Torre, I.G.D.L.; Pitts, K.R.; Matsuura, E.; Ames, P.R. Platelet thromboxane (11-dehydro-Thromboxane B2) and aspirin response in patients with diabetes and coronary artery disease. *World J. Diabetes* **2014**, *5*, 115–127. [CrossRef]
21. Johnson, R.A.; Morton, D.R.; Kinner, J.H.; Gorman, R.R.; McGuire, J.C.; Sun, F.F.; Whittaker, N.; Bunting, S.; Salmon, J.; Moncada, S.; et al. The chemical structure of prostaglandin X (prostacyclin). *Prostaglandins* **1976**, *12*, 915–928. [CrossRef]
22. Connolly, P.J.; Wetter, S.K.; Beers, K.N.; Hamel, S.C.; Chen, R.H.K.; Wachter, M.P.; Ansell, J.; Singer, M.M.; Steber, M.; Ritchie, D.M.; et al. N-Hydroxyurea and hydroxamic acid inhibitors of cyclooxygenase and 5-lipoxygenase. *Bioorganic Med. Chem. Letters* **1999**, *9*, 979–984. [CrossRef]
23. Samuelsson, B.; Dahlen, S.-E.; Lindgren, J.A.; Rouzer, C.A.; Serhan, C.N. Leukotrienes and lipoxins: Structures, biosynthesis, and biological effects. *Science* **1987**, *237*, 1171–1176. [CrossRef]
24. McMillan, R.M.; Foster, S.J. Leukotriene B4 and inflammatory disease. *Agents Actions* **1988**, *24*, 114–119. [CrossRef] [PubMed]
25. Goodman, L.; Coles, T.B.; Budsberg, S. Leukotriene inhibition in small animal medicine. *J. Vet. Pharmacol. Ther.* **2008**, *31*, 387–398. [CrossRef] [PubMed]
26. Felder, C.C.; Briley, E.M.; Axelrod, J.; Simpson, J.T.; Mackie, K.; Devane, W.A. Anandamide, an endogenous cannabimimetic eicosanoid, binds to the cloned human cannabinoid receptor and stimulates receptor-mediated signal transduction. *Proc. Natl. Acad. Sci. USA* **1993**, *90*, 7656–7660. [CrossRef]
27. Beltramo, M.; Stella, N.; Calignano, A.; Lin, S.Y.; Makriyannis, A.; Piomelli, D. Functional role of high-affinity anandamide transport, as revealed by selective inhibition. *Science* **1997**, *277*, 1094–1097. [CrossRef]
28. Sharma, C.V.; Long, J.H.; Shah, S.; Rahman, J.; Perrett, D.; Ayoub, S.S.; Mehta, V. First evidence of the conversion of paracetamol to AM404 in human cerebrospinal fluid. *J. Pain Res.* **2017**, *10*, 2703–2709. [CrossRef] [PubMed]
29. Högestätt, E.D.; Jönsson, B.A.G.; Ermund, A.; Andersson, D.A.; Björk, H.; Alexander, J.P.; Cravatt, B.F.; Basbaum, A.I.; Zygmunt, P.M. Conversion of Acetaminophen to the Bioactive N-Acylphenolamine AM404 via Fatty Acid Amide Hydrolase-dependent Arachidonic Acid Conjugation in the Nervous System. *J. Biol. Chem.* **2005**, *280*, 31405–31412. [CrossRef] [PubMed]
30. Ricciotti, E.; FitzGerald, G.A. Prostaglandins and Inflammation. *Arterioscler. Thromb. Vasc. Biol.* **2011**, *31*, 986–1000. [CrossRef] [PubMed]
31. Wang, B.; Wu, L.; Chen, J.; Dong, L.; Chen, C.; Wen, Z.; Hu, J.; Fleming, I.; Wang, D.W. Metabolism pathways of arachidonic acids: Mechanisms and potential therapeutic targets. *Signal Transduct. Target. Ther.* **2021**, *6*, 94. [CrossRef]
32. Chambers, E.; Wagrowski-Diehl, D.M.; Lu, Z.; Mazzeo, J.R. Systematic and comprehensive strategy for reducing matrix effects in LC/MS/MS analyses. *J. Chromatogr. B* **2007**, *852*, 22–34. [CrossRef]

33. Sarafian, M.H.; Gaudin, M.; Lewis, M.R.; Martin, F.-P.; Holmes, E.; Nicholson, J.K.; Dumas, M.-E. Objective Set of Criteria for Optimization of Sample Preparation Procedures for Ultra-High Throughput Untargeted Blood Plasma Lipid Profiling by Ultra Performance Liquid Chromatography–Mass Spectrometry. *Anal. Chem.* **2014**, *86*, 5766–5774. [CrossRef]
34. Cajka, T.; Fiehn, O. Toward Merging Untargeted and Targeted Methods in Mass Spectrometry-Based Metabolomics and Lipidomics. *Anal. Chem.* **2016**, *88*, 524–545. [CrossRef] [PubMed]
35. Zahra, P.W.; Simpson, N.J.K. Application of Alternative SPE Sorbents and Formats for Extraction of Urine and Plasma. In Proceedings of the 20th International Conference of Racing Analysts and Veterinarians, Mauritius, 20–27 September 2015; Volume 20, pp. 324–330.
36. Bruce, S.J.; Tavazzi, I.; Parisod, V.; Rezzi, S.; Kochhar, S.; Guy, P.A. Investigation of Human Blood Plasma Sample Preparation for Performing Metabolomics Using Ultrahigh Performance Liquid Chromatography/Mass Spectrometry. *Anal. Chem.* **2009**, *81*, 3285–3296. [CrossRef]
37. Carrier, D.J.; Bogri, T.; Cosentino, G.P.; Guse, I.; Rakhit, S.; Singh, K. HPLC studies on leukotriene A4 obtained from the hydrolysis of its methyl ester. *Prostaglandins Leukot. Essent. Fat. Acids* **1988**, *34*, 27–30. [CrossRef] [PubMed]
38. Terragno, A.; Rydzik, R.; Terragno, N.A. High performance liquid chromatography and UV detection for the separation and quantitation of prostaglandins. *Prostaglandins* **1981**, *21*, 101–112. [CrossRef] [PubMed]
39. Gandhi, A.S.; Budac, D.; Khayrullina, T.; Staal, R.; Chandrasena, G. Quantitative analysis of lipids: A higher-throughput LC-MS/MS-based method and its comparison to ELISA. *Future Sci. OA* **2017**, *3*, FSO157. [CrossRef] [PubMed]
40. Tsikas, D.; Zoerner, A.A. Analysis of eicosanoids by LC-MS/MS and GC-MS/MS: A historical retrospect and a discussion. *J. Chromatogr. B* **2014**, *964*, 79–88. [CrossRef]
41. Reis, A.; Rudnitskaya, A.; Blackburn, G.J.; Mohd Fauzi, N.; Pitt, A.R.; Spickett, C.M. A comparison of five lipid extraction solvent systems for lipidomic studies of human LDL. *J. Lipid Res.* **2013**, *54*, 1812–1824. [CrossRef]
42. Lu, X.; Zhao, X.; Bai, C.; Zhao, C.; Lu, G.; Xu, G. LC–MS-based metabonomics analysis. *J. Chromatogr. B* **2008**, *866*, 64–76. [CrossRef]
43. Kinyua, J.; Negreira, N.; Ibáñez, M.; Bijlsma, L.; Hernández, F.; Covaci, A.; van Nuijs, A.L.N. A data-independent acquisition workflow for qualitative screening of new psychoactive substances in biological samples. *Anal. Bioanal. Chem.* **2015**, *407*, 8773–8785. [CrossRef]
44. Chernushevich, I.V.; Loboda, A.V.; Thomson, B.A. An introduction to quadrupole–time-of-flight mass spectrometry. *J. Mass Spectrom.* **2001**, *36*, 849–865. [CrossRef]
45. Dass, C. Mass Analysis and Ion Detection. In *Fundamentals of Contemporary Mass Spectrometry*; John Wiley & Sons: New York, NY, USA, 2007; pp. 67–117. [CrossRef]
46. Hopfgartner, G. Can MS fully exploit the benefits of fast chromatography? *Bioanalysis* **2011**, *3*, 121–123. [CrossRef] [PubMed]
47. Fenaille, F.; Barbier Saint-Hilaire, P.; Rousseau, K.; Junot, C. Data acquisition workflows in liquid chromatography coupled to high resolution mass spectrometry-based metabolomics: Where do we stand? *J. Chromatogr. A* **2017**, *1526*, 1–12. [CrossRef] [PubMed]
48. Fernández-Costa, C.; Martínez-Bartolomé, S.; McClatchy, D.B.; Saviola, A.J.; Yu, N.-K.; Yates, J.R. Impact of the Identification Strategy on the Reproducibility of the DDA and DIA Results. *J. Proteome Res.* **2020**, *19*, 3153–3161. [CrossRef] [PubMed]
49. Xu, L.; Xu, Z.; Strashnov, I.; Liao, X. Use of information dependent acquisition mass spectra and sequential window acquisition of all theoretical fragment-ion mass spectra for fruit juices metabolomics and authentication. *Metabolomics* **2020**, *16*, 81. [CrossRef] [PubMed]
50. Drotleff, B.; Calderon Castro, C.; Cebo, M.; Dittrich, K.; Fu, X.; Laemmerhofer, M. Untargeted LC-MS Lipidomics with Data Independent Acquisition using Sequential Window Acquisition of All Theoretical Fragment-Ion Spectra. *LC GC Eur.* **2021**, *34*, 7–20.
51. Doerr, A. DIA mass spectrometry. *Nat. Methods* **2015**, *12*, 35. [CrossRef]
52. Raetz, M.; Bonner, R.; Hopfgartner, G. SWATH-MS for metabolomics and lipidomics: Critical aspects of qualitative and quantitative analysis. *Metabolomics* **2020**, *16*, 71. [CrossRef]
53. Wrona, M.; Mauriala, T.; Bateman, K.P.; Mortishire-Smith, R.J.; O'Connor, D. 'All-in-One' analysis for metabolite identification using liquid chromatography/hybrid quadrupole time-of-flight mass spectrometry with collision energy switching. *Rapid Commun. Mass Spectrom.* **2005**, *19*, 2597–2602. [CrossRef]
54. Plumb, R.S.; Johnson, K.A.; Rainville, P.; Smith, B.W.; Wilson, I.D.; Castro-Perez, J.M.; Nicholson, J.K. UPLC/MSE; a new approach for generating molecular fragment information for biomarker structure elucidation. *Rapid Commun. Mass Spectrom.* **2006**, *20*, 1989–1994. [CrossRef]
55. Koopmans, F.; Ho, J.T.C.; Smit, A.B.; Li, K.W. Comparative Analyses of Data Independent Acquisition Mass Spectrometric Approaches: DIA, WiSIM-DIA, and Untargeted DIA. *Proteomics* **2018**, *18*, 1700304. [CrossRef]
56. Tsugawa, H.; Hanada, A.; Ikeda, K.; Isobe, Y.; Senoo, Y.; Senoo, M. *High-Throughput Lipid Profiling with SWATH Aquisition and MS-Dial*; SCIEX: Framingham, MA, USA, 2019; pp. 1–5.
57. Ren, S.; Hinzman, A.A.; Kang, E.L.; Szczesniak, R.D.; Lu, L.J. Computational and statistical analysis of metabolomics data. *Metabolomics* **2015**, *11*, 1492–1513. [CrossRef]
58. Saccenti, E.; Hoefsloot, H.C.; Smilde, A.K.; Westerhuis, J.A.; Hendriks, M.M. Reflections on univariate and multivariate analysis of metabolomics data. *Metabolomics* **2014**, *10*, 361–374. [CrossRef]

59. Tsugawa, H.; Ikeda, K.; Takahashi, M.; Satoh, A.; Mori, Y.; Uchino, H.; Okahashi, N.; Yamada, Y.; Tada, I.; Bonini, P.; et al. A lipidome atlas in MS-DIAL 4. *Nat. Biotechnol.* **2020**, *38*, 1159–1163. [CrossRef] [PubMed]
60. Courant, F.; Antignac, J.-P.; Dervilly-Pinel, G.; Le Bizec, B. Basics of mass spectrometry based metabolomics. *Proteomics* **2014**, *14*, 2369–2388. [CrossRef]
61. Schymanski, E.L.; Jeon, J.; Gulde, R.; Fenner, K.; Ruff, M.; Singer, H.P.; Hollender, J. Identifying Small Molecules via High Resolution Mass Spectrometry: Communicating Confidence. *Environ. Sci. Technol.* **2014**, *48*, 2097–2098. [CrossRef]
62. Mangal, D.; Uboh, C.E.; Soma, L.R.; Liu, Y. Inhibitory effect of triamcinolone acetonide on synthesis of inflammatory mediators in the equine. *Eur. J. Pharmacol.* **2014**, *736*, 1–9. [CrossRef]
63. Knych, H.K.; Arthur, R.M.; McKemie, D.S.; Seminoff, K.; Hamamoto-Hardman, B.; Kass, P.H. Phenylbutazone blood and urine concentrations, pharmacokinetics, and effects on biomarkers of inflammation in horses following intravenous and oral administration of clinical doses. *Drug Test. Anal.* **2019**, *11*, 792–803. [CrossRef]
64. Knych, H.K.; Weiner, D.; Arthur, R.M.; Baden, R.; McKemie, D.S.; Kass, P.H. Serum concentrations, pharmacokinetic/pharmacodynamic modeling, and effects of dexamethasone on inflammatory mediators following intravenous and oral administration to exercised horses. *Drug Test. Anal.* **2020**, *12*, 1087–1101. [CrossRef]
65. Cuniberti, B.; Odore, R.; Barbero, R.; Cagnardi, P.; Badino, P.; Girardi, C.; Re, G. In vitro and ex vivo pharmacodynamics of selected non-steroidal anti-inflammatory drugs in equine whole blood. *Vet. J.* **2012**, *191*, 327–333. [CrossRef]
66. Slifer, P. *A Review of Therapeutic Drugs Used for Doping of Race Horses: NSAIDs, Acepromazine, and Furosemide*; Components, C., Ed.; Iowa State University: Ames, IA, USA, 2018; Volume 105.
67. Lees, P.; Landoni, M.F.; Giraudel, J.; Toutain, P.L. Pharmacodynamics and pharmacokinetics of nonsteroidal anti-inflammatory drugs in species of veterinary interest. *J. Vet. Pharmacol. Ther.* **2004**, *27*, 479–490. [CrossRef]
68. Murakami, M.; Naraba, H.; Tanioka, T.; Semmyo, N.; Nakatani, Y.; Kojima, F.; Ikeda, T.; Fueki, M.; Ueno, A.; Oh-ishi, S.; et al. Regulation of Prostaglandin E2 Biosynthesis by Inducible Membrane-associated Prostaglandin E2 Synthase That Acts in Concert with Cyclooxygenase-2. *J. Biol. Chem.* **2000**, *275*, 32783–32792. [CrossRef] [PubMed]
69. Knych, H.K.; Arthur, R.M.; McKemie, D.S.; Baden, R.; Oldberg, N.; Kass, P.H. Pharmacokinetics of intravenous flumetasone and effects on plasma hydrocortisone concentrations and inflammatory mediators in the horse. *Equine Vet. J.* **2019**, *51*, 238–245. [CrossRef] [PubMed]
70. Ryan, D.; McKemie, D.S.; Kass, P.H.; Puschner, B.; Knych, H.K. Pharmacokinetics and effects on arachidonic acid metabolism of low doses of cannabidiol following oral administration to horses. *Drug Test. Anal.* **2021**, *13*, 1305–1317. [CrossRef] [PubMed]
71. Cawley, A.T.; Keledjian, J. Intelligence-based anti-doping from an equine biological passport. *Drug Test. Anal.* **2017**, *9*, 1441–1447. [CrossRef]

Article

Center-of-Mass *iso*-Energetic Collision-Induced Decomposition in Tandem Triple Quadrupole Mass Spectrometry

Federico Maria Rubino

LaTMA Laboratory for Analytical Toxicology and Metabonomics, Department of Health Sciences, Università degli Studi di Milano at "Ospedale San Paolo" v. A. di Rudinì 8, I-20142 Milano, Italy; federico.rubino@unimi.it

Academic Editor: Thomas Letzel
Received: 14 April 2020; Accepted: 8 May 2020; Published: 10 May 2020

Abstract: Two scan modes of the triple quadrupole tandem mass spectrometer, namely Collision Induced Dissociation Precursor Ion scan and Neutral Loss scan, allow selectively pinpointing, in a complex mixture, compounds that feature specific chemical groups, which yield characteristic fragment ions or are lost as distinctive neutral fragments. This feature of the triple quadrupole tandem mass spectrometer allows the non-target screening of mixtures for classes of components. The effective (center-of-mass) energy to achieve specific fragmentation depends on the inter-quadrupole voltage (laboratory-frame collision energy) and on the masses of the precursor molecular ion and of the collision gas, through a non-linear relationship. Thus, in a class of homologous compounds, precursor ions activated at the same laboratory-frame collision energy face different center-of-mass collision energy, and therefore the same fragmentation channel operates with different degrees of efficiency. This article reports a linear equation to calculate the laboratory-frame collision energy necessary to operate Collision-Induced Dissociation at the same center-of-mass energy on closely related compounds with different molecular mass. A routine triple quadrupole tandem mass spectrometer can operate this novel feature (*iso*-energetic collision-induced dissociation scan; *i*-CID) to analyze mixtures of endogenous metabolites by Precursor Ion and Neutral Loss scans. The latter experiment also entails the hitherto unprecedented synchronized scanning of all three quadrupoles of the triple quadrupole tandem mass spectrometer. To exemplify the application of this technique, this article shows two proof-of-principle approaches to the determination of biological mixtures, one by Precursor Ion analysis on alpha amino acid derivatized with a popular chromophore, and the other on modified nucleosides with a Neutral Fragment Loss scan.

Keywords: amino acids; equation; HPLC; MS/MS; NTS techniques (separation, ionization, and detection); nucleosides; open access software; target gas; triple quadrupole

---

1. Introduction

The discovery of collision-induced dissociation of gas-phase ions and the introduction of tandem triple quadrupole mass spectrometers [1,2] has revolutionized the art of trace organic analysis in complex matrices in the last forty years, and has opened the way to contemporary *"omic"* measurement of the complex and dynamic composition of biological compartments [3].

In particular, the two scan modes that, although not unique to tandem triple quadrupole [4], at which this configuration of mass spectrometer performs at its best are those known as Precursor (formerly, "parent" [5]) Ion (PI) scan and Constant Neutral Loss (CNL) scan. Their analytical strength allows selectively pinpointing, in a complex mixture, only compounds that feature specific chemical groups. The compounds can be identified as giving rise, upon fragmentation of the molecular ion, to a characteristic fragment ion (precursor ion scan) or to the loss of a characteristic substructure of the molecule(s) as a neutral species (neutral loss scan). These scan modes (Scheme 1) allow:

(a) identifying families of structurally similar compounds that occur in a complex mixture;
(b) assigning molecular masses to each of the components and
(c) indicating the molecular precursors that can be characterized by a subsequent "fragment ion" scan (the earliest scan mode of tandem mass spectrometers) experiment to insight their molecular connectivity.

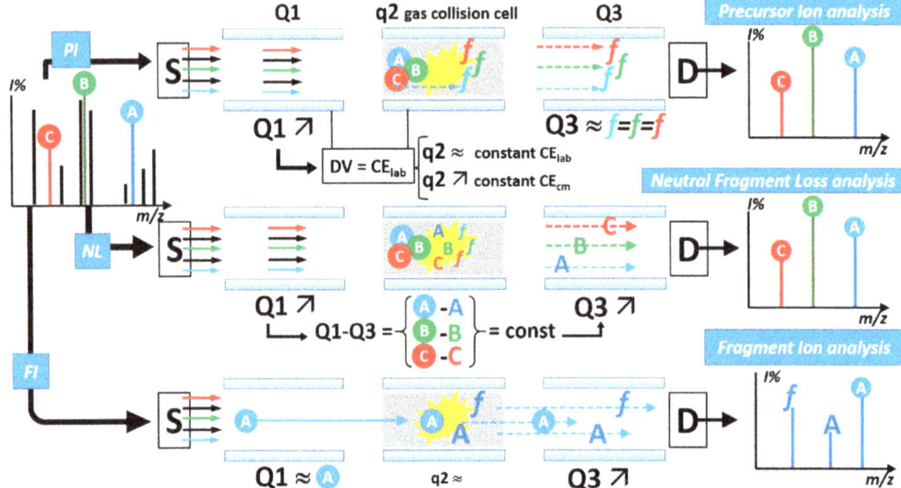

Scheme 1. Simplified operating scheme of the tandem triple quadrupole mass spectrometer in the three main scan modes: (**top**) recording of Precursor Ions (**PI**), (**middle**) Constant Neutral Fragment Loss (**NL**), (**bottom**) Fragment Ions (**FI**) of a selected precursor. The example refers to the characterization of a complex mixture that contains, apart from other components (black ions in the total source spectrum, left), of a family of (three) molecules of interest. The three molecules, each with a different molecular mass, are identified as circled **A**, **B** and **C**, blue, green and red, respectively. Each molecule is composed of a sub-structure (fragment *f*) that is common to the chemical class and of a variable part (bold **A**, **B** and **C**, blue, green and red, respectively) that adds to the individual molecular mass. The symbol ↗ indicates *scanning* of the **Q1** and **Q3** quadrupole filters; the symbol ≈ indicates "*parking*" to transmit ions of a specific m/z value. Collision-activated dissociation (CAD, yellow flash) determines fragmentation of the **PI** and occurs on a neutral collision gas (grey background of **q2**) at a collision energy value (CE_{lab}) that is determined by the voltage difference between **Q1** (reference value) and **q2**. This difference is usually fixed (**q2** ≈) but it can be varied (**q2** ↗) synchronously with and during a scan of **Q3** (Precursor Ion mode) or of both **Q1** and **Q3** (Neutral Loss mode).

This analytical strategy can be automated in Data-Dependent modes of operation of commercial triple quadrupoles. Industrial pharmacology and toxicology found a particular benefit in applying this technique systematically to identify the products that derive from biotransformation of xenobiotics, such as candidate pharmaceutical drugs, pesticides, industrial chemicals.

Several classes of conjugated xenobiotic and endogenous metabolites feature distinctive fragmentation pathways in their mass spectra, and selective scan experiments were elaborated to selectively detect the compounds that belong to each class of conjugates, such as glucuronides, sulphates, glutathione thioethers, mercapturic acids [6,7]. Specific scan modes have been developed to search for (and to discover, in the case of unexpected components) several endogenous classes of soluble metabolites, such as urinary acyl-carnitines [8], derivatized fatty acids and alcohols [9], complex lipids with a choline head-group, such as phosphatidyl-cholines and sphingomyelins [10], ceramides [4]. The Precursor Ion and Neutral Loss operation modes of the triple quadrupole tandem mass spectrometer allow the non-target screening of mixtures for structural classes, rather than for individual molecules, thus allowing the detection of new or unexpected components.

At the core of the application of tandem triple quadrupole instruments to the selective identification of chemical groups is the occurrence, in the fragmentation pattern of the molecular precursor(s), of specific decomposition channels with the following characteristics:

(a) generate a sufficiently specific charged or neutral fragment as a reporter of the considered chemical class;
(b) the considered Precursor-to-Fragment generation process ("transition") occurs in the different molecules with a sufficiently constant kinetics (reflected by a fairly constant value of collision energy for maximum production);
(c) the transition is sufficiently insensitive to the presence of other chemical arrangements in the molecule (reflected by a fairly constant value of relative intensity).

Therefore, the search for specific molecules of a defined structural class is performed by PI or CNL scan, in the triple quadrupole MS.

To operate Collision Induced Dissociation, the q2 (RF-only quadrupole collision cell) is kept at some value of the potential difference (in the instrument, the voltage drop, or collision offset voltage) with respect to that of the "first" quadrupole, considered at "ground" potential [11]. A "reasonable" value for this parameter usually corresponds to the maximum value of the production efficiency curve of the selected fragment ion that is recorded in one or more "typical" compounds as a preliminary measurement of the method setup. "Nominal" (or "laboratory frame") collision energy, CE_{lab}, is the product of the voltage drop and the number of charges of the precursor ion.

The efficiency of fragmentation, however, is not determined by the "laboratory frame" collision energy, but rather by that experienced by the isolated precursor ion undergoing collision with the stationary target gas, the "Center-of-Mass" collision energy; E_{CM}. By operating at a constant difference of the (q2-Q1) potential (the "laboratory frame" collision energy), precursor ions of different m/z experience non-linearly decreasing amounts of additional internal energy, as their m/z value increases. In addition, the increasing molecular size increases the amount of energy that is necessary to achieve fragmentation (Degree-of-Freedom, DoF effect), as has been demonstrated in a wide-purpose systematic measurement of the "characteristic energy" ($E_{50\%}$) for fragmentation of organic polymers [12].

Very recent commercial instruments can adjust collision energy for fragment ion analysis of precursors with different m/z (especially of peptides in proteomic analysis) with the use of patented "mass-related" parameters [13], such as the Normalized Collision Energy [14,15], and otherwise-named approaches used by academic scientists [14–18]. However, it is likely that the instruments' manuals do not explicitly report the equations employed to perform this task (e.g., http://proteomicsnews.blogspot.com/2014/06/normalized-collision-energy-calculation.html). In addition, the term "Normalized Collision Energy" is not even defined in the IUPAC 2013 "Definitions of Terms Relating to Mass Spectrometry" [19].

This article describes how to derive an equation to calculate the laboratory frame, or "nominal", collision energy necessary to analyze closely related compounds with different molecular mass and using different gas targets for collision-induced decomposition tandem mass spectrometry. It further demonstrates that this novel feature (*iso*-energetic collision-induced dissociation; *i*-CID) can be applied in a standard analytical triple quadrupole tandem mass spectrometer, either by continuous ramping of the (q2-Q1) potential, or with a "stepped" surrogate, if the instrument's software does not allow for continuous ramping. To suggest the utility of this application, the article reports two proof-of-principle analyses of mixtures of endogenous metabolites, one using a Precursor Ion scan, the other a Constant Neutral Loss scans. The latter experiment also entails the hitherto unprecedented synchronized scan of all three quadrupoles (Q1, q2, Q3) of the triple quadrupole tandem mass spectrometer.

2. Results

2.1. Theory and Simulations

In Collision Induced Dissociation of ions, the relation of center-of-mass, or "effective" (E_{cm}) to laboratory frame or "nominal" (E_{lab}) collision energy depends on the mass of the resting target gas in the collision cell (in most instruments Nitrogen, and occasionally Argon), (m TAR) and on that of the impinging precursor ion (m PRE), according to Equation (1) [20].

$$E_{cm} = E_{lab} \times \frac{mTAR}{(mTAR + mPRE)} \tag{1}$$

This transformation from "*laboratory frame*" to "*center-of-mass*" collision energy is widely employed to plot "breakdown curves" and to compare those obtained for precursor ion with different *m/z*.

For a fixed collision gas employed (in all described examples m TAR is Nitrogen, MW 28, or Argon, MW 40), and for each specific value of "laboratory frame" collision energy, the "center-of-mass" collision energy decreases as the *m/z* value of the precursor ions increase. The variation is not linear, neither with the *m/z* of the precursor ion, nor with the value of the "*laboratory frame*" collision energy (Figure 1).

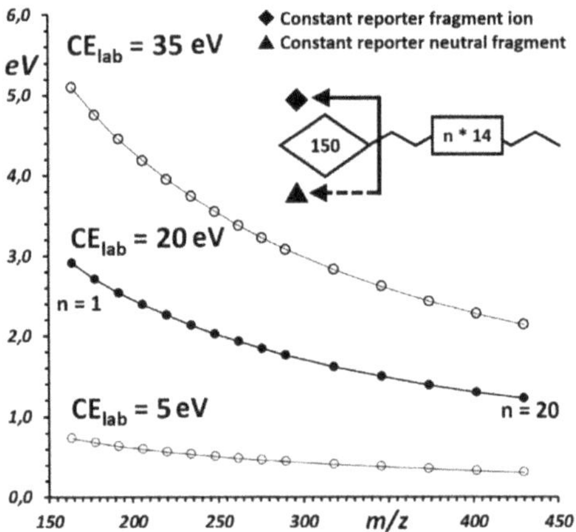

Figure 1. Center-of-mass collision energy calculated for precursor ions of different *m/z* and for different values of laboratory-frame collision energy (5, 20 and 35 eV). The *m/z* values of the precursor ions are calculated for a homologous series of metabolites constituted by a common motif of a molecular mass of 150 Da (that which gives rise to the diagnostic precursor-to-fragment transition) and of a homologous 20-methylene series. Center-of-mass collision energy is calculated according to Equation (1), with Nitrogen as the collision gas (see Section 2.1).

The simulation displayed in Figure 1 refers, as a general example, to singly charged precursors ions of molecules that contain a common portion (molecular mass 150, in the example, corresponding to fragment *f* in the example of Scheme 1) and a poly methylene variable part with one to twenty carbon atoms (this originates the different fragment ions **A**, **B** and **C**). Collision-induced dissociation occurs on a target of Nitrogen gas and is calculated for increasing values of "laboratory frame" collision energy. The ion precursor dissociates after collision according to the following pathways:

Precursor(C_n)$^{\pm}$	→	**Const$^{\pm}$ + (C_n)^0**	**Precursor ion scan**
Precursor(C_n)$^{\pm}$	→	**Const^0 + (C_n)$^{\pm}$**	**Neutral fragment loss scan**

In the example of Figure 1, precursor ion Precursor $(C_1)^{\pm}$ (m/z 164 = 150 + 14) impinges on the Nitrogen gas target at a laboratory frame collision voltage of 20 eV and experiences a center-of-mass collision energy of 2.92 eV. At the same laboratory frame collision voltage, precursor ion Precursor $(C_{20})^{\pm}$ (m/z 430 = 150 + 20 × 14) experiences a much lower center-of-mass collision energy of 1.22 eV.

The trend of the curves displayed in Figure 1 makes it possible to understand some literature results that have been reported for the measurement of molecular series by tandem mass spectrometry at constant laboratory-frame collision energy.

One such example in the literature is offered by the measurement of ten saturated, straight chain homolog esters of acyl-carnitine that span from C2 to C18, obtained by FIA-ESI and precursor ion detection of the m/z 85 fragment ($C_4H_5O_2$; protonated butadienoic acid; Scheme 2), at 20 eV nominal collision energy onto an Argon gas target [8].

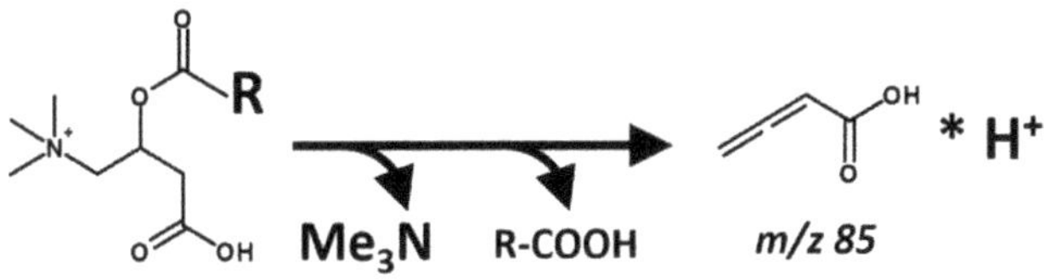

Scheme 2. Generation of the reporter fragment m/z 85 from acyl-carnitines.

When the authors infused an equimolar solution of the standard compounds, the resulting Precursor Ion spectrum (Figure 3 of ref. [8]) showed a marked and regular decrease in signal strength of the lower (C-2 to C-8) and of the higher (C-16 to C-18) homologs. Figure 2, left panel, is recalculated from the intensities in the spectrum of the cited article.

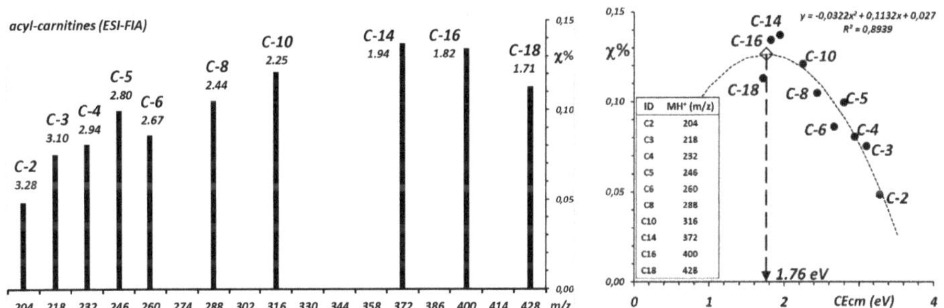

Figure 2. Re-analysis of the data from Paglia et al., 2008 (Figure 3 of ref [8]). **Left:** Intensities of the molecular precursors of an equimolecular mixture of acyl carnitines ionized by ESI-FIA and detected by a Precursor Ions scan of the fragment at m/z 85. Above each bar is the CE_{cm} calculated for the precursor ion at 40 eV of CE_{lab} on an Ar gas target, calculated from Equation (1). **Right:** Estimation of the fragmentation efficiency curve for the formation of the m/z 85 fragment by least squares fit to a parabolic curve.

In the right panel of Figure 2, the relative yields of the Precursor Ions are plotted against the center-of-mass collision energies, the values of which are in turn calculated for the respective molecular ions as corresponding to the employed laboratory frame collision energy of 20 eV against an Ar gas target, with the use of Equation (1). The bell-shaped trend can be interpolated with a parabola, the apex of which, at approximately 1.76 eV, yields an approximate estimation of the maximum of a fragmentation efficiency curve for the employed transition $M^+ \cong 85^+$. This rough calculation also implies, but does not warrant, that under FIA (i.e., at a constant composition of the solvent) the ESI ionization efficiency does not discriminate homologs with an acyl chain of a greatly different length that are simultaneously present in the droplet.

This example shows that it is useful to have a systematic method to perform analyses by collision-induced dissociation and Precursor Ion or Neutral Loss scan in the triple quadrupole over a range of *m/z*, which keeps at a constant value the collision energy in the center-of-mass frame. To achieve this, the laboratory frame collision voltage needs being systematically increased as the *m/z* value of the precursor ion (m PAR) increases.

Standard algebraic passages allow re-formulating Equation (1) to yield the linear working Equation (2):

$$E_{lab} = E_{CM} + E_{CM} \times \frac{\text{m PRE}}{\text{m TAR}} \tag{2}$$

This linear equation makes it possible to "back"-calculate the necessary laboratory frame collision voltage at each value of *m/z* of the precursor ion corresponding to a specific constant value of center-of-mass collision energy, as shown in Figure 3 for the same example of Figure 1.

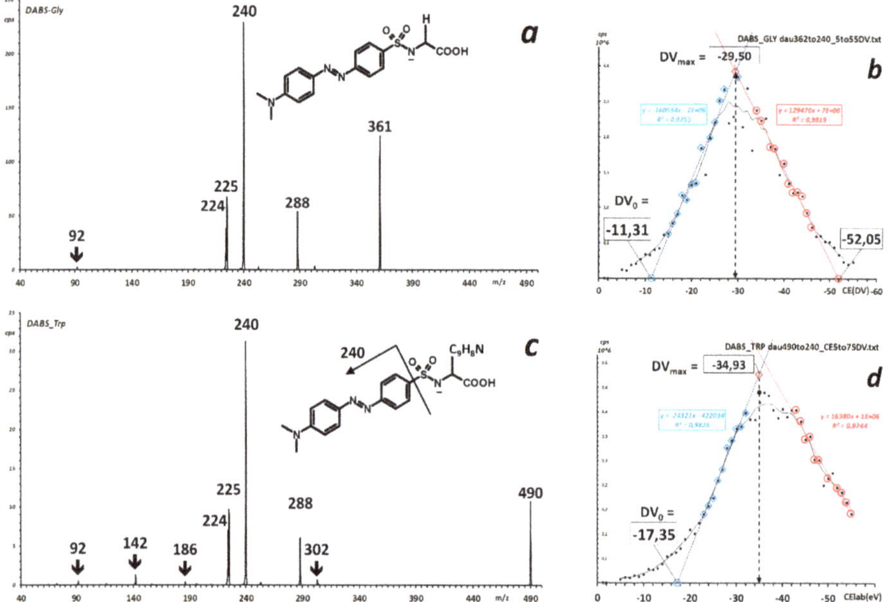

Figure 3. Integrated fragment spectra of dabsylated Glycine (**a**) and Tryptophan (**c**). The insert to each spectrum shows the relationship of the most intense fragment to molecular connectivity. The fragment generation efficiency curve of *m/z* −240 (laboratory frame collision potential) is reported, for each DABS-AA, in the right panels (**b**) and (**d**).

Equation (2) can find use in transferring the collision-induced decomposition conditions that have been optimized for one specific molecular species within a class to other homologues with a similar fragmentation pattern, and to instruments that employ a different collision gas (e.g., Nitrogen in lieu of Argon, as seen above).

2.2. Proof-of-Principle Applications

The feasibility of the *iso*-energetic scan as an analytical tool to characterize complex mixtures of metabolites was tested with some proof-of-principle examples to understand whether this strategy can be useful in organic bio-analysis of small-molecule biomarkers.

Two experimental examples of application are reported. The first employs a precursor ion scan (the reporter molecular fragment is an ion, identified by the *m*/*z* of Q3), the second a constant neutral loss scan (the reporter molecular fragment is a neutral species, identified as the fixed *m*/*z* difference between Q1 and Q3).

2.2.1. Detection of Dabsyl-Amino Acids by *iso*-Energetic CID and Precursor Ion Scan

The first proof-of-principle experiment of the use of *iso*-energetic collision-induced dissociation for the analysis of small organic molecules by Precursor Ion scan is obtained by derivatizing alpha-amino acids with a very popular chromogenic tag, dimethylamino-azobenzene (dabsyl, DABS-), linked to the amino-group as the sulphonyl amide (Scheme 3).

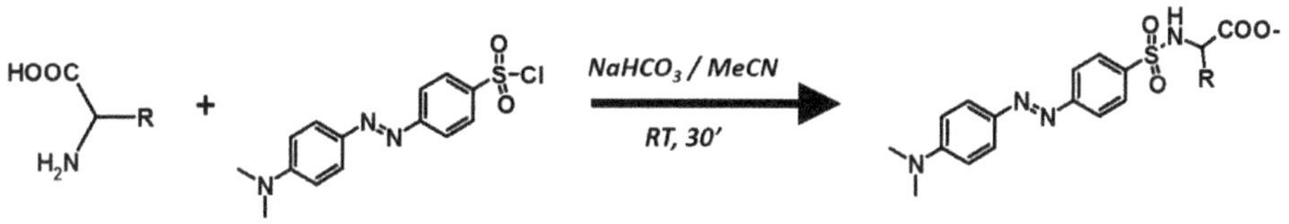

Scheme 3. Derivatization of amino acids with DABS-Cl to obtain the corresponding sulphonamides, DABS-AA.

Some amino-acid derivatives were prepared individually (list in Table S1) and the fragment spectra of their deprotonated molecules were measured in the negative ion mode over a range of collision energy up to 70 eV_{lab}. Figure 3a,c shows as examples the integrated spectra recorded for the derivatives of the lowest- and highest-mass natural amino acids, glycine and tryptophan, respectively. The most intense fragment in the spectra, at *m*/*z* 240, is generated from the appended chromophore, as indicated in the general structure of the examined derivatives, and due to its common occurrence and high intensity qualifies as reporter ion for analytical purposes.

The generation efficiency curves of *m*/*z* 240 are displayed in the corresponding right-hand panels Figure 3b,d, and show the position of the maxima calculated as described in the Materials and Methods Section 4.5. In the laboratory frame, the values of the maxima are 29.50 eV_{lab} (center-of-mass frame, 2.12 eV) for DABS-Gly, and 35.66 eV_{lab} (center-of-mass frame, 1.93 eV) for DABS-Trp. Results of the measurements in a few other DABS-AA are collected in Table 1.

Table 1. Collection of CE_{max} and corresponding $CE_{cm\text{-}max}$ measured for nine synthetic dabsylated alpha amino acids.

	Compound	PI [1]	FI [1]	FI	FI	PI	PI
		$[M-H]^-$	ArO^-	CE_{max}	$CE_{cm\text{-}max}$	CE_{max}	$CE_{cm\text{-}max}$
1	DABS-Gly	362	240	−29.2	−2.10	−29.35	−2.11
2	DABS-Val	403	240	−32.1	−2.09	−32.89	−2.14
3	DABS-Leu	417	240	−35.9	−2.26	-	-
4	DABS-Asp	419	240	−34.5	−2.16	−35.58	−2.23
5	DABS-SMC	421	240	−35.3	−2.20	−32.93	−2.05
6	DABS-Gln	432	240	-	-	−35.60	−2.17
7	DABS-Glu	433	240	−36.1	−2.20	−36.70	−2.23
8	DABS-Met	435	240	−34.0	−2.06	−33.82	−2.05
9	DABS-Trp	490	240	−40.4	−2.18	−35.66	−1.93

[1] PI Precursor ion (*m*/*z*, negative ions); FI Fragment ion (*m*/*z*, negative ions); NL Neutral Loss (Da). CE_{max} is the "nominal" value of collision energy; $CE_{cm\text{-}max}$ is the corresponding value of center-of-mass collision energy calculated with Equation (1).

Least-squares calculation of center-of-mass collision energy vs. precursor ion *m/z* affords a best-estimate value of 2.15 eV for the representative value of CE_{max} for the characteristic $[M–H]^-$ to *m/z* 240 transition (Figures S2 and S3). The obtained value corresponds to a scan-line that starts at 29.9 eV_{lab} for *m/z* 361 of deprotonated DABS-Gly and ends at 39.8 eV_{lab} for *m/z* 490 of deprotonated DABS-Trp (see parameters used in the example of Figure 1).

One interesting observation stems from the comparison of the scatter around the best-fit scan line of actual maxima measured in the individual production efficiency curves of *m/z* 240 (Figure S4). While differences as large as 3.9 eV_{lab} are found (the outlying Trp being that with the largest difference), this value should be compared to the actual span of the maxima of the round-topped curves recorded for these compounds. As can be appreciated from the right-column curves of Figure 4, the width of the 95%–100%–95% round-topped curve maxima is of 5 to 9 eV_{lab}. The uncertainty of appreciation for the apex (the intersection of the two least-squares ascending and descending lines) results in the order of fractions of eV_{lab} (0.2–0.4 eV_{lab}), less than the employed potential step in the accurate ramp-CE experiment (0.5 eV_{lab}). Therefore, the method employed to obtain the representative value of center-of-mass CE_{max} yields a reliable value for method setup.

To test whether this experiment is amenable, and useful, to real applications in bio-analysis, a DABS-derivatized urine was analyzed with an un-optimized fast gradient elution linked to a 1-s *m/z* 240 precursor ion scan from 350 to 500 and to a synchronized scan of the collision voltage (CElab, q2-Q1) from −28.4 to −39.6 eV_{lab}. Main results are displayed in Figure 4a–d.

Under the un-optimized "shotgun" chromatographic conditions of this experiment, several DABS-derivatized alpha-amino acids of urine elute as partially separated chromatographic peaks between 3.5 and 5.5 min (Figure 4a) and the integrated spectrum (merged spectra) of the corresponding chromatographic time-frame contains the molecular signals of the main physiologically expected amino acids (Figure 4b).

This un-optimized, very fast chromatographic condition does not separate the isomers of Leucine, nor isobaric hydroxy-proline, at *m/z* 417. The mono-DABS derivatized histidine signal is observed (*m/z* 441), and the signal at *m/z* 455 may correspond to the unseparated, isomeric methyl-histidines. Methionine can be identified from the co-occurrence of the molecular species at m/z 435 (32S) and −437 (^{34}S). By analogy, a weaker signal at *m/z* 421 can be attributed to the trace amino acid S-Methyl-cysteine (SMC) [21], based on the appearance of a corresponding signal at *m/z* 423 of the corresponding ^{34}S-species. This observation, in particular, opens the way to a non-target screening of cysteine thioethers in adductomics applications. No signals are observed, nor were likely expected, for those that yield a doubly dabsylated derivative, Cysteine and Lysine. Only one chromatographic peak is observed in the XIC of *m/z* 432 (Glutamine), and a very weak signal at *m/z* 460 is observed for mono-DABS-Arginine. Precursor Ion spectra extracted corresponding to the XIC for the main identified amino acids feature the expected $[M-H]^-$ molecular signal expected for the individual eluting species (Figure 4c,d).

Identification of the amino acids by Data-Dependent Fragment Ion analysis was not performed, since all DABS-derivatives yield the same main fragment ions, all of which deriving from the appended chromophore (see Figure 4a,c).

2.2.2. Detection of Nucleosides by *Iso*-Energetic CID and Constant Neutral Fragment Loss Scan

The second proof-of-principle example employs nucleosides, measured in the positive ion mode, to demonstrate that iso-energetic CID is compatible with Constant Neutral fragment Loss Scan (CNL) experiments in the triple quadrupole. In this experiment, all three quadrupoles scan synchronously; in fact, the q2-Q1 voltage difference is "ramped" simultaneously with the scan time of both Q1 and Q3.

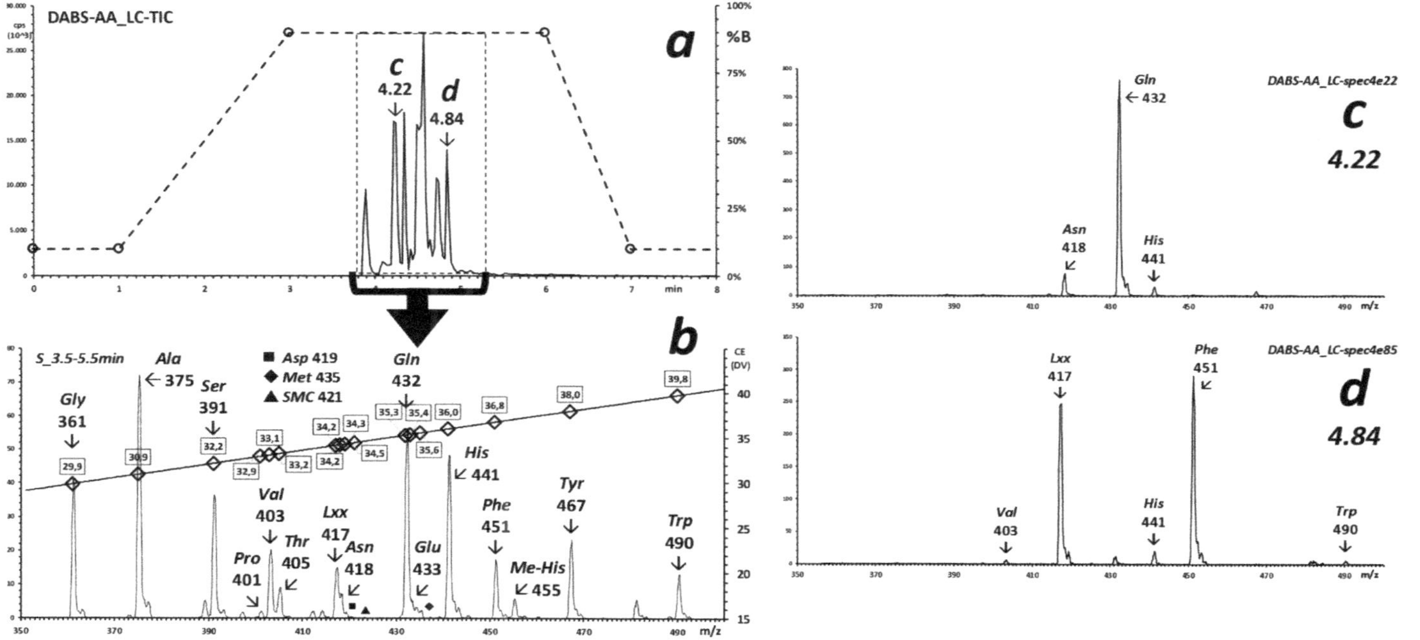

Figure 4. (**a**) XIC of the main deprotonated DABS-AAs detected in a derivatized urine from a healthy laboratory worker. (**b**) Integrated Precursor Ion spectrum of *m*/*z* 240 of deprotonated DABS-AAs with *m*/*z* from *m*/*z* 350 to 500 (negative ions), eluted from the chromatographic column between 3.5 and 5.5 min. (**c**,**d**) Precursor Ion spectra recorded across the chromatographic peaks eluting at 4.22 and 4.84 min (see panel (**a**)), respectively.

Protonated nucleosides (listed in Table S4) undergo collision-induced fragmentation through a main pathway that involves loss of the N-linked sugar as neutral (loss of 116 Da from the 2-deoxy-ribosides and of 132 Da from the ribosides) and formation of the protonated nucleobase (Scheme 4). A very minor pathway generates the complementary positively charged sugar species (*m/z* 117 and 133, respectively, for 2-deoxy-ribose and ribose) but their intensity is lower by more than two orders of magnitude with respect to that of the protonated nucleobases, therefore making a Precursor Ion scan utterly unsuitable for analysis.

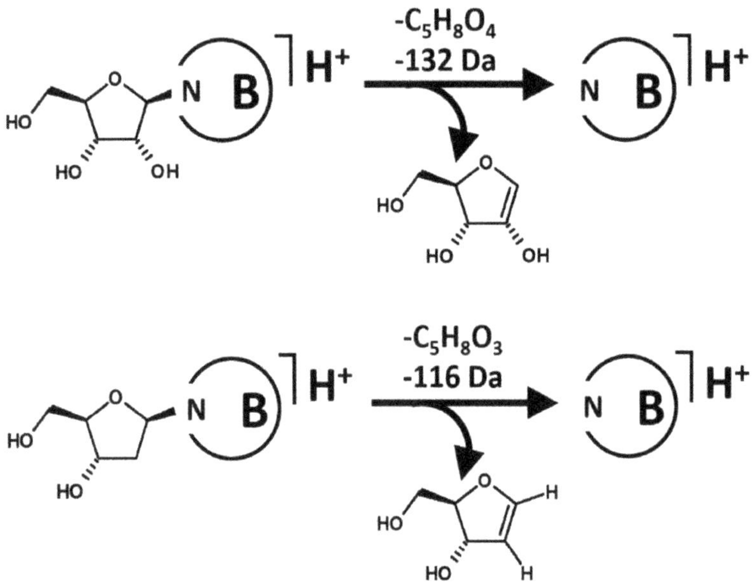

Scheme 4. Fragmentation of protonated nucleosides with (above) ribose and (below) 2-deoxy-ribose to afford the respective protonated nucleobases.

As in the example before, the collision energy corresponding to the maxima of the production efficiency curves of the protonated nucleobases was measured from the fragment ion spectra (measurements in Table 2 and results of calculation of the representative CE_{max} in Figures S6 and S7). The maxima are between approximately 1.5 and 2.0 eV_{CM}, with no analytically relevant difference between the 2-deoxy-nucleosides (loss of 116 Da; median 1.60 eV) and the nucleosides (loss of 132 Da; median 1.67 eV).

From the perspective of fast method setup, it is appealing to observe that the values of CE_{max} measured from the fragment intensity curves of the Fragment Ion spectra and those measured from the "reverse" experiment (CNL, in the case of nucleotides) are sufficiently close for the analytical application. As an example, Figure S7 shows the curves for Guanosine (*m/z* 284 > 152, Fragment Ion spectrum) and (CNL132, *m/z* 284), both recorded from a mixture to ensure that calculation results derive from measurements of the same (noisy) quality. Estimated CE_{max} is 17.86 and 17.89 eV_{lab}, respectively.

That such close results can be obtained is advantageous when a few standard compounds are available. In this case, one single "reverse" experiment, rather than several recordings of individual Fragment Ion spectra, may still yield sufficiently reliable data to calculate a starting representative eV_{max} value for method optimization with the spreadsheet module of Appendix A.

To demonstrate the selectivity of the two CNL scans, Figure 5 shows the source spectrum of a mixture that contains the four DNA nucleosides (dA, dC, dG, dT) and two RNA nucleosides (only C and G; A was deliberately omitted, since it is isobaric to dG). The two inserts display the results of consecutive alternate 1-s CNL-scans of 116 Da (selective for deoxy-ribonucleosides) and 132 Da (selective for ribonucleosides) at 17 eVlab (1.6 eVCM for precursor *m/z* 270).

Table 2. Collection of CE_{max} and corresponding $CE_{cm\text{-}max}$ measured for four DNA nucleosides, three RNA nucleosides and six modified adenine ribosides.

ID	Compound	PI [1]	FI [1]	NL [1]	FI	FI	NL	NL
		MH^+	BH^+	x	CE_{max} [1]	$CE_{cm\text{-}max}$	CE_{max}	$CE_{cm\text{-}max}$
1	2-deoxy-cytidine	228	112	116	14.60	1.60	15.28	1.67
2	2-deoxy-thymidine	243	131	116	-	-	14.80	1.53
3	2-deoxy-adenosine	252	136	116	18.87	1.89	19.26	1.93
4	2-deoxy-guanosine	268	152	116	15.63	1.48	14.09	1.33
5	cytosine	244	112	132	17.18	1.77	16.76	1.73
6	adenosine	268	136	132			21.52	2.04
7	guanosine	284	152	132	17.87	1.60	17.89	1.61
8	N6-isopentenyl_Ade	336	204	132	22.04	1.70	22.21	1.71
9	kinetin-riboside	348	216	132	22.87	1.70	23.28	1.73
10	trans-zeatine-riboside	352	220	132	23.39	1.72	24.37	1.80
11	N6-Bn-Ade	358	226	132	24.63	1.79	25.81	1.87
12	N6-(4-OH-Bn)-Ade	374	242	132	24.32	1.69	24.04	1.67
13	3,4-diOH-PhEt-Ade	404	272	132	27.09	1.76	28.79	1.87

[1] PI Precursor ion (*m/z*, positive ions); FI Fragment ion (*m/z*, positive ions); NL Neutral Loss (Da). CE_{max} is the "nominal" value of collision energy; $CE_{cm\text{-}max}$ is the corresponding value of center-of-mass collision energy calculated with Equation (1).

All expected signals that correspond to the protonated molecules of the six compounds are observed in the respective NL spectra. Adenosine was deliberately omitted from this mixture to confirm that the NL116 scan is specific for deoxy-ribonucleosides, since the precursor at *m/z* 268 from isobaric A is absent from the NL132 spectrum, but that from dG appears in the NL116 scan.

A more complex mixture that contains the four DNA nucleosides (dA, dC, dG, dT), three RNA nucleosides (A, C, G), and six 6N-substituted derivatives of adenine (listed in Table S4; complete ESI source spectrum in Figure S8) was further employed to investigate the potentiality of the iso-energetic CID scan for applications in the field of nucleoside analysis. In this mixture, the ratio of the isobaric dG:A is approximately 1:4.

The possibility of screening the sample for modified nucleobases is exemplified by six 6N-substituted derivatives of adenine (8–13) [22,23] with masses between 350 and 400 Da, and thus 30% to 50% higher than that of the generating riboside, adenosine. Their fragment spectra (one example is reported in Figure S9) feature loss of the ribose, and the first-generation protonated adenine further loses the N6-adducted substituent to yield the *m/z* 136 $AdeH^+$ fragment, and often one from the appended molecular unit.

The experiment reported in Figure 6a–d is a concept with prospective real applications, such as the identification of natural epigenetic modifications of DNA nucleobases, of xenobiotic DNA adducts, of regulatory t-RNA modified bases, and of synthetic nucleobase tools for research in molecular biology [24]. Its aim is to evaluate whether the *i*-CID approach is feasible in a routine triple quadrupole instrument without compromising mass resolution and at a scan speed compatible with coupling to liquid chromatography. To test whether the use of a CE ramp improves the accuracy in the determination of nucleoside mixtures, the peak intensities of the non-isobaric nucleosides obtained in the fixed-CE and in the ramp-CE Neutral Loss scans were compared to those measured in the ESI source spectra of the same mixture. The cycle of measurement follows the pattern and conditions reported in Table S10. In the same sample, a consecutive NL116_rampCE experiment, selective for 2-deoxy-ribonucleotides was also performed, with essentially similar results (data not shown).

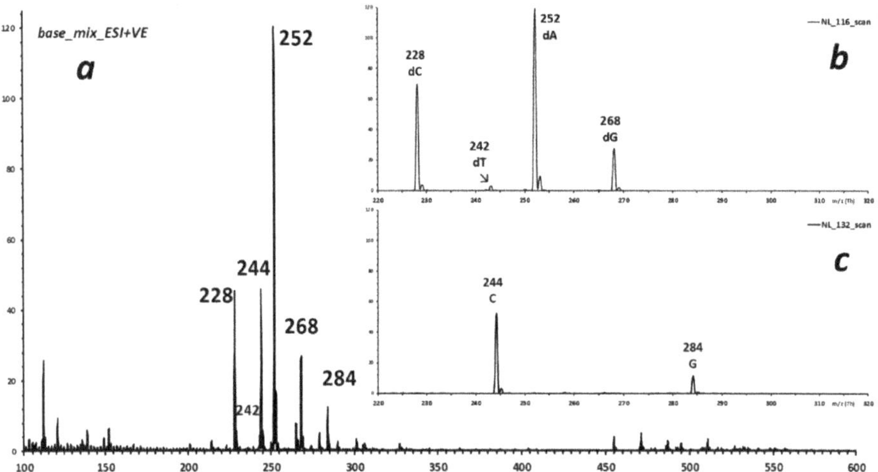

Figure 5. Source spectrum of a mixture of dA, dC, dG, dT, C and G (mean of 20 1-s scans, infusion mode). Inserts: consecutive 1-s alternate NL-scans of 116 Da (upper, selective for deoxy-ribonucleosides) and 132 Da (lower, selective for ribonucleosides).

As is apparent from the ion profiles (insert in Figure S12), no deterioration of resolution occurs when the q2 voltage is ramped synchronously with the Q1 and Q3 scans to achieve the collision energy scan. The fastest conditions chosen in the employed routine triple quadrupole instrument (infusion mode, CNL scan with synchronized q2 ramping) were a scan speed of 2.5 seconds/1.000 Da (*m*/*z* 220–420 in 0.5 seconds) and a collision voltage ramp speed of 23.6 eV_{lab}/s (14.6–26.4 eV_{lab} in 0.5 seconds). To hint at the analytical potential of this experiment in bio-analysis, the relative signal intensity of the considered ribo-nucleosides was compared between the different analytical conditions, as summarized in Figures S13 and S14. In particular, the relative signal abundances of the substituted adenines were closer to those measured in the source spectra when the CEscan mode was employed, rather than at any fixed CE value (Figure S14).

In the employed routine triple quadrupole instrument (infusion mode), scan speed of the Neutral Loss scan with synchronized q2 ramping can be as fast as 0.5 s in a *m*/*z* range of 200 Da, without any consistent signal instability or loss of resolution (insert in Figure S12), and allowing for two scan modes to alternate (*i*-NL116 and *i*-NL132).

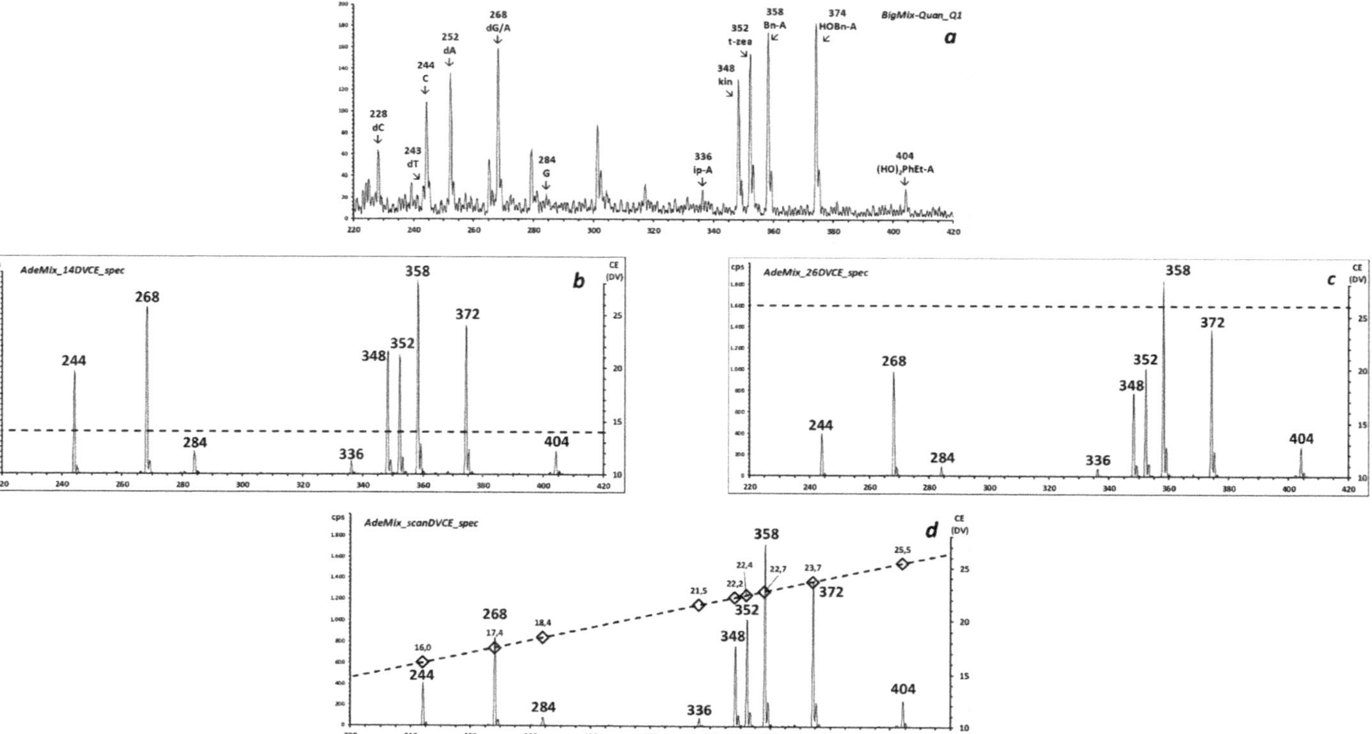

Figure 6. (**a**) Partial source spectrum (positive ions) of a mixture of dA, dC, dG, dT, C, A, G, and six N6-substituted adenosine compounds (mean of 20 1-s scans, infusion mode). (**b**,**c**) Consecutive 1-s NL-scans of 132 Da (selective for ribonucleosides) at fixed laboratory CE_{lab} of 14 and 26 eV, respectively. (**d**) CID is performed with a synchronized scan of CE_{lab} from 14.6 eV (*m/z* 220) to 26.4 eV (*m/z* 420).

3. Discussion

The work described in this article shows that continuous variation of the collision voltage in the cell of a routine triple quadrupole instrument during a fast scan of the mass filter(s) is technically feasible both in the Precursor Ion and in the Constant Neutral Loss modes.

This unprecedented operation of the TSQ can be used without degrading sensitivity and mass resolution, with respect to holding the collision cell at any intermediate constant value. This modification to the more common fixed-voltage practice allows collision-induced dissociation of molecules of a common structure but with different molecular mass to occur with each precursor ion facing the same value of effective (center-of-mass) energy deposition, and thus with the same, or close, relative kinetics. Therefore, when analyzing homogeneous series of chemical species, such as homologous series of metabolites, the intensity of corresponding fragments in different compounds will be more similar. Thus, the detection efficiency of the different homologs within a series improves, by overcoming the variation in collision efficiency that occurs when compounds over a wide range of m/z values are analyzed at a "compromise" value of laboratory frame collision voltage.

The convenience of this approach over the use of a fixed, "compromise", value of collision voltage is more apparent when the "variable" part of the structure contributes more to the increase of molecular mass than the "fixed" one. In this case, the relative increase of molecular mass of the target compound with that of the "variable" part is larger, and so is the associated interval of collision voltages between the smallest and the larger molecules of the class. The other factor is the value of center-of-mass collision energy for the considered, analytically useful transition: a higher Center-of-Mass collision energy (slope of the CE scan line) leads to a higher interval of collision voltages for the same mass range.

This article examined two molecular classes with a different type of fragmentation as proof-of-principle, among the many different examples of possible applications. One class, dabsyl-derivatized amino-acids, feature a common reporter fragment ion in the spectrum, and can be detected by employing a precursor ion scan. The derivatization approach is still widely popular in many bio-analytical applications [25], and several molecular classes of natural compounds, such as complex lipids, are similarly featured, with a common charge-carrying substructure and variable modifications that increase the molecular mass [4]. Another class of examined molecules, ribosylated nucleotides, features a common decomposition pathway that eliminates as a neutral the common carbohydrate sub-structure, and yields the variable nucleobase as an ionic fragment, so a Constant Neutral Loss scan is the most adequate tandem-MS experiment to highlight this class of molecules.

4. Materials and Methods

4.1. Reagents and Standards

All solvents, reagents and standards were of analytical purity (Prodotti Gianni, Milano, Italy), and were available in the laboratory from previous or current analytical and spectroscopic studies. The Supplementary Information section (Tables S1, DABS-amino acids and S4, nucleosides) reports the structures and relevant characteristics of the analyzed compounds.

4.2. Instrumentation

A SCIEX API 3000-LTQ mass spectrometer (AB Sciex Framingham, MA, USA) with the standard ESI source and data system is employed for this study, and operated according to the manufacturer's indications. The instrument employs Nitrogen from the purification system both as the nebulizer and as auxiliary gas in the ion source and as the collision gas in the q2 quadrupole.

4.3. Measurements

The infusion mode was used throughout this work to measure spectroscopic parameters and to analyze standards and samples. The final dilution of solutions for infusion was in 1:1 *v/v* water-methanol containing 0.1% formic acid. Flow rate of the built-in syringe was set to 10 microL $\times$ min^{-1}.

The concentration of standard compounds in the infused solutions varied between 0.1 and 10 microM. All experiments were carried out in the centroid mode. All compounds used in this study were characterized by means of a three-step procedure, which is described below:

(a) The source spectrum of the compound or mixture was acquired, and the source conditions were optimized for maximum signal stability and intensity. This was accomplished by regulating spray gas flows and curtain plate temperature, and tuning the Declustering Potential to the value that yields the most intense signal of the desired ion species with the use of the ramp function of the instrument's operating software.
(b) For each compound, a series of fragment ion spectra was acquired in the Collision Energy ramp function mode of the instrument's operating software, within a range of laboratory collision energies between 0 and approximately 50 to 70 V. The maximum value of collision energy depends on the molecular mass of the compound, and corresponds to a maximum of 4–6 eV_{CM}, as dictated by the compound's structure and yield of structurally informative fragments. Collision gas pressure was at the lowest setting (Low, corresponding to an ion gauge reading of 1.2×10^{-5} Torr (1.60×10^{-3} Pascal), vs. 0.8×10^{-5} Torr (1.07×10^{-3} Pascal) in the absence of collision gas).
(c) The collision efficiency curves for the transitions of analytical interest were recorded both in the "forward" (Fragment Ion Scan, *see* above) and in the "backwards" (Precursor Ion Scan or Neutral Loss Scan) modes, according to the intended use of the transition, to ensure that no overlapping fragmentation processes occur.

4.4. Data Elaboration

Mass spectrometry measurements were extracted as spectra (ion intensity vs. *m*/*z*), XIC (ion intensity vs. time), and as Ramp curves (ion intensity vs. the relevant modified instrument voltage), and exported as .txt files from the instrument's data system for further elaboration. Custom developed Microsoft Excel spreadsheets were used for all post-acquisition elaborations [26,27]. A few measurements were replicated across several months, in between other analytical work, to gauge robustness of the determined parameters.

4.5. Calculation of the Collision Energy Maximum in the Fragmentation Efficiency Curve

The value of collision energy corresponding to the maximum yield of the analytically useful fragment (or transition) was calculated, in the intensity vs. collision energy (either as laboratory, or Center-of-Mass scale) plot, as the intersection of two (least-square) straight lines, each one approximating the raising and the decreasing portion of the curve, respectively ("*teepee* method"). The interval of approximately 25% to 75% of the maximum intensity was usually selected as the linear portion for optimal least-squares calculation based on maximizing the R^2 coefficient. The intersection point of the two straight lines was calculated from their intercept and slope values by standard analytical geometry equations. The uncertainty of the calculation was calculated from the span, on the x-axis, of the diamond-shaped intersection of the 95% confidence limit stripes of the two straight lines (Figure 4).

4.6. Application in the Triple Quadrupole Mass Spectrometer

The obtained value of the maximum of the collision efficiency curve for the analytically relevant fragment ion is converted to the Center-of-Mass scale (eV), and used to infer the corresponding "laboratory frame" value of collision energy for other compounds that belong to the same chemical class with the use of Equation (2). A custom Microsoft Excel spreadsheet calculates the parameters to run the *iso*-energetic Precursor Ion (*i*-PI) and Neutral Loss (*i*-NL) scans with the use of Equations (1) and (2). The layout of Scheme 5 is set to match that of the instrument's data system, and the operator copy-and-pastes calculation results from the blue background cells directly for use.

	A	B	C	D	E	F	G
1	mTAR	CE_{CM}(eV)	m/z START	m/z STOP	dwell (s)	CE START	CE STOP
2	28	-2.13	-350	-500	1.0	24.5	35.8
3	N2 28	PAR_ref mz	PAR_ref mz	PAR_CE(DV)	eV		eV = DV *(mTAR/)mTAR +mPAR))
4	Ar 40	input >	421	-34.1	-2.13		eV + eV *(mTAR/PAR_ref)

Scheme 5. Calculation module of the custom spreadsheet for the calculation of collision energy ramp.

The actual values of the parameters reported are those of the first proof-of-principle example, see above (Section 2.2.1). The value of mTAR (cell A2) is defaulted at 28 (Nitrogen collision gas). Reference variables are typed in cells C4 (*m*/*z* of Precursor ion) and D4 (collision voltage for max. yield of fragment, interpolated from a ramp-CE experiment). The corresponding value of collision energy is calculated in E4 (formula is reported in G3). To operate the *i*-CID experiment, the desired scanning mass interval is input in cells C2 (min, START) and D2 (max, STOP), scan speed in cell E2, and collision voltage is calculated in cells F2 and G2 for CE START and CE STOP, respectively with formula in G4. Cells C2-G2 are thus directly ready for export. That reported in Scheme 2 only uses one reference compound to infer the value of CE_{max} for a series of homolog compounds. To improve accuracy, and to test whether the assumption holds for a wider series of available compounds, this spreadsheet is linked to a module (further lines below line 5 of the displayed section of the Excel spreadsheet) that calculates a more representative value of CE_{max} from results of a ramp-CE linked to the PI or CNL group scan.

A tabular form of the complete spreadsheet with instructions is presented in Appendix A and an Excel file is available from the Author upon request. A further module of the spreadsheet (Appendix B) allows the adaption of the *i*-CID technique in case the instrument software does not allow the modulation of the (q2-Q1) collision potential synchronously with the scan of the Q1 and Q3 mass filters. In this case, a "segmented pseudo-scan" of the quadrupole filters Q1 and Q3, e.g., by 14-u "steps" can accommodate a "stepped" variation of the (q2-Q1) collision potential, with only minimal difference of collision energy from that of a continuous variation (Figure S13). Since we did not need to implement this alternative in our instrument, Appendix B reports the calculation module with the fictitious example of Figures 1 and 3 (an Excel file is available from the Author upon request).

4.7. Derivatization of Alpha-Amino Acids as Dimethylamino-Azobenzene-Sulphonyl (Dabsyl) Amides

A mixture of alpha-amino acids, each approximately 1 μM is prepared in deionized water. A 100-μL aliquot was mixed to 10 microliters of 1M sodium bicarbonate and to 100 microliters of a 10 μmol/mL (3.3 mg/mL) of dabsyl chloride in acetonitrile [23]. The mixture was vortexed for 30 s and left standing at room temperature for 2 h. An appropriate aliquot was diluted 1:100 *v*/*v* with the ESI infusion solvent for infusion experiments, or 10 microliters were directly injected into the HPLC-MS system for analysis. Extemporary urine from a healthy laboratory volunteer (100 μL) was treated in the same way.

4.8. Shotgun Separation by Liquid Chromatography of Alpha-Amino Acids Derivatized as Dabsyl Amides

A chromatographic column (Phenomenex Gemini C18, 100 × 2mm i.d., 3 mm particle size, 110 A porosity) was employed for separation. A 10 min mobile phase gradient of A (10 mM ammonium formate, 0.1% formic acid, pH 4) and B (acetonitrile) starts at 10% B for 2 min, was linearly increased to 90% B in 7 min and brought back to the initial conditions in 1 min. The column was held at 40 °C and 10 microliters were injected with the instrument's auto-sampler. Detection by tandem mass spectrometry was achieved in the negative ion mode, by a precursor ion scan of *m*/*z* 240, scanning the m/z interval between 320 and 440 in 1 s, at the Low q2 pressure setting (approximately reading 1,2 mTorr; 1.60×10^{-3} Pascal). Collision energy ramp was synchronized to the Q1 scan, with a continuous variation between 22 eV_{lab} at *m*/*z* 320 and 44 eV_{lab} at *m*/*z* 440.

5. Conclusions

A possible advantage of the *iso*-energetic CID approach in comprehensive, or "*omic*", bio-analyses is that more comparable fragment spectra can be obtained for closely related analytes, in which reporter fragments have closer relative intensity. Thus, quantification of target metabolites or biomarkers may improve, since the signal-to-concentration relationship will be much closer for the different terms of a series of compounds. In addition, the semi-quantitative appreciation of the level of unexpected compounds, for which calibration curves cannot be established in advance, would also be more reliable.

A recently published application of this strategy allowed characterizing some unusual ceramide glycosides that occur in pistachio and almond nuts [28]. Briefly, the sequential use of alternate *i*-PI and *i*-NL scans highlighted the presence, in the nut fatty extracts, of ceramides with a modified sphingosine linked to a hexose at their C-1 hydroxyl group. Fragment ion analysis highlighted that saturated and hydroxylated fatty acids are linked through an amide bond. This application also confirmed that the technique is compatible with scan speed used in the coupling of the triple quadrupole tandem mass spectrometer with conventional narrow bore liquid chromatography.

Several classes of natural substances, including Phase 2 xenobiotic metabolites [7] and covalent protein adducts [29,30], are similarly featured, and yield either or both common structure-specific ("*pivot*") charged fragments (therefore prompting for the Precursor Ion approach) and neutral fragments (thus calling for the Neutral Loss approach). Given the easy availability of the additional feature of *iso*-energetic CID in the triple quadrupole tandem mass spectrometer, one prospective aim of this work is preparing a catalog of structure-specific transitions and optimized collision energies for an increasingly wide choice of organic and bio-organic classes of compounds. This information is the basis for the non-target screening and measurement of important classes of biomolecules in complex mixtures and for the discovery of new structurally related chemical entities, to incorporate into metabolomics protocols.

Supplementary Materials: The following are available online. Table S1. Structures of the examined dabsyl-amino acids, Figures S2 and S3. Measurement of representative CE_{max} and calculation of the scan line (DABS-AA), Table S4. Structures of the examined nucleosides, Figure S5, S6. Measurement of representative CEmax of riboside transition $MH^+ \cong BH^+$ and calculation of the scan line, Figure S7. Comparison of the generation efficiency curves of the BH^+ fragment of protonated guanosine in Fragment Ion and in Neutral Loss spectra. Figure S8. ESI source spectrum of a mixture of nucleosides. Figure S9. Fragment ion spectrum of a protonated N6-subsituted adenosine (12), Table S10. Experiments for the measurement of nucleosides. Figure S11. Relative abundance of nucleosides in different conditions. Figure S12. Stability of signal in fast-scan i-CE Neutral Loss spectra. Figure S13. Comparison of CE_{lab} in a continuous and stepped scan of collision energy.

Funding: This research received no external funding.

Acknowledgments: I am grateful to Prof. Enzo Santaniello (retired, formerly Università degli Studi di Milano) for the generous gift of compounds 8-13 (Table S4) and to Dr. Jacopo Antognetti for generous assistance in the measurement of DABS-AA mass spectra.

Conflicts of Interest: The author declares no conflict of interest.

Appendix A Calculation Module with Instructions

A tabular form of the complete calculation spreadsheet with formulae and instructions (an Excel file is available from the Author upon request).

INSTRUCTIONS
Column headings and constants
input in green cells (column heading has same color of input cells)
Calculated values
copy blue cells to instrument ***(column heading has same color of input cells)***

(a) For basic module (lines 1-4, Scheme 2 of main article), follow instructions (*b*) to (*f*)

(b) Record Ramp-CE fragment spectrum of reference compound (***PAR_ref mz***) with target fragment ion

(c) Identify (approximate) value for max. intensity of target fragment ion ***Dau_CE(DV)***

(d) Type (***PAR_ref mz***) ***DauCE(DV$_{max}$)*** in the green cells C4 and D4: eV will appear in orange cells E4 and B2

(e) Type ***m/z STARTm/z STOP*** in **D2** and **E2** complete with scan time (***dwell (s)***)

(f) Copy-and-paste (results) **D2 to H2** from spreadsheet to instrument's SW

(g) For optimization module, follow instructions (*b*), (*c*), then go to (*h*)

	A	B	C	D	E	F	G
1	*mTAR*	*CE$_{CM}$(eV)*	*m/z START*	*m/z STOP*	*dwell (s)*	*CE START*	*CE STOP*
2	28	**−2.13**	−350	−500	1.0	24.5	35.8
3	*N2 28*	*PAR_ref mz*	***PAR_ref mz***	***DauCE(DV$_m$***	***eV***		
4	*Ar 40*	***input >***	**421**	**−34.1**	**−2.13**		*eV + eV *(mTAR/PAR_ref)*
5				*CE_ref (DV)*	*eV*	*eV_mean*	*optimization_module*
6				*intercept*	***−1.94***	***−2.16***	
7				*slope*	***−0.00052***		
8				*R2*	***0.0743***		
9	*DABS-AA*	*compound_1*	*PAR mz*	*DauCE(DV$_m$*	*eVmax*	*eV$_{max}$_rec*	*DVmax*
10	*min_mz*	*Precursor_362*	*362*	−29.3	−2.10	−2.12	−30.0
11		*Precursor_403*	*403*	−32.5	−2.11	−2.15	−33.2
12		*Precursor_417*	*417*	−35.9	−2.26	−2.15	−34.3
13		*Precursor_419*	*419*	−35.0	−2.19	−2.15	−34.4
14		*Precursor_421*	*421*	−34.1	−2.13	−2.16	−34.6
15		*Precursor_432*	*432*	−35.6	−2.17	−2.16	−35.4
16		*Precursor_433*	*433*	−36.4	−2.21	−2.16	−35.5
17		*Precursor_435*	*435*	−33.9	−2.05	−2.16	−35.7
18	*max_mz*	*Precursor_490*	*490*	−40.4	−2.18	−2.19	−39.9

(h) Import *m/z* of Precursor ion (***PAR mz***) and ***DauCE(DV$_{max}$)*** in columns D and E, lines 11 downwards, respectively

(i) The optimized value of ***eV_mean*** will appear in orange cells **G7** and **B2**

(j) Like *e*: Type ***m/z START m/z STOP*** in **D2** and **E2** complete with scan time (***dwell (s)***)

(k) Like *f*: Copy-and-paste (results) **D2 to H2** from spreadsheet to instrument's SW

Notes. Values reported as example in cols. D and E correspond to Table S3 (values in col. E correspond to the mean of cols. 4 and 6 of Table S1.

Cell(s)	Formula	
B2, F4, F10 – F18	*eV = DV × (mPAR/(mTAR + mPAR)*	
G2, H2	*eV + eV × (mTAR/PAR_ref)*	
F7:intercept	*Intercept (E10:E18; D10:D18)*	*eV$_{(mz = 0)}$*
F8: slope	*Slope(E10:E18; D10:D18)*	*Closest to 0: horizontal line*
F9: R2	*RQ(E10:E18; D10:D18)*	*Closest to 0: all y values (eV) approximately constant*
G10 – G18		*eV recalculated from F8, F9, D11 – D18*
G6	*Mean of G10 – G18*	*Working value (eV) for B2 (alternative to F4)*

Appendix B Calculation Module with Instructions for Stepped *i*-CID Precursor Ion and Constant Neutral Loss Scans

A tabular form of the complete spreadsheet with formulae and instructions (an Excel file is available from the Author upon request).

INSTRUCTIONS
Column headings and constants
input in green cells (column heading has same color of input cells)
Calculated values
copy blue cells to instrument *(column heading has same color of input cells)*

Calculation and graph are for the example of Figures 1 and 3 of the main article.

a. Type ***constant part (m/z)*** in the green cell **D2** and minimum number of methylene groups in chain in **A4**. Remaining values will be calculated and fill-in.

b. Type collision energy (eV) in **K2**

c. Type scan time (***dwell (s)*** in **H4** (constant scan time for all m/z windows)

d. Check whether total scan time (cell **H2**), inclusive of hold time between segments (cell **G2**) is acceptable for application. In case, try a wider mass window (28u) in cell **G4**

	A	B	D	E	F	G	H	I	J	K	H	L
1	1	Co150	MW	eV	N2	28	scan_tot	n			scan_tot	
2	CH2]n	constant part	150		hold (s)	0.005	1.05	10	eV =	2.00	1.05	scan line
3	n	Cpd_ID	PAR	DAU	center	width	dwell (s)	CE	mz START	mz STOP	dwell (s)	CE(DV)
4	11	Co150-CH2(11)	304	150	304	13.9	0.1	23.7	297.1	311.0	0.1	23.7
5	12	Co150-CH2(12)	318	150	318	13.9	0.1	24.7	311.1	325.0	0.1	24.7
6	13	Co150-CH2(13)	332	150	332	13.9	0.1	25.7	325.1	339.0	0.1	25.7
7	14	Co150-CH2(14)	346	150	346	13.9	0.1	26.7	339.1	353.0	0.1	26.7
8	15	Co150-CH2(15)	360	150	360	13.9	0.1	27.7	353.1	367.0	0.1	27.7
9	16	Co150-CH2(16)	374	150	374	13.9	0.1	28.7	367.1	381.0	0.1	28.7
10	17	Co150-CH2(17)	388	150	388	13.9	0.1	29.7	381.1	395.0	0.1	29.7
11	18	Co150-CH2(18)	402	150	402	13.9	0.1	30.7	395.1	409.0	0.1	30.7
12	19	Co150-CH2(19)	416	150	416	13.9	0.1	31.7	409.1	423.0	0.1	31.7
13	20	Co150-CH2(20)	430	150	430	13.9	0.1	32.7	423.1	437.0	0.1	32.7

e. To use center/width for *m/z* input: copy-and-paste (results) **F4** to **I13** from spreadsheet to instrument's SW

f. To use ***m/z START m/z STOP*** for m/z input: copy-and-paste (results) **J4** to **L13** from spreadsheet to instrument's SW

References

1. Yost, R.A.; Enke, C.G. Triple quadrupole mass spectrometry for direct mixture analysis and structure elucidation. *Anal. Chem.* **1979**, *51*, 1251–1264. [CrossRef] [PubMed]
2. Schwartz, J.C.; Wade, A.P.; Enke, C.G.; Cooks, R.G. Systematic delineation of scan modes in multidimensional mass spectrometry. *Anal. Chem.* **1990**, *62*, 1809–1818. [CrossRef] [PubMed]
3. Griffiths, W.J.; Wang, Y. Mass spectrometry: from proteomics to metabolomics and lipidomics. *Chem. Soc. Rev.* **2009**, *38*, 1882. [CrossRef] [PubMed]
4. Rubino, F.M.; Zecca, L.; Sonnino, S. Characterization of Ceramide Mixtures by Fast Atom Bombardment and Precursor and Fragment Ion Analysis Tandem Mass Spectrometry. *Biol. Mass Spectrom.* **1994**, *23*, 82–90. [CrossRef]
5. Adams, J. Letters to the editor. *J. Am. Soc. Mass Spectrom.* **1992**, *3*, 473. [CrossRef]
6. Holčapek, M.; Kolářová, L.; Nobilis, M. High-performance liquid chromatography–tandem mass spectrometry in the identification and determination of phase I and phase II drug metabolites. *Anal. Bioanal. Chem.* **2008**, *391*, 59–78. [CrossRef]
7. Scholz, K.; Dekant, W.; Völkel, W.; Pähler, A. Rapid detection and identification of N-acetyl-L-cysteine thioethers using constant neutral loss and theoretical multiple reaction monitoring combined with enhanced product-ion scans on a linear ion trap mass spectrometer. *J. Am. Soc. Mass Spectrom.* **2005**, *16*, 1976–1984. [CrossRef]
8. Paglia, G.; D'Apolito, O.; Corso, G. Precursor ion scan profiles of acylcarnitines by atmospheric pressure thermal desorption chemical ionization tandem mass spectrometry. *Rapid Commun. Mass Spectrom.* **2008**, *22*, 3809–3815. [CrossRef]

9. Rubino, F.M.; Zecca, L. Application of triple quadrupole tandem mass spectrometry to the analysis of pyridine-containing derivatives of long-chain acids and alcohols. *J. Chromatogr. B: Biomed. Sci. Appl.* **1992**, *579*, 1–12. [CrossRef]
10. Kerwin, J.L.; Tuininga, A.R.; Ericsson, L.H. Identification of molecular species of glycerophospholipids and sphingomyelin using electrospray mass spectrometry. *J. Lipid Res.* **1994**, *35*, 1102–1114.
11. Sleno, L.; Volmer, D.A. Ion activation methods for tandem mass spectrometry. *J. Mass Spectrom.* **2004**, *39*, 1091–1112.
12. Memboeuf, A.; Nasioudis, A.; Indelicato, S.; Pollreisz, F.; Kuki, A.; Kéki, S.; Van den Brink, O.F.; Vékey, K.; Drahos, L. Size Effect on Fragmentation in Tandem Mass Spectrometry. *Anal. Chem.* **2010**, *82*, 2294–2302. [CrossRef] [PubMed]
13. Schwartz Jae, C.; Gilroy, P.; Remes, M. Methods for calibration of usable fragmentation energy in mass spectrometry. US Patent USOO827862OB2, 2 October 2012.
14. Diedrich, J.K.; Pinto, A.F.M.; Yates, I.J.R. Energy Dependence of HCD on Peptide Fragmentation: Stepped Collisional Energy Finds the Sweet Spot. *J. Am. Soc. Mass Spectrom.* **2013**, *24*, 1690–1699. [CrossRef] [PubMed]
15. Facchini, L.; Losito, I.; Cataldi, T.R.; Palmisano, F. Seasonal variations in the profile of main phospholipids in Mytilus galloprovincialis mussels: A study by hydrophilic interaction liquid chromatography-electrospray ionization Fourier transform mass spectrometry. *J. Mass Spectrom.* **2017**, *53*, 1–20. [CrossRef]
16. Qu, J.; Liang, Q.; Luo, G.; Wang, Y. Screening and Identification of Glycosides in Biological Samples Using Energy-Gradient Neutral Loss Scan and Liquid Chromatography Tandem Mass Spectrometry. *Anal. Chem.* **2004**, *76*, 2239–2247. [CrossRef]
17. Madala, N.E.; Steenkamp, P.; Piater, L.A.; Dubery, I.A. Collision energy alteration during mass spectrometric acquisition is essential to ensure unbiased metabolomic analysis. *Anal. Bioanal. Chem.* **2012**, *404*, 367–372. [CrossRef]
18. Qiao, X.; Lin, X.-H.; Ji, S.; Zhang, Z.-X.; Bo, T.; Guo, D.-A.; Ye, M. Global Profiling and Novel Structure Discovery Using Multiple Neutral Loss/Precursor Ion Scanning Combined with Substructure Recognition and Statistical Analysis (MNPSS): Characterization of Terpene-Conjugated Curcuminoids in Curcuma longa as a Case Study. *Anal. Chem.* **2015**, *88*, 703–710. [CrossRef]
19. Murray, K.K.; Boyd, R.K.; Eberlin, M.N.; Langley, G.; Li, L.; Naito, Y. Definitions of terms relating to mass spectrometry (IUPAC Recommendations 2013). *Pure Appl. Chem.* **2013**, *85*, 1515–1609. [CrossRef]
20. Busch, K.L.; Glish, G.L.; McLuckey, S.A.; Monaghan, J.J. Mass spectrometry/mass spectrometry: techniques and applications of tandem mass spectrometry. *Anal. Chim. Acta* **1990**, *237*, 509. [CrossRef]
21. Rubino, F.M.; Pitton, M.; Di Fabio, D.; Meroni, G.; Santaniello, E.; Caneva, E.; Pappini, M.; Colombi, A. Measurement of S-methylcysteine and S-methyl-mercapturic acid in human urine by alkyl-chloroformate extractive derivatization and isotope-dilution gas chromatography-mass spectrometry. *Biomed. Chromatogr.* **2011**, *25*, 330–343. [CrossRef]
22. Colombo, F.; Falvella, F.S.; De Cecco, L.; Tortoreto, M.; Pratesi, G.; Ciuffreda, P.; Ottria, R.; Santaniello, E.; Cicatiello, L.; Weisz, A.; et al. Pharmacogenomics and analogues of the antitumour agent N6-isopentenyladenosine. *Int. J. Cancer* **2009**, *124*, 2179–2185. [CrossRef] [PubMed]
23. Ottria, R.; Casati, S.; Baldoli, E.; Maier, J.A.; Ciuffreda, P. N^6-Alkyladenosines: Synthesis and evaluation of in vitro anticancer activity. *Bioorg Med. Chem.* **2010**, *18*, 8396–8402. [CrossRef] [PubMed]
24. Guo, J.; Villalta, P.W.; Turesky, R.J. Data-Independent Mass Spectrometry Approach for Screening and Identification of DNA Adducts. *Anal. Chem.* **2017**, *89*, 11728–11736. [CrossRef] [PubMed]
25. Lin, J.-K.; Chang, J.-Y. Chromophoric labeling of amino acids with 4-dimethylaminoazobenzene-4′-sulfonyl chloride. *Anal. Chem.* **1975**, *47*, 1634–1638. [CrossRef] [PubMed]
26. Rubino, F.M.; Pitton, M.; Brambilla, G.; Colombi, A. A study of the glutathione metaboloma peptides by energy-resolved mass spectrometry as a tool to investigate into the interference of toxic heavy metals with their metabolic processes. *J. Mass Spectrom.* **2006**, *41*, 1578–1593. [CrossRef]
27. Rubino, F.M.; Pitton, M.; Caneva, E.; Pappini, M.; Colombi, A. Thiol-disulfide redox equilibria of glutathione metaboloma compounds investigated by tandem mass spectrometry. *Rapid Commun. Mass Spectrom.* **2008**, *22*, 3935–3948. [CrossRef]

28. Rubino, F.M.; Cas, M.D.; Bignotto, M.; Ghidoni, R.; Iriti, M.; Paroni, R. Discovery of Unexpected Sphingolipids in Almonds and Pistachios with an Innovative Use of Triple Quadrupole Tandem Mass Spectrometry. *Foods* **2020**, *9*, 110. [CrossRef]
29. Rubino, F.M.; Pitton, M.; Di Fabio, D.; Colombi, A. Toward an "omic" physiopathology of reactive chemicals: Thirty years of mass spectrometric study of the protein adducts with endogenous and xenobiotic compounds. *Mass Spectrom. Rev.* **2009**, *28*, 725–784. [CrossRef]
30. Grigoryan, H.; Edmands, W.; Lu, S.S.; Yano, Y.; Regazzoni, L.; Iavarone, A.T.; Williams, E.R.; Rappaport, S.M. Adductomics Pipeline for Untargeted Analysis of Modifications to Cys34 of Human Serum Albumin. *Anal. Chem.* **2016**, *88*, 10504–10512. [CrossRef]

Sample Availability: Samples of the compounds are not available from the authors.

MDPI

Article

Collision Cross Section Prediction with Molecular Fingerprint Using Machine Learning

Fan Yang [1,*], Denice van Herwerden [2], Hugues Preud'homme [1,*] and Saer Samanipour [2,3,*]

[1] Institut des Sciences Analytiques et de Physico-Chimie Pour l'Environnement et les Materiaux (IPREM-UMR5254), E2S UPPA, CNRS, 64000 Pau, France
[2] Van 't Hoff Institute for Molecular Sciences (HIMS), University of Amsterdam, Science Park 904, 1098 XH Amsterdam, The Netherlands
[3] UvA Data Science Center, University of Amsterdam, 1098 XH Amsterdam, The Netherlands
* Correspondence: fyang@univ-pau.fr (F.Y.); hugues.preudhomme@univ-pau.fr (H.P.); s.samanipour@uva.nl (S.S.)

Abstract: High-resolution mass spectrometry is a promising technique in non-target screening (NTS) to monitor contaminants of emerging concern in complex samples. Current chemical identification strategies in NTS experiments typically depend on spectral libraries, chemical databases, and in silico fragmentation tools. However, small molecule identification remains challenging due to the lack of orthogonal sources of information (e.g., unique fragments). Collision cross section (CCS) values measured by ion mobility spectrometry (IMS) offer an additional identification dimension to increase the confidence level. Thanks to the advances in analytical instrumentation, an increasing application of IMS hybrid with high-resolution mass spectrometry (HRMS) in NTS has been reported in the recent decades. Several CCS prediction tools have been developed. However, limited CCS prediction methods were based on a large scale of chemical classes and cross-platform CCS measurements. We successfully developed two prediction models using a random forest machine learning algorithm. One of the approaches was based on chemicals' super classes; the other model was direct CCS prediction using molecular fingerprint. Over 13,324 CCS values from six different laboratories and PubChem using a variety of ion-mobility separation techniques were used for training and testing the models. The test accuracy for all the prediction models was over 0.85, and the median of relative residual was around 2.2%. The models can be applied to different IMS platforms to eliminate false positives in small molecule identification.

Keywords: collision cross section; ion mobility spectrometry; non-target screening; machine learning

Citation: Yang, F.; van Herwerden, D., Preud'homme, H.; Samanipour, S. Collision Cross Section Prediction with Molecular Fingerprint Using Machine Learning. *Molecules* **2022**, *27*, 6424. https://doi.org/10.3390/molecules27196424

Academic Editor: Thomas Letzel

Received: 30 August 2022
Accepted: 19 September 2022
Published: 29 September 2022

Publisher's Note: MDPI stays neutral with regard to jurisdictional claims in published maps and institutional affiliations.

1. Introduction

A large number of chemicals have been released into the environment by human activities, such as agriculture, industrial productions, and their relative byproducts. Once these chemicals enter the environment, transformation products (TPs) can be produced through hydrolysis, photosynthesis, and biological metabolism [1–6]. Most of these chemicals and their TPs are missing molecular and/or structure information. Thus, these chemicals' human and environmental risk assessments remain an open question [6–12]. Although most legacy pollutants have been banned for decades in many countries, they can still be detected at trace-level in the environment [2,13–15]. The known pollution is only the tip of the iceberg compared to the number of environmental hazards [1,13,14].

Non-target screening/analysis (NTS) is considered as an appropriate methodology to identify a variety of chemicals, especially for the unknown unknowns, such as contaminants of emerging concern (CECs) [16–18]. High-resolution mass spectrometry (HRMS) coupled with gas or liquid chromatography (GC or LC) is the most commonly used analytical technique in human health and environmental assessments. Thanks to the advance of HRMS, it has been increasingly applied in NTS studies in the last decades [17,19–21]. HRMS (i.e.,

Time-of-flight (TOF) and Orbitrap) maintains a high mass accuracy within ±5 mDa *m/z* error, and it can be acquired in full scan MS data or plus MS/MS data [10,21–24]. The accurate mass of the parent ion and the fragments are used to identify unknowns [17,19,21]. The isotopic pattern is one of the additional criteria which can help determine the presence of hetero-elements in non-target analysis [25]. However, mass spectral information is not enough for highly confident structural elucidation [22,25,26]. Therefore, inclusion of orthogonal sources of information such as measured or predicted retention time and/or retention time indices is necessary [21,27,28]. Such measurements are complex to perform and require particular experimental conditions [29–31].

Collision cross section (CCS) is a platform-independent measure of chemical structure in the gas phase and the three-dimensional space [32–34]. Studies have demonstrated that the inter-laboratory CCS biases are within 2% for the same IMS technique [35,36]. Moreover, cross-platform biases are below 3% for over 98% of the chemicals included in their studies [37,38]. Drift tube ion mobility (DTIM) and traveling wave ion mobility (TWIM) are two of the most used IMS techniques to measure the CCS value or drift time [37,39]. CCS value and drift time have been employed in NTS as an additional source of information, to increase confidence level in structural elucidation [40–42]. In addition to experimentally defined CCS values, CCS values can be estimated/predicted via theoretical calculations or Machine Learning (ML) [43,44]. ML CCS predictions take advantage of large datasets of the experimentally defined CCS values to train, validate, and test the regression models [44]. Zhou et al. [45] reported the first CCS prediction tool using the support vector regression (SVR) ML algorithm for metabolites. Plante et al. [46] published a deep neural networks CCS prediction strategy for cross-platform CCS measurement. The currently available CCS prediction tools rely on molecular descriptors or the combination of the chemical class and the *m/z* value of the parent compound [44–52]. Molecular fingerprints, which are more accurate and representative of the structure of a molecule [53], have not been used for the prediction of CCS values due to the difficulties associated with variable selection.

This study proposes a novel approach for CCS prediction using molecular topology fingerprints instead of molecular descriptors. First, we built a classification model to predict the chemical super classes based on their fingerprints. This model was used to classify chemical super classes. Then, CCS prediction models were developed for each super class. Additionally, all 13,324 chemicals were combined and to build a direct CCS prediction model. We also evaluated the impact of the chemical classes on the model accuracy.

2. Materials and Methods

2.1. Datasets

Experimental CCS databases and chemical information were collected from Zenodo, PubChem, and published articles as referenced in Table 1. Firstly, we retrieved all the missing SMILES notations from PubChem by PubChem CID using the Python PubChemPy library [54]. All the datasets were concatenated, and molecular fingerprints were generated by RDKit [55] (Open-source cheminformatics https://www.rdkit.org) (accessed on 10 April 2022) modules in Python. Hence, a dataset containing PubChem CID, SMILES [56,57], and empirical CCS value was saved as a csv file ready for model development and validation. The datasets and the source codes are available at https://github.com/fyang22/CCS-Prediction-Publish (accessed on 10 April 2022). Additional details about model optimization and construction are available in the Supplementary Materials.

The merged dataset included 13,324 unique empirical CCS values from 108.4 to 450.6 Å^2, measured by TWIM and DTIM. The merged dataset of 3313 chemicals was categorized into 43 super classes, including POPs, lipids, sugars, metabolites, hormones, drugs, etc. This dataset was then used for a classification model training and testing. Topological torsion (TT) fingerprints were chosen as features to encode chemical structure. TT fingerprints were first introduced by Nilakantan et al. [58], which describe the atom type, the topological distance between two atoms within four bonds, and torsion angles [59].

Four examples of molecular substructures are shown in Figure 1. The SMILES were converted to 1024 bit-strings fingerprints (FPs) by the implemented module in RDKit. The FPs were used to calculate molecular similarity, then visualized by principal component analysis (PCA) and fit machine learning models.

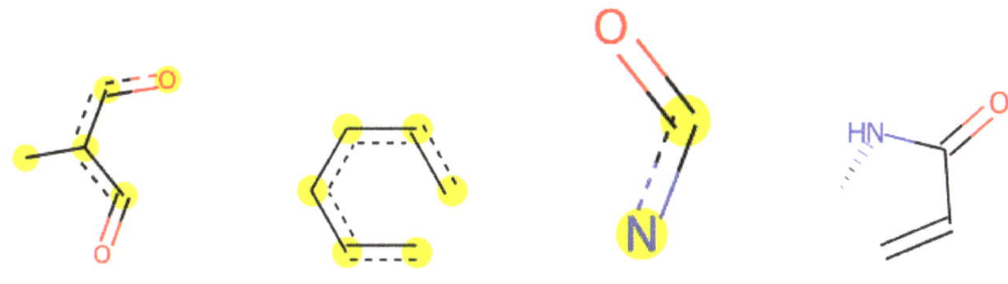

(**a**) bit-string 194 (**b**) bit-string 792 (**c**) bit-string 842 (**d**) bit-string 588

Figure 1. Examples of a substructure of a bit string. The most relevant features in the prediction models.

2.2. Overall Workflow

This study consists of two major parts and three models, and the workflow is summarized in Figure 2. Firstly, we developed a classification model to categorize chemicals into five groups, so-called "super class", based on their FPs similarity. The number of the "super class" was selected to create a balanced distribution of chemicals in each class. Five class-based CCS prediction models were developed using the optimized predicted category. Meanwhile, a direct CCS prediction model was built with the complete dataset without considering chemical categories. We also compared the two strategies to assess the prediction accuracy of these two modeling approaches. Finally, we applied the models to NORMAN SusDat (i.e., 101,684 chemicals) and carried out the direct and class-based prediction of the CCS values for SusDat.

2.2.1. Dataset for Classification Model

The dataset consisted of the identified chemical super classes which were merged from three CCS libraries [60–62]. This split dataset was used for chemical classification model training, validation, and testing. Initially, 43 super classes were defined, where most super classes contained less than 20 chemicals. To avoid overfitting of the classification model, we merged different super classes based on the calculated similarity scores of the chemicals. This enabled a more balanced distribution of chemicals in each super class. First, we calculated pair-wised fingerprint similarity by the Tanimoto similarity using RDKit. Tanimoto coefficient is a way to calculate the distance metric using molecular fingerprints [53,63]. Based on the distribution of the chemicals, super classes, and the similarity scores (plotted in Figure 3a), we kept the 5 super classes with the highest population of chemicals (listed in Figure 3b) and used them as ground truth. Chemicals in other super classes were assigned to one of the referred classes based on their similarity with a minimum similarity threshold of 0.6 since around 97% of pair-wise similarities were under 0.6 (shown in Figure 3a). Chemicals (n = 118) not meeting the similarity score criteria were manually assigned to a new super class (5 super classes) based on their characterized functional groups. Meanwhile, we kept the chemicals from the same given class (43 super classes from the raw dataset) in the same new super class. The final dataset consisted of 5 super classes having around 1000 unique chemicals in each class (in Supplementary Table S1), the classification of chemicals is visualized in Figure 3b. This dataset was used for random forest classification modeling. The final dataset for classification included fingerprints with 1024 bit-strings and the assigned super classes. Our super-class reassignment strategy effectively differentiated chemical classes from each other. For example, Organic acid and derivatives (in blue) and Benzenoid (in green) are two separate clusters in the middle left and in the bottom left.

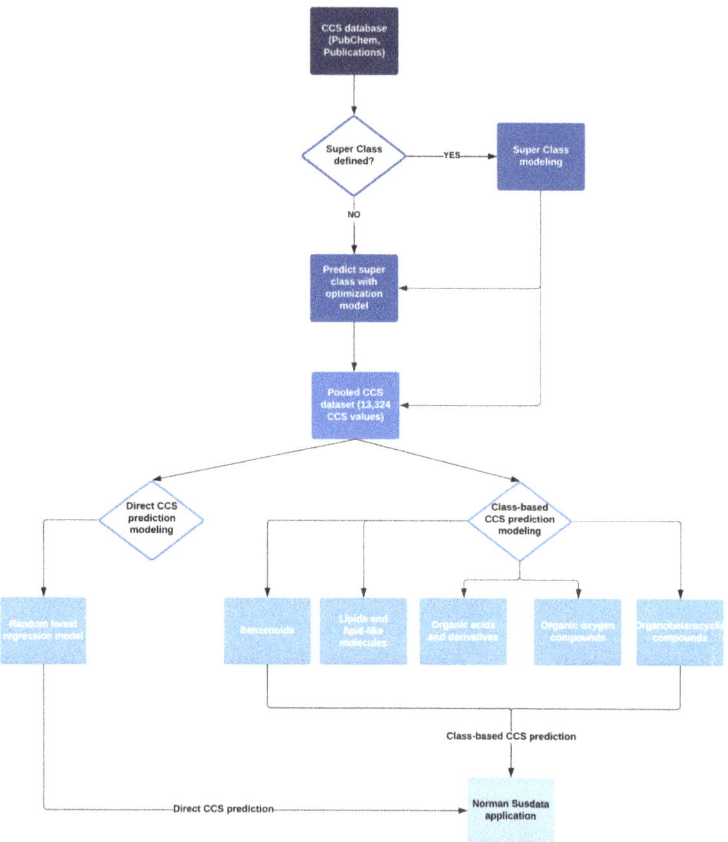

Figure 2. Modeling workflow: CCS empirical databases were collected from 6 different laboratories and PubChem. Two CCS prediction approaches were developed and validated. One model was class-based CCS prediction, and 5 super classes were defined for modeling. Another was a direct CCS prediction model. In the end, both prediction approaches were applied to the Norman Susdat list.

2.2.2. Dataset for Regression

For CCS regression modeling, we only considered protonated ions (8620 chemicals of $[M + H]^+$), deprotonated ions (4589 chemicals of $[M - H]^-$) and radical ions (115 chemicals of $[M]^{\cdot}$). Then, all the replications were removed by the SMILES, adduct ion and CCS values. Meanwhile, we calculated the standard deviation of CCS values for the same chemicals (same SMILES and adduct ion). In the training and test datasets, 642 chemicals have replications with different measured CCS values. The median of relative standard deviation (RSD) was about 1.4% (shown in Supplementary Figure S1) for both positive and negative ionization mode, and studies from multiple laboratories, which are consistent with the results reported by Hinnenkamp et al. [37] and Feuerstein et al. [38] Aspartame resulted in RSD of 12.5%, Picache et al. [60] recorded a CCS value of 127.4 $Å^2$ for Aspartame $[M + H]^+$, which is 40 $Å^2$ lower than the one measured in other references. Different Aspartame CCS values are also recorded in https://pubchem.ncbi.nlm.nih.gov/compound/134601

#section=Collision-Cross-Section (accessed on 1 June 2022). Hence, this dataset, collected from different laboratories and measured by different IM-MS platforms, was appropriate for CCS prediction. The entire dataset contained 13,324 unique empirical CCS values ranging from 108.4 to 450.6 Å^2, covering metabolites, drugs, lipids, etc., and it is available in Supplementary Table S2.

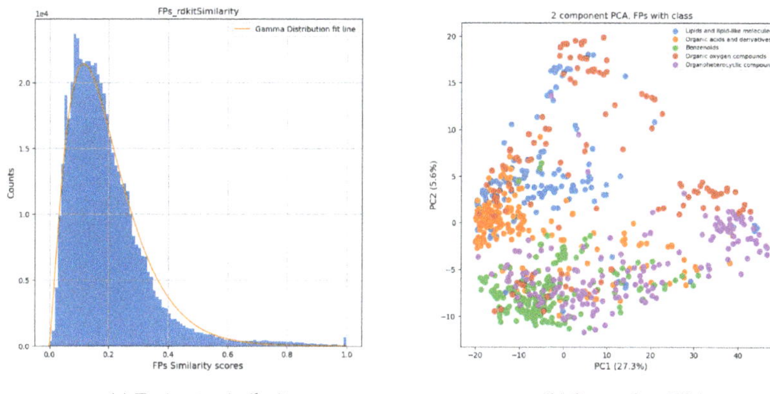

(**a**) Tanimoto similarity (**b**) Super class PCA

Figure 3. Super class distribution: A histogram of pair-wise fingerprints similarity is plotted in (**a**), and a normalized gamma distribution was fitted to the data and is shown as a red line. Based on the gamma distribution curve, similarity ≥ 0.6 was chosen to arrange the dataset. In (**b**), a 2D-scatter plot of PCA is generated by fingerprints.

Table 1. Summary of the dataset used in CCS prediction model optimization.

Reference	Number of Chemicals	Instrument *
Picache et al. [60]	1195	Agilent 6560 IM-QTOF MS
Hines et al. [64]	1304	Waters Synapt G2-Si HDMS
Celma et al. [40]	631	Waters VION IMS-QTOF MS
Zheng et al. [61,62]	891	Agilent 6560 IM-QTOF MS
Belova et al. [65]	145	Agilent 6560 IM-QTOF MS
Bijlsma et al. [51]	193	Waters VION IMS-QTOF MS
PubChem [66]	8965	

* Agilent: Drift tube ion mobility (DTIM), Waters: Traveling wave ion mobility (TWIM).

2.3. Modeling

In this study, we optimized three models: (1) Class prediction, (2) Class-based CCS regression model, and (3) a direct CCS regression model. A super class prediction model was first optimized using random forest classification. This model was used to assign the super class (i.e., five classes) of the whole dataset. Then, a regression model was built for each super class to predict the CCS values based on the FPs. Finally, we developed a model using only molecular FPs for CCS prediction. We compared the pros and cons of two CCS prediction approaches. All the modelings were performed using a 5-fold cross-validation by GridSearchCV build-in functionality in Scikit-learn. The details of each modeling strategy are provided below.

2.3.1. Class Prediction

The Class prediction model was first optimized using the random forest classification algorithm. The dataset was split into a training set (80%, n = 836) and a test set (20%, n = 210) with even distribution by super classes. In the random forest classifier, different hyper-parameters impact the model accuracy differently [67]. In this study, we focused on the number of trees in the random forest (n_estimators) and the minimum number of samples required at each leaf node (min_samples_leaf). These two parameters appeared to have the highest impact on the balance between the model robustness and accuracy. We generated a grid with 25 candidates for the number of trees ranging from 100 to 200 and 2 to 15 for minimum sample leaf. For each model, we performed 5 folds of cross-validation to assess the model accuracy. The model with the highest cross-validation accuracy was chosen as the optimized classification model, and the GridSearchCV scores are plotted in Supplementary Figure S2. The accuracy and F1 scores of each class are listed in Table 2.

Table 2. Results of super-class prediction modeling.

Super Class	Training	Test	F1 Score	Accuracy
Benzenoids	181	46	0.905	0.935
Lipids and lipid-like molecules	189	47	0.909	0.889
Organic acids and derivatives	184	46	0.848	0.813
Organic oxygen compounds	142	36	0.861	0.861
Organoheterocyclic compounds	140	35	0.822	0.857

2.3.2. Class-Based CCS Regression

For class-based regression modeling, we applied the optimized classification model (mentioned above) to the entire dataset, and the results are shown in Supplementary Table S2 and Figure S3. We independently performed the CCS prediction modeling for 5 data splits based on this classification, using the random forest regression algorithm. A total of 80% of the datasets were trained and tested by the rest. Similarly, we generated a grid with 50 candidates and the number of tree fits of 100 to 500. To avoid overfitting, the minimum sample leaf was set from 5 to 20. For each model and each class, 5 folds of cross-validation were evaluated to assess the model accuracy (Supplementary Figure S4a–e).

2.3.3. Direct CCS Regression

For comparison, we developed and tested a direct CCS prediction model for the entire dataset (13,324 compounds). A total of 80% of the data was used to train the model, and 20% of the data to test with 5-fold cross-validation (Supplementary Figure S4f). Similarly to the class-based CCS prediction model, n_estimators, and min_samples_leaf were optimized. The hyper-parameter optimization followed the same steps as class-based modeling (mentioned above). The model details and accuracy are listed in Table 3.

Table 3. Results of CCS prediction modeling.

	Training		Test		
Dataset	Data	R^2	Data	R^2	MRE (%)
All	10,659	0.972	2665	0.958	2.20
Benzenoids	1930	0.942	483	0.869	1.89
Lipids and lipid-like molecules	3675	0.940	919	0.932	2.33
Organic acids and derivatives	1392	0.950	348	0.901	2.21
Organic oxygen compounds	754	0.925	189	0.860	2.33
Organoheterocyclic compounds	2907	0.960	724	0.933	1.96

3. Results

3.1. Random Forest Classifier and Regression Prediction Model

Random forest is a suitable supervised machine learning algorithm for categorical and nonlinear data. We used a random forest classifier model to divide chemicals into 5 super classes by their molecular fingerprints. Then, we developed two CCS prediction strategies using molecular fingerprints. One is based on molecular super classes and molecular fingerprints, and another is a direct prediction by molecular fingerprints. As a CCS value is related to the chemical structure, we described each chemical structure by 1024 bit-strings molecular fingerprints, which were used as the prediction features. Each bit represents a substructure of a chemical, and some refer to a characteristic chemical substructure. These bits build up sets of nodes and leaves, then a decision tree.

A collection of decision trees results in a random forest model (decision trees files are available in Supplementary Materials). In order to obtain a generalized CCS prediction model, we merged 7 CCS libraries containing 13,324 unique CCS values (108.4 to 450.6 Å^2) measured by TWIM and DTIM platforms from multiple laboratories. Additionally, using a merged dataset for modeling allowed us to understand the variation of CCS measurement.

3.2. Evaluation of Classification Model

We obtained a classification model to separate 5 super classes with a global test accuracy (R^2) ≥ 0.871. In the classification model, it is crucial to have sufficient examples and similar training weights for each class. For example, if the dataset is randomly split to 80% of the training set that contains 50 organoheterocyclic compounds but over 200 chemicals of other classes, it would lead to insufficient training for organoheterocylic compounds and an overfitting problem, which can impact the overall performance of the classifier prediction. As shown in Table 2, the training and test sets were evenly distributed by super classes before modeling. The F1 score was over 0.9 for two classes and over 0.82 for the other three, indicating that the training data were balanced between classes. To further evaluate the classification model, we also generated a confusion matrix (Figure 4).

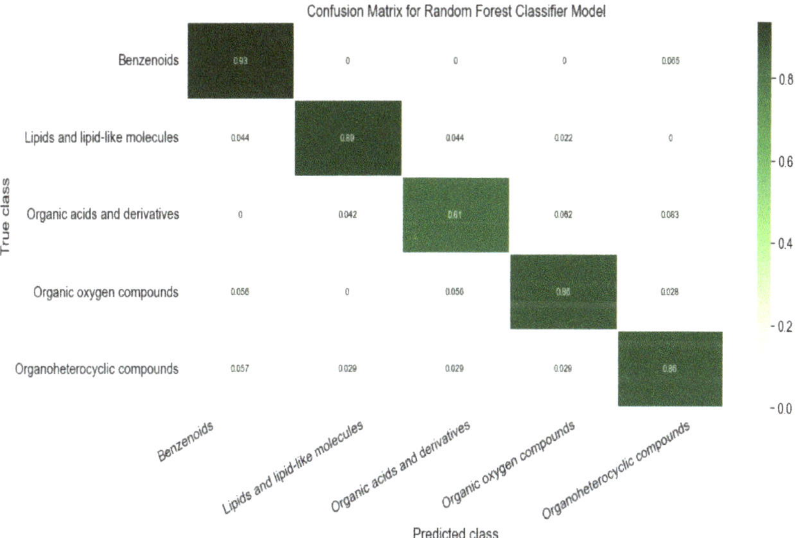

Figure 4. Confusion matrix of classification model.

Our model correctly predicted the super class of around 87% of the chemicals while around 8% of Organic acids and derivatives were classified as Organoheterocyclic compounds or Organic oxygen compounds. We noticed that errors frequently occurred in carboxylic acid compounds with phosphate esters or peptides. We randomly selected 3 incorrectly classified chemicals in each class. For instance, sulfadimethoxine (Figure 5a) was defined as Benzenoids due to an aniline. Nevertheless, it also contains pyrimidine, which was predicted as an Organoheterocyclic compound. Similarly, 3-Methyloxindole (Figure 5b) is an oxinole derivative consisting of a benzene ring and a heterocyclic with nitrogen. It was assigned to Organoheterocyclic (indole) in the collected dataset but went to Benzenoids compounds by prediction. We further investigated these incorrect classifications by examining the feature importance, shown in Supplementary Figure S5. Figure 1 shows a possible substructure of the most relevant bit-strings. For example, bit 792 (Figure 1b) would define whether a compound is classified as a Benzenoid or Organoheterocyclic compound. On the other hand, the bit-string 842 (Figure 1c) was used to decide whether a chemical should go to Organic oxygen compounds. None of the bit-strings displayed significant importance from others, indicating that the "incorrect" classification mainly has to do with which functional groups were given the higher priority when the original training set was being compiled.

(**a**) Sulfadimethoxine (**b**) 3-Methyloxindole

Figure 5. Random examples of "incorrect" predicted chemical.

3.3. Evaluation of Regression Models

In class-based modeling, the prediction R^2 was from 0.860 to 0.933, and the median relative error (MRE) of prediction was from 1.89% to 2.33% (Table 3). Direct CCS prediction, on the other hand, reached an R^2 of 0.95 and MRE of 2.2%, showing a good performance. Although we dropped replicated chemicals having the same CCS values before generating the modeling, considering that this dataset was merged by inter-laboratory studies, some chemicals might have been seen during training. Thus it can affect the prediction accuracy. Chemicals with less measurement deviation will increase the accuracy. On the contrary, those who have a significant deviation will bias prediction performance. We confirmed that for the direct prediction model, only 2% of the chemicals were common over 2665 test samples. The dataset was split by category in the class-based prediction, and the replications percentage was varied by chemical class. About 10% chemicals in the test set of Organic oxygen compounds were used in training before prediction, and less than 5% for other classes. Furthermore, except for a few outliers, the deviation of replications was under 6%. Therefore, we considered that the impact of replicated chemicals was negligible.

Additionally, we compared the performance of class-based models. Organic oxygen compound model obtained the lowest accuracy due to the lack of training data. Moreover, in its test split, the relative error $\geq$10% only occurred to macromolecules (e.g., maltodecaose ($C_{60}H_{102}O_{51}$)), contributing 15% to the test split, which resulted in poor prediction accuracy. Since we could not remeasure outliers' CCS values, we hypothesize that the error is associated with the compact and complex chemical structure. For instance, IMS measures the rotational-average surface of the maltodecaose ion. While a 1024 bit fingerprint is not enough to represent its complex chemical structure, resulting in a relative prediction error of 41.9% (true CCS at 390.3 while predicted 226.6 Å^2). Another possible reason can be the training weight. The dataset size of Organic oxygen compounds were almost 5 times less than Lipids and lipid-like molecules dataset, and glucose was the minority in the Organic oxygen compounds dataset. The model cannot properly generate the chemical rarely present during training. Therefore, higher accuracy was reached by Lipids and lipid-like molecules model and the direct prediction model. Outliers of other models were further investigated (shown in Supplementary Figure S6), and Figures 6 and 7a compare the predicted results of class-based models and direct prediction model. Four error cases have occurred to macromolecules (e.g., Diphenyl phosphate ($C_{39}H_{34}O_8P_2$)), which can be explained by the same hypothesis as maltodecaose (mentioned above). Metronidazole ($C_6H_9N_3O_3$) has 6 empirical CCS values measured with Waters TWIM, 5 were between 124 to 133 Å^2, while 200 Å^2 was measured by Picache et al. [60], leading a -61 Å^2 residual error (predicted CCS = 139.3 Å^2). L-tenuazonic acid ($C_{10}H_{15}NO_3$) was predicted to have a twice higher CCS than the measured one by the class-based model (35% higher by the direct prediction model). It might result from an inappropriate prediction by certain important features. Predicted CCS of vinyl acetate ($C_4H_6O_2$) was 127.4 Å^2 through the class-based model, and 147.9 Å^2 by direct prediction, while the empirical one was 227.2 Å^2. We hypothesize that vinyl acetate might be polymerized leading to higher measured CCS values. Benefiting from datasets from multiple sources, class-based and direct prediction models can verify experimental CCS and evaluate the inter-laboratory and inter-platform deviation.

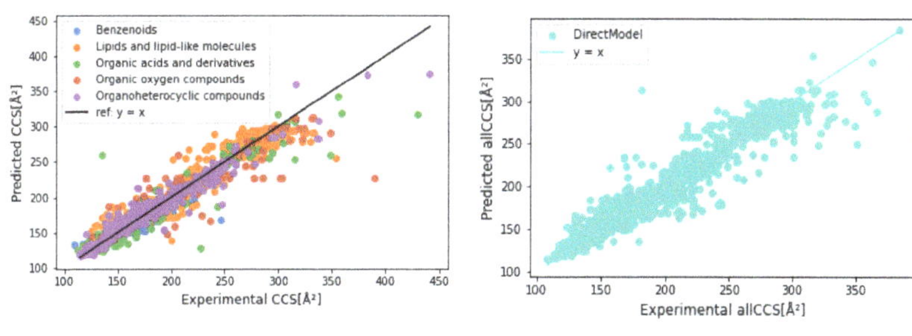

(**a**) Class-based Models (**b**) Direct prediction Model

Figure 6. Precision comparison between each predictive class and direct prediction without class. Class-based models lead to a better precision from 150 to 300 Å^2, while giving more bias by the small and macro molecules. The direct prediction model is less affected by the extreme cases.

Figure 6 compares empirical and predicted CCS values generated by different models. We noticed that the direct prediction model was less biased by chemical class and structure, small and/or macro molecules, leading to higher prediction accuracy than the class-based prediction results. Although class-based models generated lower MREs (Table 3), a higher residual error was obtained in vinyl acetate and macro molecules resulting in lower R^2. As we can see in Figure 7c, over 25% of the test dataset obtained relative residual lower than 1%, and class-based models gained slightly higher, at 26.6%. All prediction models were further evaluated by feature importance (shown in Supplementary Figures S7 and S8). In both prediction approaches, the most relevant features divided chemicals into relative low CCS and high CCS. In other words, the decision tree was made of different CCS ranges based on certain substructures. For example, the most relevant feature in Organic acids and derivatives CCS prediction model was bit 588 (Figure 1d). If a chemical has its represented substructure, this chemical will be considered as CCS > 150 Å^2, which might yield the prediction error for l-tenuazonic acid. Overall, the direct CCS model generated the best prediction performance, and a more extensive dataset ensured a more robust model.

MetCCS was a support vector regression (SVR) based on a prediction method only for metabolites. It achieved an excellent $R^2 > 0.96$ with the intra-laboratory and inter-laboratory measurements, relative residual was within 5% [45]. Bijlsma et al. [51] developed an artificial neural network (ANN)-based CCS prediction tool and [52] published an multivariate adaptive regression splines (MARS) CCS prediction model. Both were trained by TWIM data, and the relative error was within 6% for 95% of the chemicals. Belova et al. [68] compared experimental DTIM measured CCS values to predicted CCS values by the ANN-based and MARS-based predictors. A total of 95% of the protonated and deprotonated ions observed the relative error under 6.7%. However, only 56 compounds with 108 DTIM measured CCS values were compared in their study. We obtained comparable results by direct and class-based models, 87% of predicted results obtained the relative error within 7% (Figure 7c). DeepCCS is a more generalized CCS prediction model generated by SMILES with the deep neural network. R^2 was greater than 0.97, and MRE was below 2.6% [46]. However, only 1637 datasets were initially used to train the model, and the prediction power might be declined by chemical class [49,50]. We achieved a comparable accuracy for a wider scope of chemicals by direct prediction model (R^2 over 0.95, MRE within 2.2%). AllCCS and CCSbase generated better accuracy, with R^2 over 0.98 and MRE below 2%, since both tools used a larger and more diverse training set than DeepCCS and MetCCS.

More structural-related features were emphasized in their studies. Considering our models, we reached comparable MREs with other tools and over 90% of the chemicals predicted within 8% relative residual. The results are satisfied with the CCS measurement bias via different IMS instrumentation and techniques [37].

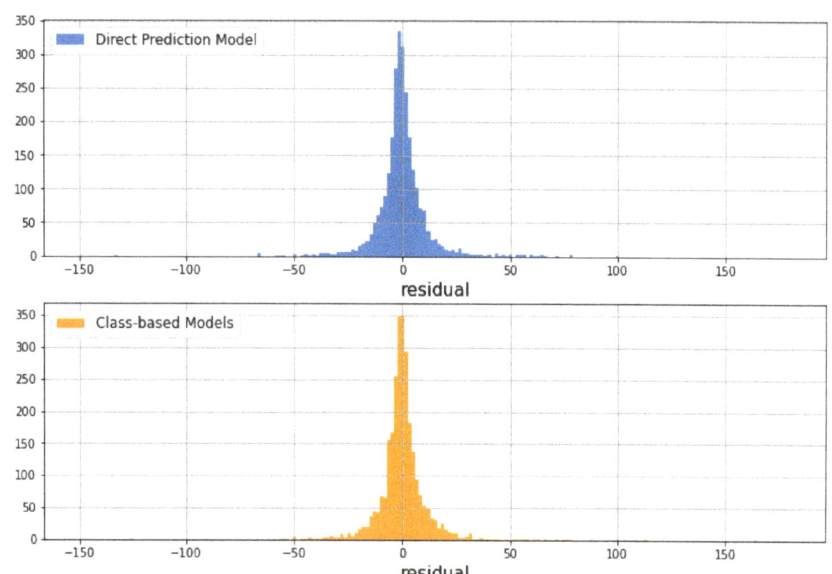

(**a**) Residuals of class-based and direct model

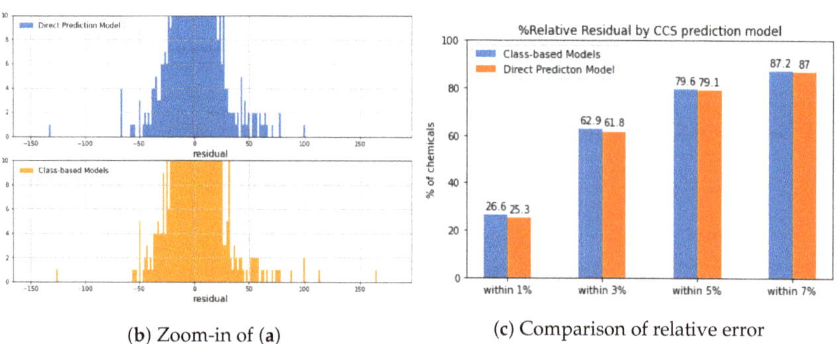

(**b**) Zoom-in of (**a**)

(**c**) Comparison of relative error

Figure 7. (**a**) compared the residuals of predicted CCS from class-based CCS prediction model and direct CCS prediction model. (**b**) is a zoomed-in of (**a**). Both approaches generate a good prediction power. A total of 98% of chemicals has a predicted difference within 25 Å^2. (**c**) Comparison of relative error in the testing set between the two approaches within 1%, 3%, 5%, and 7%.

3.4. Application on SusDat

NORMAN SusDat database contains over 111,000 environmentally relevant chemicals, with SMILES, accurate mass, and physiochemical properties [69]. We applied direct CCS prediction and class-based CCS prediction to the SusDat database, which contains chemicals that have never been seen during training and test, such as antibiotics and transformation products. A total of 96 % of the chemicals have a predicted difference within 25 Å^2 by two approaches (shown in Figure 8). The lack of true CCS values in SusDat, thus, by comparing the differences in predicted results generated by two approaches, demonstrates the robustness of models, and the direct prediction model can discriminate different chemical classes.

Predicted CCS values are provided in Supplementary Table S3 for use in non-target screening or retrospective analysis. Moreover, these predicted CCS values can be compared to the measured CCS values by standard inter-laboratory evaluation and inter-platform deviation and improve the performances of our models.

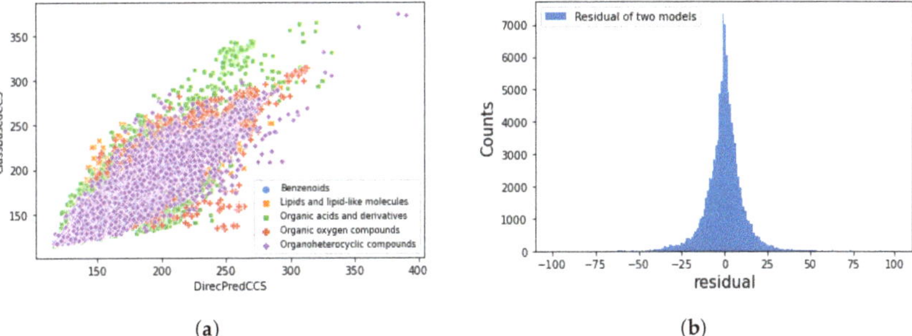

(a) (b)

Figure 8. Comparison of direct and class-based CCS prediction model using Norman Susdat. (**a**) Scatter plot of class-based predicted CCS value against direct predicted CCS value. (**b**) Difference of predicted CCS values between class-based and direct prediction models. A total of 96 % of the chemicals have a predicted difference within 25 Å^2.

4. Discussion

In this study, we introduced topological fingerprints to categorize chemicals and generate CCS prediction models using the random forest algorithm. Our methods are generalized to TWIM and DTIM measured CCS data collected from seven sources. Prediction models were developed for five super classes of chemicals (Benzenoids, Lipids and lipid-like molecules, Organic acids and derivatives, Organic oxygen compounds, and Organoheterocyclic compounds) and the entire dataset. The test prediction accuracy was 0.958 by the direct prediction approach, 3 class-based prediction models more than 0.9, and over 0.86 for the remaining two classes. The MRE was between 1.89% to 2.33%. Additionally, models only required SMILES to encode fingerprints. A significant predicted variation was observed in macro molecules and vinyl acetate, with over 100 Å^2 residual. We noticed that the residuals were reduced through the direct prediction model due to an extensive training set and a higher presence of macro molecules in the dataset. The prediction performances are highly dependent on the collected CCS libraries. Therefore, it is emphasized that multiple and accurate empirical CCS libraries with a broad scope of chemicals are crucial to CCS machine learning studies. Moreover, this bias indicated a limited prediction performance for chemicals with unique structures. A better classification model or other structural importance features might improve the prediction accuracy. Since fingerprint was the only input feature for prediction, adduct ions (e.g., $[M + Na]^+$) were eliminated in this study. Other features can be introduced in the models to generate more ion types. Moreover, fingerprints offer a novel aspect in CCS prediction using machine learning. The generated feature importance of 1024 bits was directly related to the structures and thus easier to interpret chemically.

Supplementary Materials: The following supporting information can be downloaded at: https://www.mdpi.com/article/10.3390/molecules27196424/s1, Table S1: SuperClassModeling.csv; Table S2: CCSPredictedData.csv; Table S3: SusDatCCSprediction.csv; Decision trees files: DecisionTrees.zip. Figure S1: RSD of replicated chemicals; Figure S2: Classification GridSearchCV scores; Figure S3: Distribution of predicted super classes; Figure S4: Regression modeling scores of hyper-parameters optimization by GridSearchCV; Figure S5: Feature Importance of classification model; Figure S6: Outliers with predicted Class, predicted CCS by Class-based and direct models; Figure S7: Regression modeling Feature Importances; Figure S8: Example of most relevant features.

Author Contributions: F.Y., D.v.H. and S.S. designed the study. F.Y. performed the experiments and wrote the original draft. H.P. reviewed the manuscript. S.S. edited, reviewed the manuscript and supervised this project. All authors have read and agreed to the published version of the manuscript.

Funding: This study was funded by SETASAR PhD Project from Région Nouvelle Aquitaine and LPL. The exchange project was support by E2S UPPA.

Institutional Review Board Statement: Not applicable.

Informed Consent Statement: Not applicable.

Data Availability Statement: The datasets and the source codes can be found at https://github.com/fyang22/CCS-Prediction-Publish (accessed on 10 April 2022) and https://www.mdpi.com/article/10.3390/molecules27196424/s1.

Acknowledgments: The authors thank Région Nouvelle Aquitaine and LPL for their financial support of SETASAR PhD Project, E2S UPPA for the exchange grant in the University of Amsterdam. S. Samanipour is grateful to UvA Data Science center and ChemistryNL for their financial support, projects EDIFIED and SCOPE.

Conflicts of Interest: The authors declare no conflicts of interest.

References

1. Muir, D.C.; Howard, P.H. Are there other persistent organic pollutants? A challenge for environmental chemists. *Environ. Sci. Technol.* **2006**, *40*, 7157–7166. [CrossRef] [PubMed]
2. Howard, P.H.; Muir, D.C. Identifying new persistent and bioaccumulative organics among chemicals in commerce II: Pharmaceuticals. *Environ. Sci. Technol.* **2011**, *45*, 6938–6946. [CrossRef] [PubMed]
3. Escher, B.I.; Stapleton, H.M.; Schymanski, E.L. Tracking complex mixtures of chemicals in our changing environment. *Science* **2020**, *367*, 388–392. [CrossRef] [PubMed]
4. Newton, S.R.; McMahen, R.L.; Sobus, J.R.; Mansouri, K.; Williams, A.J.; McEachran, A.D.; Strynar, M.J. Suspect screening and non-targeted analysis of drinking water using point-of-use filters. *Environ. Pollut.* **2018**, *234*, 297–306. [CrossRef] [PubMed]
5. Shi, Q.; Xiong, Y.; Kaur, P.; Sy, N.D.; Gan, J. Contaminants of emerging concerns in recycled water: Fate and risks in agroecosystems. *Sci. Total Environ.* **2022**, *814*, 152527. [CrossRef] [PubMed]
6. Rizzo, L.; Gernjak, W.; Krzeminski, P.; Malato, S.; McArdell, C.S.; Perez, J.A.S.; Schaar, H.; Fatta-Kassinos, D. Best available technologies and treatment trains to address current challenges in urban wastewater reuse for irrigation of crops in EU countries. *Sci. Total Environ.* **2020**, *710*, 136312. [CrossRef]
7. Manaia, C.M. Assessing the risk of antibiotic resistance transmission from the environment to humans: Non-direct proportionality between abundance and risk. *Trends Microbiol.* **2017**, *25*, 173–181. [CrossRef]
8. López-Pacheco, I.Y.; Silva-Núñez, A.; Salinas-Salazar, C.; Arévalo-Gallegos, A.; Lizarazo-Holguin, L.A.; Barceló, D.; Iqbal, H.M.; Parra-Saldívar, R. Anthropogenic contaminants of high concern: Existence in water resources and their adverse effects. *Sci. Total Environ.* **2019**, *690*, 1068–1088. [CrossRef]
9. Ma, Y.; He, X.; Qi, K.; Wang, T.; Qi, Y.; Cui, L.; Wang, F.; Song, M. Effects of environmental contaminants on fertility and reproductive health. *J. Environ. Sci.* **2019**, *77*, 210–217. [CrossRef]
10. Alygizakis, N.A.; Samanipour, S.; Hollender, J.; Ibáñez, M.; Kaserzon, S.; Kokkali, V.; Van Leerdam, J.A.; Mueller, J.F.; Pijnappels, M.; Reid, M.J.; et al. Exploring the potential of a global emerging contaminant early warning network through the use of retrospective suspect screening with high-resolution mass spectrometry. *Environ. Sci. Technol.* **2018**, *52*, 5135–5144. [CrossRef]
11. Pedrazzani, R.; Bertanza, G.; Brnardić, I.; Cetecioglu, Z.; Dries, J.; Dvarionienė, J.; García-Fernández, A.J.; Langenhoff, A.; Libralato, G.; Lofrano, G.; et al. Opinion paper about organic trace pollutants in wastewater: Toxicity assessment in a European perspective. *Sci. Total Environ.* **2019**, *651*, 3202–3221. [CrossRef] [PubMed]
12. Rueda-Ruzafa, L.; Cruz, F.; Roman, P.; Cardona, D. Gut microbiota and neurological effects of glyphosate. *Neurotoxicology* **2019**, *75*, 1–8. [CrossRef] [PubMed]
13. Lohmann, R.; Breivik, K.; Dachs, J.; Muir, D. Global fate of POPs: Current and future research directions. *Environ. Pollut.* **2007**, *150*, 150–165. [CrossRef] [PubMed]
14. Samanipour, S.; Martin, J.W.; Lamoree, M.H.; Reid, M.J.; Thomas, K.V. Optimism for nontarget analysis in environmental chemistry. *Environ. Sci. Technol.* **2019**, *53*, 5529–5530. [CrossRef] [PubMed]
15. Vermeulen, R.; Schymanski, E.L.; Barabási, A.L.; Miller, G.W. The exposome and health: Where chemistry meets biology. *Science* **2020**, *367*, 392–396. [CrossRef]
16. Schymanski, E.L.; Jeon, J.; Gulde, R.; Fenner, K.; Ruff, M.; Singer, H.P.; Hollender, J. Identifying small molecules via high resolution mass spectrometry: communicating confidence. *Environ. Sci. Technol.* **2014**, *48*, 2097–2098. [CrossRef]
17. Schulze, B.; Jeon, Y.; Kaserzon, S.; Heffernan, A.L.; Dewapriya, P.; O'Brien, J.; Ramos, M.J.G.; Gorji, S.G.; Mueller, J.F.; Thomas, K.V.; et al. An assessment of quality assurance/quality control efforts in high resolution mass spectrometry non-target workflows for analysis of environmental samples. *TrAC Trends Anal. Chem.* **2020**, *133*, 116063. [CrossRef]

18. Pérez-Lemus, N.; López-Serna, R.; Pérez-Elvira, S.I.; Barrado, E. Analytical methodologies for the determination of pharmaceuticals and personal care products (PPCPs) in sewage sludge: A critical review. *Anal. Chim. Acta* **2019**, *1083*, 19–40. [CrossRef]
19. Hollender, J.; Schymanski, E.L.; Singer, H.P.; Ferguson, P.L. Nontarget screening with high resolution mass spectrometry in the environment: Ready to go? *Environ. Sci. Technol.* **2017**, *51*, 11505–11512. [CrossRef]
20. Guo, Z.; Huang, S.; Wang, J.; Feng, Y.L. Recent advances in non-targeted screening analysis using liquid chromatography—High resolution mass spectrometry to explore new biomarkers for human exposure. *Talanta* **2020**, *219*, 121339. [CrossRef]
21. Hollender, J.; Van Bavel, B.; Dulio, V.; Farmen, E.; Furtmann, K.; Koschorreck, J.; Kunkel, U.; Krauss, M.; Munthe, J.; Schlabach, M.; et al. High resolution mass spectrometry-based non-target screening can support regulatory environmental monitoring and chemicals management. *Environ. Sci. Eur.* **2019**, *31*, 42. [CrossRef]
22. Knolhoff, A.M.; Callahan, J.H.; Croley, T.R. Mass accuracy and isotopic abundance measurements for HR-MS instrumentation: Capabilities for non-targeted analyses. *J. Am. Soc. Mass Spectrom.* **2014**, *25*, 1285–1294. [CrossRef] [PubMed]
23. Hernandez, F.; Sancho, J.V.; Ibáñez, M.; Abad, E.; Portolés, T.; Mattioli, L. Current use of high-resolution mass spectrometry in the environmental sciences. *Anal. Bioanal. Chem.* **2012**, *403*, 1251–1264. [CrossRef] [PubMed]
24. Kaufmann, A. The current role of high-resolution mass spectrometry in food analysis. *Anal. Bioanal. Chem.* **2012**, *403*, 1233–1249. [CrossRef] [PubMed]
25. Knolhoff, A.M.; Croley, T.R. Non-targeted screening approaches for contaminants and adulterants in food using liquid chromatography hyphenated to high resolution mass spectrometry. *J. Chromatogr. A* **2016**, *1428*, 86–96. [CrossRef] [PubMed]
26. Kind, T.; Fiehn, O. Metabolomic database annotations via query of elemental compositions: Mass accuracy is insufficient even at less than 1 ppm. *BMC Bioinform.* **2006**, *7*, 234. [CrossRef]
27. d'Atri, V.; Causon, T.; Hernandez-Alba, O.; Mutabazi, A.; Veuthey, J.L.; Cianferani, S.; Guillarme, D. Adding a new separation dimension to MS and LC–MS: What is the utility of ion mobility spectrometry? *J. Sep. Sci.* **2018**, *41*, 20–67. [CrossRef]
28. Boelrijk, J.; van Herwerden, D.; Ensing, B.; Forré, P.; and Samanipour, S. Predicting RP-LC retention indices of structurally unknown chemicals from mass spectrometry data. *ChemRxiv* **2022**. [CrossRef]
29. Celma, A.; Ahrens, L.; Gago-Ferrero, P.; Hernández, F.; López, F.; Lundqvist, J.; Pitarch, E.; Sancho, J.V.; Wiberg, K.; Bijlsma, L. The relevant role of ion mobility separation in LC-HRMS based screening strategies for contaminants of emerging concern in the aquatic environment. *Chemosphere* **2021**, *280*, 130799. [CrossRef]
30. Mairinger, T.; Causon, T.J.; Hann, S. The potential of ion mobility–mass spectrometry for non-targeted metabolomics. *Curr. Opin. Chem. Biol.* **2018**, *42*, 9–15. [CrossRef]
31. Goscinny, S.; Joly, L.; De Pauw, E.; Hanot, V.; Eppe, G. Travelling-wave ion mobility time-of-flight mass spectrometry as an alternative strategy for screening of multi-class pesticides in fruits and vegetables. *J. Chromatogr. A* **2015**, *1405*, 85–93. [CrossRef] [PubMed]
32. Hill, H.H., Jr.; Siems, W.F.; St. Louis, R.H. Ion mobility spectrometry. *Anal. Chem.* **1990**, *62*, 1201A–1209A. [CrossRef] [PubMed]
33. Borsdorf, H.; Eiceman, G.A. Ion mobility spectrometry: Principles and applications. *Appl. Spectrosc. Rev.* **2006**, *41*, 323–375. [CrossRef]
34. Eiceman, G.A.; Karpas, Z. *Ion Mobility Spectrometry*; CRC Press: Boca Raton, FL, USA, 2005.
35. Hernández-Mesa, M.; D'atri, V.; Barknowitz, G.; Fanuel, M.; Pezzatti, J.; Dreolin, N.; Ropartz, D.; Monteau, F.; Vigneau, E.; Rudaz, S.; et al. Interlaboratory and interplatform study of steroids collision cross section by traveling wave ion mobility spectrometry. *Anal. Chem.* **2020**, *92*, 5013–5022. [CrossRef]
36. Stow, S.M.; Causon, T.J.; Zheng, X.; Kurulugama, R.T.; Mairinger, T.; May, J.C.; Rennie, E.E.; Baker, E.S.; Smith, R.D.; McLean, J.A.; et al. An interlaboratory evaluation of drift tube ion mobility–mass spectrometry collision cross section measurements. *Anal. Chem.* **2017**, *89*, 9048–9055. [CrossRef] [PubMed]
37. Hinnenkamp, V.; Klein, J.; Meckelmann, S.W.; Balsaa, P.; Schmidt, T.C.; Schmitz, O.J. Comparison of CCS values determined by traveling wave ion mobility mass spectrometry and drift tube ion mobility mass spectrometry. *Anal. Chem.* **2018**, *90*, 12042–12050. [CrossRef] [PubMed]
38. Feuerstein, M.L.; Hernández-Mesa, M.; Kiehne, A.; Le Bizec, B.; Hann, S.; Dervilly, G.; Causon, T. Comparability of Steroid Collision Cross Sections Using Three Different IM-HRMS Technologies: An Interplatform Study. *J. Am. Soc. Mass Spectrom.* **2022**. [CrossRef]
39. Borsdorf, H.; Mayer, T.; Zarejousheghani, M.; Eiceman, G.A. Recent developments in ion mobility spectrometry. *Appl. Spectrosc. Rev.* **2011**, *46*, 472–521. [CrossRef]
40. Celma, A.; Sancho, J.V.; Schymanski, E.L.; Fabregat-Safont, D.; Ibanez, M.; Goshawk, J.; Barknowitz, G.; Hernandez, F.; Bijlsma, L. Improving target and suspect screening high-resolution mass spectrometry workflows in environmental analysis by ion mobility separation. *Environ. Sci. Technol.* **2020**, *54*, 15120–15131. [CrossRef]
41. Menger, F.; Celma, A.; Schymanski, E.L.; Lai, F.Y.; Bijlsma, L.; Wiberg, K.; Hernández, F.; Sancho, J.V.; Lutz, A. Enhancing Spectral Quality in Complex Environmental Matrices: Supporting Suspect and Non-Target Screening in Zebra Mussels with Ion Mobility. *SSRN Electron. J.* **2022**. [CrossRef]
42. Izquierdo-Sandoval, D.; Fabregat-Safont, D.; Lacalle-Bergeron, L.; Sancho, J.V.; Hernández, F.; Portoles, T. Benefits of Ion Mobility Separation in GC-APCI-HRMS Screening: From the Construction of a CCS Library to the Application to Real-World Samples. *Anal. Chem.* **2022**, *94*, 9040–9047. [CrossRef] [PubMed]

43. Gabelica, V.; Marklund, E. Fundamentals of ion mobility spectrometry. *Curr. Opin. Chem. Biol.* **2018**, *42*, 51–59. [CrossRef] [PubMed]
44. Ross, D.H.; Xu, L. Determination of drugs and drug metabolites by ion mobility-mass spectrometry: A review. *Anal. Chim. Acta* **2021**, *1154*, 338270. [CrossRef] [PubMed]
45. Zhou, Z.; Shen, X.; Tu, J.; Zhu, Z.J. Large-Scale Prediction of Collision Cross-Section Values for Metabolites in Ion Mobility-Mass Spectrometry. *Anal. Chem.* **2016**, *88*, 11084–11091. [CrossRef] [PubMed]
46. Plante, P.L.; Francovic-Fontaine, É.; May, J.C.; McLean, J.A.; Baker, E.S.; Laviolette, F.; Marchand, M.; Corbeil, J. Predicting Ion Mobility Collision Cross-Sections Using a Deep Neural Network: DeepCCS. *Anal. Chem.* **2019**, *91*, 5191–5199. [CrossRef]
47. Zhou, Z.; Tu, J.; Zhu, Z.J. Advancing the large-scale CCS database for metabolomics and lipidomics at the machine-learning era. *Curr. Opin. Chem. Biol.* **2018**, *42*, 34–41. [CrossRef]
48. Mollerup, C.B.; Mardal, M.; Dalsgaard, P.W.; Linnet, K.; Barron, L.P. Prediction of collision cross section and retention time for broad scope screening in gradient reversed-phase liquid chromatography-ion mobility-high resolution accurate mass spectrometry. *J. Chromatogr. A* **2018**, *1542*, 82–88. [CrossRef]
49. Zhou, Z.; Luo, M.; Chen, X.; Yin, Y.; Xiong, X.; Wang, R.; Zhu, Z.J. Ion mobility collision cross-section atlas for known and unknown metabolite annotation in untargeted metabolomics. *Nat. Commun.* **2020**, *11*, 4334. [CrossRef]
50. Ross, D.H.; Cho, J.H.; Xu, L. Breaking down structural diversity for comprehensive prediction of ion-neutral collision cross sections. *Anal. Chem.* **2020**, *92*, 4548–4557. [CrossRef]
51. Bijlsma, L.; Bade, R.; Celma, A.; Mullin, L.; Cleland, G.; Stead, S.; Hernandez, F.; Sancho, J.V. Prediction of collision cross-section values for small molecules: Application to pesticide residue analysis. *Anal. Chem.* **2017**, *89*, 6583–6589. [CrossRef]
52. Celma, A.; Bade, R.; Sancho, J.V.; Hernández, F.; Humpries, M.; Bijslma, L. Prediction of Retention Time and Collision Cross Section (CCSH+, CCSH-and CCSNa+) of Emerging Contaminants Using Multiple Adaptive Regression Splines. 2022. Available online: https://doi.org/10.21203/rs.3.rs-1249834/v1 (accessed on 13 January 2022).
53. Cereto-Massagué, A.; Ojeda, M.J.; Valls, C.; Mulero, M.; Garcia-Vallvé, S.; Pujadas, G. Molecular fingerprint similarity search in virtual screening. *Methods* **2015**, *71*, 58–63. [CrossRef] [PubMed]
54. Swain, M. PubChemPy: A Way to Interact with PubChem in Python. 2014. Available online: https://pubchempy.readthedocs.io/en/latest/ (accessed on 13 January 2022).
55. Landrum, G. RDKit: Open-Source Cheminformatics. 2006. Available online: https://doi.org/10.5281/zenodo.3732262 (accessed on 13 January 2022).
56. Weininger, D. SMILES, a chemical language and information system. 1. Introduction to methodology and encoding rules. *J. Chem. Inf. Comput. Sci.* **1988**, *28*, 31–36. [CrossRef]
57. Weininger, D.; Weininger, A.; Weininger, J.L. SMILES. 2. Algorithm for generation of unique SMILES notation. *J. Chem. Inf. Comput. Sci.* **1989**, *29*, 97–101. [CrossRef]
58. Nilakantan, R.; Bauman, N.; Dixon, J.S.; Venkataraghavan, R. Topological torsion: A new molecular descriptor for SAR applications. Comparison with other descriptors. *J. Chem. Inf. Comput. Sci.* **1987**, *27*, 82–85. [CrossRef]
59. Capecchi, A.; Probst, D.; Reymond, J.L. One molecular fingerprint to rule them all: Drugs, biomolecules, and the metabolome. *J. Cheminformatics* **2020**, *12*, 1–15. [CrossRef]
60. Picache, J.A.; Rose, B.S.; Balinski, A.; Leaptrot, K.L.; Sherrod, S.D.; May, J.C.; McLean, J.A. Collision cross section compendium to annotate and predict multi-omic compound identities. *Chem. Sci.* **2019**, *10*, 983–993. [CrossRef]
61. Zheng, X.; Aly, A.N.; Zhou, Y.; Dupuis, K.T.; Bilbao, A.; Paurus, V.L.; Orton, D.J.; Wilson, R.; Payne, S.H.; Smith, R.D.; et al. A structural examination and collision cross section database for over 500 metabolites and xenobiotics using drift tube ion mobility spectrometry. *Chem. Sci.* **2017**, *8*, 7724–7736. [CrossRef]
62. Zheng, X.; Dupuis, K.T.; Aly, N.A.; Zhou, Y.; Smith, F.B.; Tang, K.; Smith, R.D.; Baker, E.S. Utilizing ion mobility spectrometry and mass spectrometry for the analysis of polycyclic aromatic hydrocarbons, polychlorinated biphenyls, polybrominated diphenyl ethers and their metabolites. *Anal. Chim. Acta* **2018**, *1037*, 265–273. [CrossRef]
63. Bajusz, D.; Rácz, A.; Héberger, K. Why is Tanimoto index an appropriate choice for fingerprint-based similarity calculations? *J. Cheminform.* **2015**, *7*, 20. [CrossRef]
64. Hines, K.M.; Ross, D.H.; Davidson, K.L.; Bush, M.F.; Xu, L. Large-Scale Structural Characterization of Drug and Drug-Like Compounds by High-Throughput Ion Mobility-Mass Spectrometry. *Anal. Chem.* **2017**, *89*, 9023–9030. [CrossRef]
65. Belova, L.; Caballero-Casero, N.; van Nuijs, A.L.N.; Covaci, A. Ion Mobility-High-Resolution Mass Spectrometry (IM-HRMS) for the Analysis of Contaminants of Emerging Concern (CECs): Database Compilation and Application to Urine Samples. *Anal. Chem.* **2021**, *93*, 6428–6436. [CrossRef] [PubMed]
66. Schymanski, E.; Zhang, J.; Thiessen, P.; Bolton, E. Experimental CCS Values in Pubchem. *Zenodo* **2022**. [CrossRef]
67. Svetnik, V.; Liaw, A.; Tong, C.; Culberson, J.C.; Sheridan, R.P.; Feuston, B.P. Random forest: A classification and regression tool for compound classification and QSAR modeling. *J. Chem. Inf. Comput. Sci.* **2003**, *43*, 1947–1958. [CrossRef] [PubMed]

68. Belova, L.; Celma, A.; Van Haesendonck, G.; Lemière, F.; Sancho, J.V.; Covaci, A.; van Nuijs, A.L.; Bijlsma, L. Revealing the differences in collision cross section values of small organic molecules acquired by different instrumental designs and prediction models. *Anal. Chim. Acta* **2022**, *1229*, 340361. [CrossRef] [PubMed]
69. Dulio, V.; Koschorreck, J.; Van Bavel, B.; Van den Brink, P.; Hollender, J.; Munthe, J.; Schlabach, M.; Aalizadeh, R.; Agerstrand, M.; Ahrens, L.; et al. The NORMAN association and the European partnership for chemicals risk assessment (PARC): Let's cooperate! *Environ. Sci. Eur.* **2020**, *32*, 100. [CrossRef]

MDPI

Article

Evaluation of Sample Preparation Methods for Non-Target Screening of Organic Micropollutants in Urban Waters Using High-Resolution Mass Spectrometry

Nina Huynh [1,2], Emilie Caupos [1,2,3], Caroline Soares Peirera [1,2], Julien Le Roux [1,2,*], Adèle Bressy [1,2] and Régis Moilleron [1,2]

[1] LEESU, Université Paris-Est Créteil, F-94010 Creteil, France; tinh-nghi-nina.huynh@u-pec.fr (N.H.); emilie.caupos@u-pec.fr (E.C.); soarespereiracaroline@gmail.com (C.S.P.); adele.bressy@enpc.fr (A.B.); moilleron@u-pec.fr (R.M.)

[2] LEESU, Ecole des Ponts, F-77455 Marne-la-Vallee, France

[3] OSU-EFLUVE, Université Paris-Est Créteil, F-94010 Creteil, France

* Correspondence: julien.le-roux@u-pec.fr

Citation: Huynh, N.; Caupos, E.; Soares Peirera, C.; Le Roux, J.; Bressy, A.; Moilleron, R. Evaluation of Sample Preparation Methods for Non-Target Screening of Organic Micropollutants in Urban Waters Using High-Resolution Mass Spectrometry. *Molecules* **2021**, *26*, 7064. https://doi.org/10.3390/molecules26237064

Academic Editor: Thomas Letzel

Received: 2 November 2021
Accepted: 19 November 2021
Published: 23 November 2021

Abstract: Non-target screening (NTS) has gained interest in recent years for environmental monitoring purposes because it enables the analysis of a large number of pollutants without predefined lists of molecules. However, sample preparation methods are diverse, and few have been systematically compared in terms of the amount and relevance of the information obtained by subsequent NTS analysis. The goal of this work was to compare a large number of sample extraction methods for the unknown screening of urban waters. Various phases were tested for the solid-phase extraction of micropollutants from these waters. The evaluation of the different phases was assessed by statistical analysis based on the number of detected molecules, their range, and physicochemical properties (molecular weight, standard recoveries, polarity, and optical properties). Though each cartridge provided its own advantages, a multilayer cartridge combining several phases gathered more information in one single extraction by benefiting from the specificity of each one of its layers.

Keywords: emerging contaminants; high-resolution mass spectrometry; micropollutant fingerprint; non-target screening; solid phase extraction; statistical analysis; urban waters

1. Introduction

The presence of organic micropollutants in the aquatic environment is a major issue in many countries due to their potential ecotoxicological impact on aquatic organisms as well as their potential risks for human health. European directives concerning the quality of surface waters advocate better control of the presence of micropollutants, but they deal with priority organic molecules that are well known by scientists for their high environmental impact [1]. However, a lot of emerging molecules (e.g., substitutes of forbidden molecules, metabolites, and degradation byproducts) enter the environment due to human activities through wastewater effluents. The survey of these unknown molecules using non-target screening (NTS) enables us to anticipate further regulation on water quality [2]. This analytical approach has undergone high development in the last decade thanks to the spreading of high-resolution mass spectrometry, but it requires powerful analytical systems and computational software as well as non-specific sample preparation methods [3].

The increased resolving power of high-resolution mass spectrometry such as time-of-flight mass spectrometry (TOF-MS) or Orbitrap-MS allows the accurate mass measurement and isotopic pattern identification [4]. Therefore, these techniques are suitable for the development of three main screening strategies based on accurate-mass databases [5,6]: multi-residue target screening [7], suspect screening [8–10], and NTS [11,12]. Contrary to target and suspect screening which aim to evaluate the presence of molecules based on a preliminary list, NTS is the characterization of substances present in a sample without any prior information, based on full scan analysis that enables the acquisition of all peaks comprised within a range of masses. It thus requires the subsequent selection (e.g., by statistical analysis) of features of interest (i.e., specific signal characterized by its retention time and m/z ratio) followed by their tentative identification with the use of databases and literature information [13,14].

In combination with high-resolution mass spectrometry acquisition methods, the computational technique for data processing is of primary interest. Molecular identification depends on high-resolution data precision [13] but also on the software possibilities to deconvolute, filter, re-align signals [15], associate raw formulas to features of interest and finally compare with existing spectra libraries for identification [16,17]. NTS studies can be performed using proprietary software (e.g., Agilent Masshunter and MPP, Waters Unifi, Thermo Compound Discoverer, ABSCiex MarkerView, SpectralWorks AnalyzerPro, associated with ChemSpider, SciFinder, METLIN, PubChem, or MassBank libraries) as well as free ones (e.g., MZmine, R, EnviMass, and MetFrag) [18] for the identification of molecules, using the information on the fragments. The qualification of a single sample requires a lot of computational treatment, and statistical studies are relevant as a complementary approach to assess organic contamination by comparing different samples. Statistical approaches in NTS have been used to differentiate samples in time [19] and space [20] or to highlight variations during processes [11]. Binary comparison, principal component analysis (PCA), and Venn diagrams are examples of methods that can be applied to a sample set in order to isolate features of interest (e.g., molecules in common between samples or specific to groups of samples) [21]. Discriminated samples can be further characterized, and molecules of interest can then be identified.

Data analysis and statistical evaluation of samples being time-consuming, the quality of acquired data must be guaranteed. Sample preparation is essential to ensure the detection of as many molecules as possible and thus needs to be as unselective as possible. Different strategies and methods can be used in order to reach this objective and also to prevent samples from contamination. Direct injection of water samples can be employed to avoid potential contaminations or losses during sample preparation. However, the low concentration of molecules and interferences with other constituents of the sample (e.g., solid particles, colloids, or organic matter) can prevent the detection of trace organic compounds. Preparation techniques such as filtration [22], concentration [23], and solid-phase extraction [24] are widely used for targeted quantitative analysis and for NTS to decrease interferences and to concentrate the samples.

Most NTS studies employ a sample preparation step that is considered to be as "exhaustive" as possible, and little information is available about the efficiency of SPE for NTS purposes. For target screening purposes, knowing the properties of the targeted molecules makes it easier to select the most appropriate cartridge and phase. Concerning NTS studies, the choice of a cartridge is trickier given that, for example, the choice of a cartridge that is recommended for polar compounds will only retain polar compounds and thus, will leave the non-polar compounds un-retained. As a result, most non-target studies employ *universal* phases such as Oasis HLB to retain as many compounds as possible. Validating the extraction and analysis steps is also a challenge for NTS. The choice of standards to assess cartridge efficiency or to evaluate and correct matrix effects are tricky since the compounds studied in NTS are supposed to be unknown. Therefore, a mix of internal standards composed of a large variety of molecules (in terms of molecular weight, polarity, acidity, functional groups, etc.) is often used to assess the efficiency of the method in recovering and analyzing a large panel of compounds [25–28]. In addition to the injection of internal standards, various methods can be used to check the stability of the instrument or to correct the standard deviation of each feature, such as the injection of a pooled sample (consisting of a mix of all samples to be analyzed), regularly during the sequence. These normalization methods were recently reviewed for NTS applications in lipidomics [29]. Finally, using replicated injections to discard signals that are not repeatable and that could lead to false-positive identifications is a commonly used step in NTS studies [25,28,30–32]. Still, only a few studies compared the efficiency of different preparation methods for NTS purposes. One study showed that non-target strategies clearly discriminated the signals in terms of number and type of features obtained with several preparation methods (liquid-liquid extraction and solid-phase extraction on two cartridges) while the extraction yields of targeted molecules were not significantly different between those techniques [33]. In all cases, the properties of the cartridge, the pH of the sample, as well as elution solvents, and conditioning solvents will play a major role in the selection of molecules finally detected in the analyzed sample. A literature review revealed that numerous phases (e.g., HLB, ENV+, C18, X-CW, X-AW, X-C, and X-A, individually or in series), different sample pH, and different conditions of elution were used in NTS studies for environmental applications (Table 1). Several studies used a multilayer cartridge (composed of HLB, ENV+, X-AW, and X-CW) developed to extract a wide range of micropollutants [27].

Table 1. Examples of SPE cartridges, pH of samples, and elution solvents used for sample preparation in environmental studies using non-target screening for the analysis of water samples.

Matrix	pH	SPE Cartridge	Elution Solvents	Ref
Surface water and wastewater (influent and effluent)	N.A. [a]	HLB	N.A. [a]	[8]
Surface water, groundwater and Drinking water	2	HLB, MCX	Acetonitrile (HLB) Acetonitrile, Acetonitrile + 5% ammonia (MCX)	[34]
Surface water, wastewater influent and effluent	6.7	Multilayer (HLB, ENV+, X-AW, X-CW)	Methanol/Ethyl acetate (50:50, v/v) + 2% ammonia and methanol/ Ethyl acetate (50:50, v/v) + 1.7% formic acid	[11,26,27,35–37]
Landfill leachate and groundwater	7 and 3	ENV+	Methanol	[21]
Wastewater effluent	2	MCX and Strata X in series	Methanol + 5% ammonia	[38]
Wastewater (influent and effluent)	N.A. [a]	HLB	Methanol	[39]
Surface water	N.A. [a]	Multilayer (HR-X, HR-XAW, HR-XCW)	Ethyl acetate; methanol; methanol + 2% ammonia and methanol + 1% formic acid	[5]
Wastewater effluent	N.A. [a]	MAX and MCX in series	Methanol/ethyl acetate/formic acid (69:29:2, v/v) (MAX) methanol/ethyl acetate/ammonia solution (67.5:27.5:5, v/v) (MCX)	[10]
Riverbank filtration system	N.A. [a]	HLB	Methanol	[32]

[a] N.A. = Not Available.

The overall objective of this work was to assess the influence of the extraction method, in particular the type of SPE phase/cartridge, on the non-targeted HRMS analysis of organic contaminants in urban water. To reach this objective, we pursued the following specific aims: (i) to test different nature of phases to extract a large range of micropollutants in urban waters, based on a panel of 9 commercially-available stationary phases and designed for a variety of applications (i.e., from non-polar/moderately polar to more polar compounds and ionic species); (ii) to develop a strategy based on relevant indicators to compare the results, such as the number of detected molecules, their range, and characteristics; and (iii) to apply the optimized SPE method and the developed data analysis strategy to urban water samples.

2. Results and Discussion

2.1. Comparison of the Cartridge Retention Ability

In this section, the retention abilities of several cartridges (ENV+, X-A, X-AW, X-CW, HLB pH 2, HLB pH 6, Multilayer, X-C, C18 ENV+, SDBL, and C18) were compared on the basis of different indicators.

2.1.1. Discrimination of Cartridges Based on Optical Properties

To get a preliminary overview of the diversity of retained organic materials, optical properties (UV absorbance at 254 nm and 3D-spectrofluorescence spectra) of one sample from a river located in an urbanized area (Marne River) were measured before and after extraction (i.e., on the water phase during the SPE loading step). These optical properties are commonly used to characterize dissolved organic matter (DOM). As DOM is composed of molecules of various sizes and polarities with which micropollutants can interact, the behavior of DOM can be related to the behavior of micropollutants. Based on these interactions (e.g., hydrogen bonding, hydrophobic, van der Waals, or dipole-dipole interactions), fluorescence can be used for a better understanding of the fate of micropollutants, for example, in advanced wastewater treatment processes such as adsorption onto activated carbon [40]. DOM retention on SPE cartridges was evaluated to characterize their ability to retain a large variety of organic materials but also to determine if it can hinder the detection of micropollutants. DOM is indeed known to sometimes reduce the adsorption of target compounds on SPE cartridges (competition effect) and to limit the ionization of molecules for their detection in mass spectrometry [41,42]. The good retention of DOM on SPE cartridges could thus be detrimental to the detection of organic molecules in non-target analyses. Figure 1 presents the retention of DOM on each SPE cartridge based on the percentage change of UV absorbance at 254 nm before and after extraction. The highest recoveries were obtained for X-A, X-AW, X-C, and the Multilayer phases (recoveries $\geq$ 90%). Other cartridges exhibited significantly lower retention of organic materials (<60%). Especially, extractions with the HLB cartridge retained fewer organic materials (~60%). SDBL, X-CW, and HLB (pH 6) do not seem to be appropriate phases for a large screening of organic material.

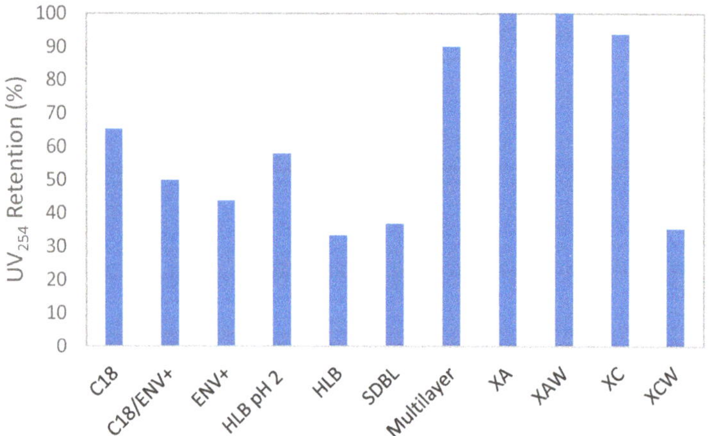

Figure 1. Retention of DOM based on UV absorbance at 254 nm (UV_{254}) for different SPE cartridges. UV_{254} retention (%) was calculated as the percentage change between UV_{254} of the Marne River sample and UV_{254} measured at the outlet of each cartridge (during the loading phase).

For a better evaluation of the quality of DOM retained on the cartridges, 3D-fluorescence spectra were acquired from the Marne River sample before and after extractions. Fluorescence regional integration (FRI) was performed [40,43], calculating the regional fluorescence intensities for each excitation-emission matrix (EEM) spectrum. The EEM spectrum of the Marne River sample before extraction (Supplementary Materials Figure S1) exhibited mainly two fluorophores: one in the region III of FRI and the other in region V. They correspond to large molecules like polysaccharides and humic-like substances [23,44,45]. Other fluorophores (regions I and II) were also observed, representing more hydrophilic and smaller molecules, proteins, and aromatic amino acids [46]. Table 2 presents fluorophores retention depending on the phase of SPE used. Similar results were obtained with other indexes derived from [43] (data not shown).

Table 2. DOM retention (%) based on fluorophores for different SPE cartridges (indexes FRI derived from [43]).

Index	C18	C18 ENV+	ENV+	HLB pH 2	HLB	Multilayer	SDBL	XA	XAW	XC	XCW
Region I FRI	84	85	64	70	58	71	58	94	89	55	49
Region II FRI	76	69	67	68	58	86	48	96	90	59	47
Region III FRI	72	46	46	55	55	73	27	92	86	50	28
Region IV FRI	68	62	64	65	45	80	41	93	86	53	45
Region V FRI	70	47	50	56	53	55	28	93	86	48	28

As expected, the cartridges retained aromatic proteins and humic-like materials (regions I, III, and V [37,40]). Higher retention was observed for XA and XAW ($\geq$85%). Like the results obtained with UV absorbance, the extractions on SDBL and XCW phases led to poor DOM retention (retentions generally lower than 50%). It is interesting to note that regions II and IV were well retained by XA, XAW, and Multilayer ($\geq$80%). They can be associated with the retention of simpler, more polar, and nitrogenous compounds [47]. Based on EEM spectra, these three cartridges seem to be the most efficient to retain hydrophobic materials as well as smaller and more polar compounds for global screening of the molecules present in the sample.

2.1.2. Discrimination of Cartridges Based on HRMS Features

Samples were then analyzed by HRMS, and differences between SPE cartridges were investigated through criteria classically used in omics studies. A PCA of the HRMS features found in the different extracts of the Marne River sample was performed to quickly identify similarities and differences between the feature sets obtained from each cartridge (Figure 2).

All cartridges were clearly discriminated by the first two components of the PCA (covering 45.3% of the total variance), which indicated that each type of SPE phase extracted specific sets of HRMS features. The first component (PC 1) was mostly correlated with the features derived from the Multilayer cartridge and explained 30% of the dataset variability. The second component (PC 2) explained 15% of the variability and discriminated all individual cartridges. Polymeric cartridges (ENV+, SDBL, and C18/ENV+) were all clustered near the center of the PCA, meaning that they all extracted similar features. Cartridges for which sample loading was performed at pH 2 (HLB pH 2 and XA) also exhibited similar behavior. Features from anionic exchange cartridges (XA and XAW) were almost identical.

To better understand the repartition of cartridges obtained by the PCA (Figure 2), fingerprints (i.e., bubble plots of all detected features) were compared between the most different cartridges (XCW, ENV+, HLB, and the Multilayer cartridge) for the Marne river sample (Figure 3).

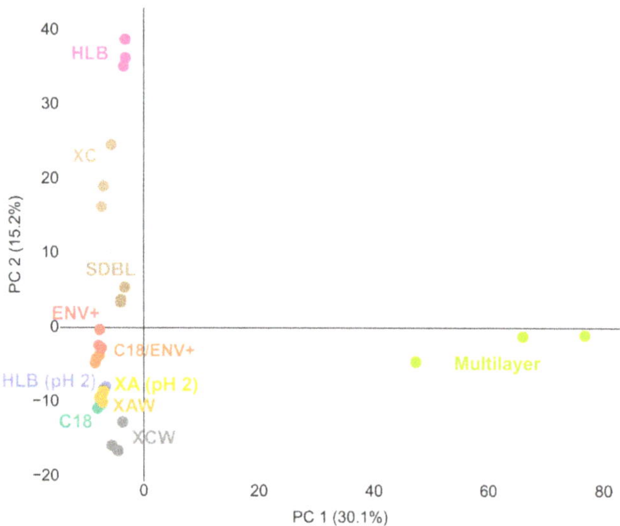

Figure 2. PCA (graph of individuals) of Marne sample extracts on different SPE cartridges.

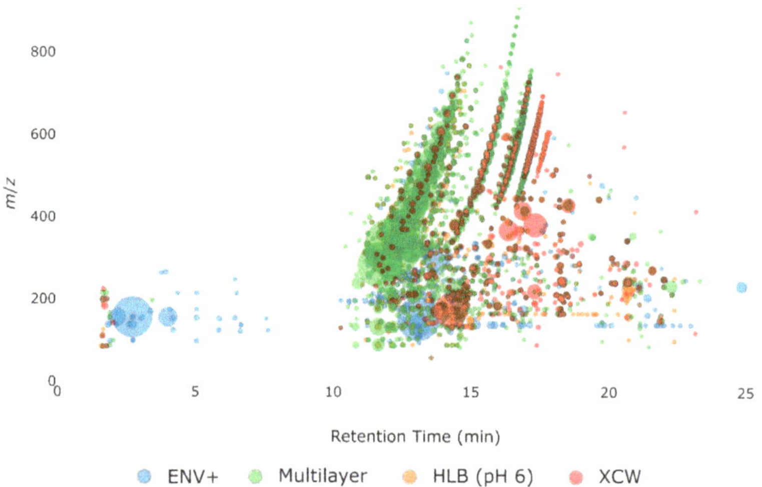

Figure 3. Fingerprints of detected features from the Marne River sample after SPE on ENV+ (blue), HLB (orange), Multilayer (green), and XCW (red) cartridges. The size of bubbles is proportional to the area of the feature.

The Multilayer cartridge displayed the most intense features on the entire m/z range, whereas ENV+ displayed more intense features for $m/z < 400$, and HLB retained more intense features with $m/z > 400$. XCW showed more intense signals at retention times greater than 15 min (i.e., less polar compounds), whereas most intense signals on the two other individual cartridges were obtained at retention times before 15 min. Thus, the PC 2 component of the PCA may describe the polarity of the retained features on each cartridge. This hypothesis is supported by the fact that the C18 cartridge, which is designed for less polar compounds, was close to the XCW one on the PCA. Finally, a comparison of common features retained by the cartridges was performed (Figure 4).

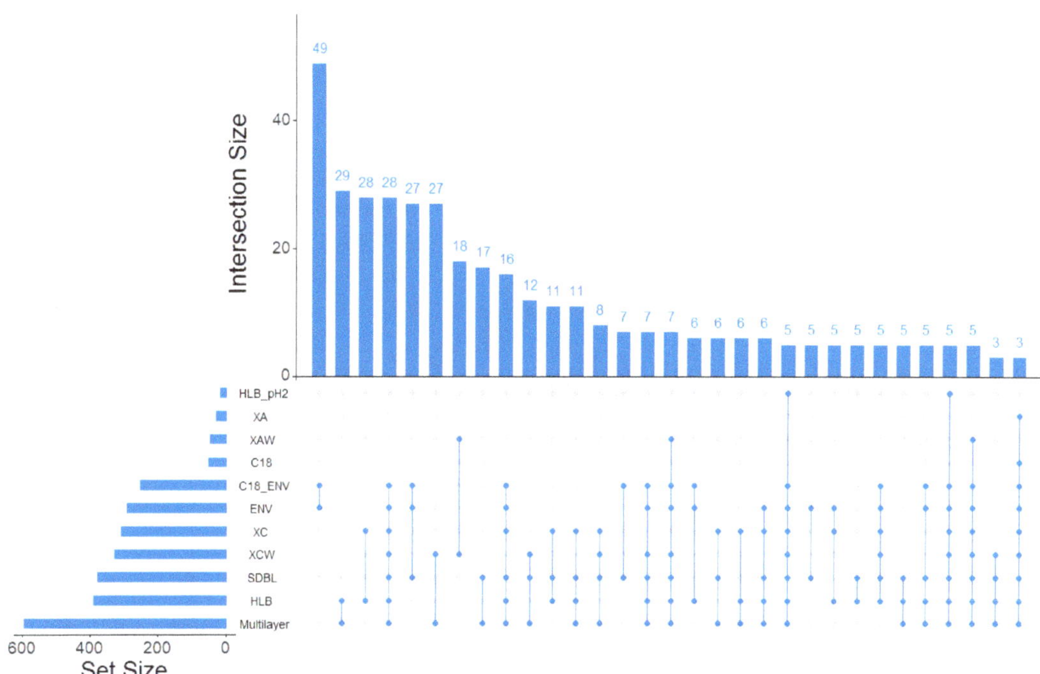

Figure 4. Upset diagram, representing the intersection of the different cartridges in terms of retained features. The "set size" represents the total number of features from each cartridge, the "Intersection size" represents the size of the intersection (i.e., number of features in the intersection designated by the dots), the dots represent the cartridges intersected (i.e., number of intersections).

The first thing to notice is that the Multilayer cartridge had the biggest set size among the studied cartridges, meaning that it retained the highest number of features. The two closest cartridges in terms of retained features were the ENV+ and C18/ENV+ cartridges with 49 common features, which is not surprising given the similar nature of the phases. The cartridge involved in the highest number of intersections was the SDBL (23 intersections for 210 features intersected); however, the cartridge with the highest number of features intersected was the Multilayer one (20 intersections for 212 features intersected). Therefore, the Multilayer was the cartridge with the highest number of specific features (323 features only found in this cartridge), but also the one with the highest number of intersected features, making it the most interesting cartridge for non-target studies by retaining the most extensive set of features.

Compared to the results obtained when considering the optical properties (Section 2.1.1), the cartridges that retained more fluorescent organic materials exhibited a lower number of HRMS features (e.g., XA, XAW). On the contrary, SDBL and XCW showed low retention of fluorescing materials but a high number of HRMS features. This could be due to the competition of DOM with organic compounds for the adsorption on the SPE cartridge [41,48] or to matrix effects in the HRMS analysis. Interestingly, this behavior can also be observed for the same cartridge used at different pH. Extraction on HLB at pH 6 gave a high number of features in HRMS and low retention of fluorescing organic materials (fluorophores), contrary to the extraction on HLB at pH 2. Notably, the Multilayer cartridge exhibited both retention of fluorophores and a large number of HRMS features; this cartridge thus seemed less affected by organic content.

2.2. Characterization of the Cartridge Retention Ability with Different Matrices

Based on the previous observations (HRMS fingerprints and DOM retention), five cartridges were retained for further investigation with different matrices: the Multilayer cartridge (highest number of retained features and highest intensities), HLB at pH 6 (universal cartridge, used in many studies), ENV+ (universal cartridge, representing the polymeric cartridges clustered in the PCA), XAW (high retention of fluorophores) and XCW (retaining more apolar compounds). Three types of samples were extracted and compared, with the increasing complexity of matrix (i.e., presence of organic and inorganic constituents): Ultrapure water (Milli-Q), surface water (Marne River), and wastewater effluent (WWe).

2.2.1. Recoveries of Internal Standards

To compare the different SPE cartridges and evaluate the corresponding matrix effects, the recovery of internal standards was first calculated (Figure 5).

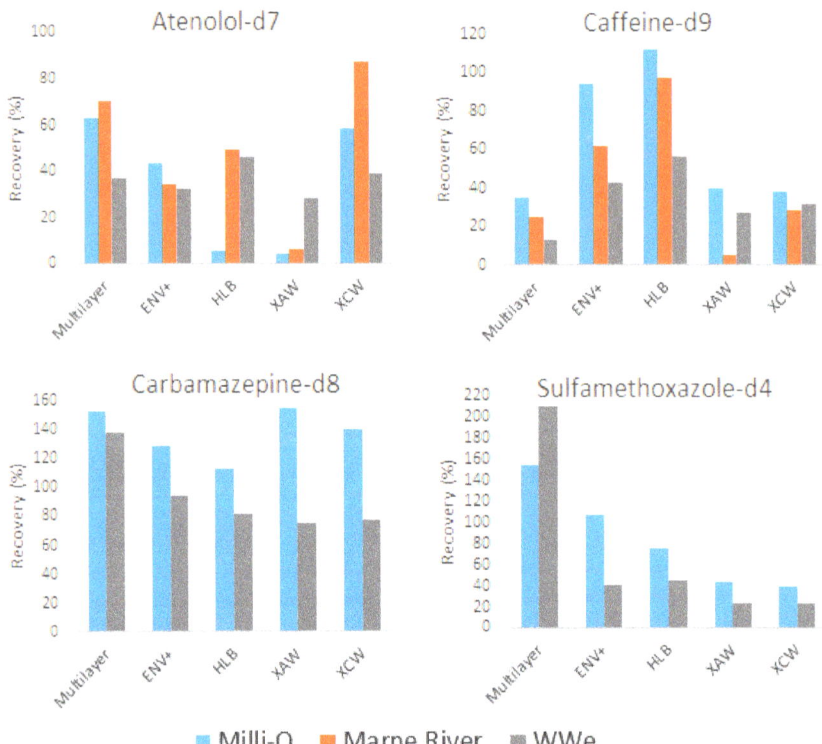

Figure 5. Recovery of internal standards after extraction by the cartridges for different matrices.

In Milli-Q, recoveries ranged from 4% to 63% for atenolol-d7, from 35% to 112% for caffeine-d9, from 113% to 154% for carbamazepine-d8 and from 40% to 155% for sulfamethoxazole-d4. Carbamazepine-d8 exhibited an overall great recovery with all cartridges and showed the least differences between cartridges. On the contrary, atenolol-d7 gave lower recoveries, and sulfamethoxazole-d4 and caffeine-d9 exhibited large differences in recoveries between cartridges. Carbamazepine-d8 has the highest log Kow among the four molecules; its better recovery on all cartridges could be explained by its lower hydrophilicity. Extractions were all carried out at pH = 6.5–7, meaning that, for atenolol-d7, caffeine-d9 and carbamazepine-d8, the acidic form was predominant, whereas, for sulfamethoxazole-d4, its basic form is predominant. The only molecule showing a significant difference between XAW and XCW cartridges (designed for compounds with pKa < 5 and pKa > 8, respectively) was atenolol-d7 (pKa = 9.6), with a better recovery on XCW as expected. The strong acidity (pKa of sulfamethoxazole-d4 = 1.6) or basicity (pKa of caffeine-d9 and carbamazepine-d8 are 14 and 13.9, respectively) of the three other compounds may explain their less contrasted retention of those two cartridges. Overall, the Multilayer cartridge gave the highest recoveries of those standards (despite a quite poor recovery of caffeine-d9).

Lower recoveries of caffeine-d9 were recorded with the Marne River and WWe samples for all five cartridges as compared to the Milli-Q water sample. The three universal cartridges (Multilayer, HLB, and ENV+) were more affected by complex matrices. The ionic XCW cartridge showed the least difference in recoveries for the three matrices. However, the XAW cartridge exhibited a very low recovery of most internal standards from the Marne River sample, while the two other samples (Milli-Q and WWe) showed similar recoveries. Better retention of caffeine-d9 was expected on the XCW cartridge because of its high pKa value of 14. Compared to other cartridges, recoveries of caffeine-d9 on XCW was low but more consistent across the different types of water. Its lower recovery can be explained by the employed elution step that was recommended by the cartridge supplier for specific compounds (Alprenolol, Acetaminophen, and Clomipramine). The elution step could be optimized with this cartridge to obtain better recoveries for internal standards such as caffeine-d9 as well as for the total number of features retained.

An unexpected high signal was obtained for carbamazepine-d8 and sulfamethoxazole-d4 in the Marne River sample after SPE, leading to excessive recovery values (>1000%), which were therefore not included in Figure 5. The extraction of those two components on WWe, led to signal suppression as compared to the Milli-Q sample. Such matrix effects were previously reported in wastewater effluent with ion suppression of 34% for sulfamethoxazole, 5.4% for caffeine, and 23% for carbamazepine on Strata-X cartridges [49]. Atenolol-d7 exhibited various behaviors, with a signal enhancement for HLB and XAW, signal suppression for ENV+, and both signal suppression (WWe sample) and signal enhancement (Marne River sample) for the Multilayer cartridge and XCW.

Matrix effects (i.e., signal suppression or enhancement) or adsorption competition on the cartridge were observed for all detected internal standards. A high suppression was especially observed for the XAW cartridge, in accordance with the impact of adsorbed organic materials already described in Section 2.1.2.

2.2.2. Range and Properties of Retained Features

The number of features retained on each cartridge was compared for the WWe and the Marne River sample (Table 3). As mentioned previously, the Multilayer cartridge displayed the highest number of features compared to the other cartridges, which explains its widespread use in non-target screening studies [27,35,36]. HLB and ENV+, commonly considered as "universal cartridges", were also quite efficient for the Marne River sample, retaining 32% and 47% fewer features than the Multilayer cartridge, respectively. They also retained the highest number of features from the WWe sample, ENV+ being the most efficient. XCW retained 11% more features from the Marne River sample than ENV+, which suggests the presence of cationic substances in this sample. Table 3 displays the properties of the features retained on each cartridge for both samples.

The Multilayer cartridge retained globally bigger molecules than other cartridges, as seen from both average m/z and weighted average m/z. For the two types of samples, HLB also retained features with larger molecular sizes and ENV+ smaller ones. When those values were not weighted by the feature area, no clear trend could be observed. The polarity of retained features (as described by the average retention) was also not significantly different between the cartridges. However, it can be noted that the Multilayer cartridge average retention of features from the Marne River sample (46% ACN) was equal to the mean of the average retention of all four other cartridges (i.e., the four phases used in the

Multilayer cartridge). For the WWe sample, the total area of the detected features followed the same ranking order as the total number of features (i.e., ENV > HLB > XAW > XCW). However, this was not the case for the Marne River sample, for which ENV+ exhibited 15% less detected features compared to HLB but a slightly more intense total signal. This observation indicates that the molecules retained on ENV+ were more ionizable, therefore different from HLB, or better retained, as it was demonstrated by the difference in their fingerprints (Figure 3). The Multilayer cartridge showed the highest total intensity among all cartridges, which demonstrates its ability to retain a large number of compounds with a high signal. This property is clearly beneficial for NTS studies to obtain clean mass spectra and thus to allow easier identifications.

Table 3. Properties of features retained on various SPE cartridges. The average retention is given as the percentage of acetonitrile in the mobile phase (ACN) needed to elute a given feature. The weighted values correspond to the average parameter, weighted by the intensity of each individual marker.

	Number of Features	Sum of Detected Areas	Average *m/z*	Weighted Average *m/z*	Average Retention (% ACN)	Weighted Average Retention (% ACN)
				Marne River		
ENV	315	4.25×10^6	306.6403	300.0973	43	41
HLB	403	4.20×10^6	367.9942	379.1581	44	43
XAW	49	8.69×10^5	295.4186	323.9237	52	52
XCW	350	3.43×10^6	404.9278	366.7695	48	46
Multilayer	594	5.84×10^6	470.6467	441.9003	46	44
				WWe		
ENV	301	2.72×10^7	306.7059	339.2997	37	33
HLB	201	1.75×10^7	337.8578	311.8444	35	27
XAW	149	9.03×10^7	345.6214	384.1570	34	22
XCW	100	8.32×10^6	297.6594	367.0206	28	23
Multilayer	7515	4.89×10^7	461.3586	416.1775	47	35

The distribution of m/z values was visualized for both types of the sample matrix and for each cartridge (Figure 6). To the authors' knowledge, this representation has not yet been used for the characterization of HRMS data, although it is useful for identifying differences between samples. Comparison of the cartridges shows that ENV+ was less effective in retaining molecules with $m/z > 600$, whereas HLB covered a larger range of m/z. It is interesting to note that the Multilayer cartridge combined the ranges of m/z distribution of ENV+, HLB, XAW, and XCW.

2.3. Application: Evaluating a Disinfection Treatment by Performic Acid

The above-mentioned indicators showed that the Multilayer cartridge was the most efficient among the different tested cartridges and that it is the most suitable to be used as a universal cartridge. It was therefore tested in a real-case application to characterize the evolution of organic compounds during a disinfection treatment of wastewater effluents by performic acid (PFA). Raw wastewater effluent and the same effluent treated with 30 ppm of PFA (contact time of 10 min) were extracted with the Multilayer cartridge and analyzed by HRMS (Figure 7).

The Multilayer cartridge allowed the detection of a wide number of compounds covering the whole range of m/z values at low and high retention times. Sample after treatment showed a significant decrease in the area of the markers with a retention time between 17 and 21 min and an increase for the markers between 5 and 7 min compared to the non-treated wastewater.

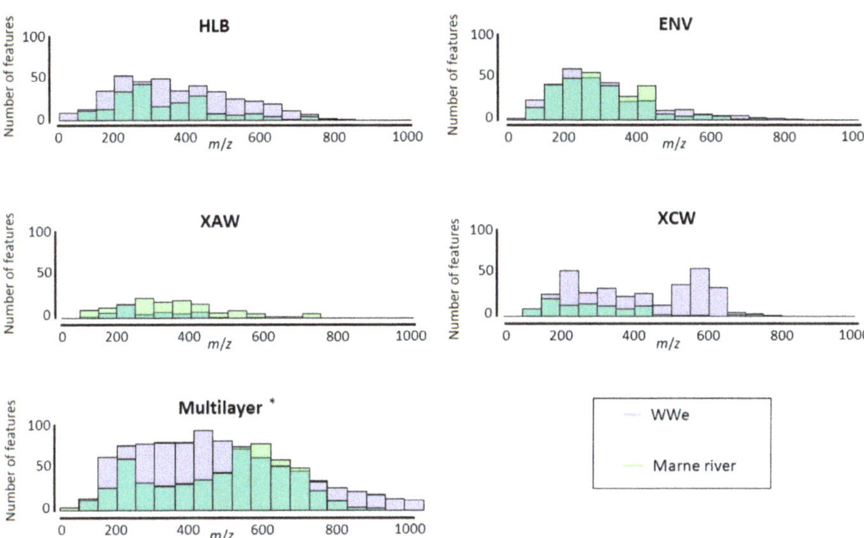

Figure 6. m/z distribution of the retained features on the different cartridges. * Number of features for Multilayer WWe were divided by eight.

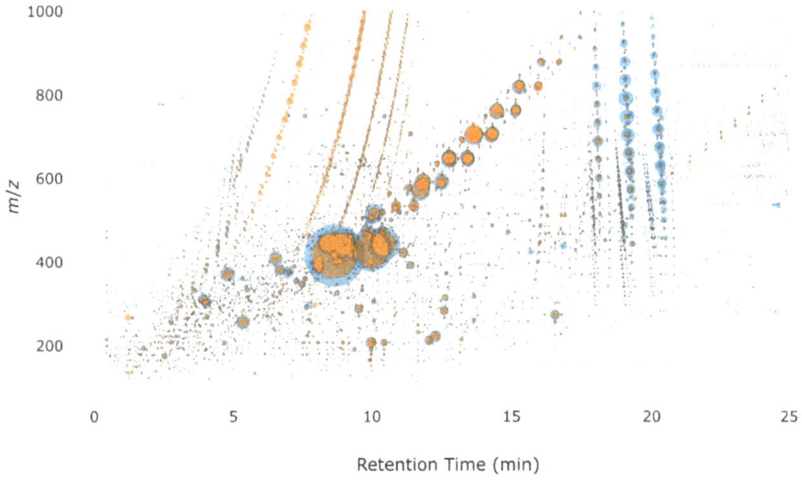

Figure 7. Bubble plot of WWe (blue) and WWe after 30 ppm PFA treatment during 10 min (orange) extracted on a Multilayer cartridge.

This observation suggests that compounds with lower polarity (i.e., higher retention time) were transformed into more polar ones by the treatment. Table 4 displays the distribution of the number of markers before and after treatment in different zones defined by the m/z and retention time ranges. Zone 2 (bigger and more polar molecules) exhibited a significant increase in both the number of markers and their total area. Zone 1 and zone 3 (molecules with $m/z < 500$) showed a decrease by almost a third of the number of markers after treatment, and a decrease in their total intensity as well. Finally, PFA treatment slightly increased the number of markers in zone 4 (bigger and less polar molecules), but the overall intensity of signals decreased in that zone. Globally, these results show that PFA treatment formed more polar molecules and removed a larger proportion of small molecules.

Table 4. Distribution of markers before and after PFA treatment.

	Before PFA		After 30 ppm PFA	
	Number of Markers	Sum of Detected Areas	Number of Markers	Sum of Detected Areas
Zone 1 m/z: 0–500 Retention Time: 0–10 min	4861	2.51×10^8	3781	1.85×10^8
Zone 2 m/z: 500–1000 Retention Time: 0–10 min	996	0.35×10^8	2301	1.36×10^8
Zone 3 m/z: 0–500 Retention Time: 10–25 min	2795	1.43×10^8	1728	0.75×10^8
Zone 4 m/z: 500–1000 Retention Time: 10–25 min	2770	2.93×10^7	3557	1.95×10^8

3. Materials and Methods

3.1. Chemicals and Standards

All standards were obtained from Sigma-Aldrich (Dr. Ehrenstorfer, Augsburg, Germany, purity > 99%). Methanol (MeOH), ammonia (35%), hydrochloric acid (37%), and formic acid (98%) were LCMS grade and purchased from Fischer Scientific (Illkirch Cedex, France). Ethyl acetate (EtAc), dichloromethane (DCM), and acetone were obtained from Merck (Darmstadt, Germany), and acetonitrile (ACN) was purchased from Biosolve (Dieuze, France), at analytical grade. HPLC water (Milli-Q) was produced from deionized water using a Millipore Milli-Q system (IQ 7000, Merck, Darmstadt, Germany) equipped with an LC-pak polisher (Merck, Darmstadt, Germany).

Deuterated standards were used to evaluate matrix effects from different samples, to evaluate the extraction efficiency on the various SPE phases, and to correct the retention time of the chromatograms. A mixed solution of 4 deuterated compounds (atenolol-d7, caffeine-d9, carbamazepine-d8, sulfamethoxazole-d4) was prepared in MeOH at a concentration of 10 mg/L for further injection in samples.

Before sampling and analysis, glassware was washed with TFD4 (Franklab, Montigny-le-Bretonneux, France), rinsed with deionized water, and calcined at 500 °C to remove any trace of organic contamination.

3.2. Sample Collection and Preliminary Preparation

Various types of samples were used: Milli-Q water produced in the laboratory, surface water under anthropic pressure (Marne river, France), and wastewater effluent (WWe).

Thirty liters of water from the Marne River were collected from a bridge at Chennevières-sur-Marne (GPS coordinates: 48.79071188130754, 2.5215935793518462) downstream the wastewater treatment plant of *Marne Aval* (Noisy-le-Grand, France). The sample was collected in 3 × 10 L amber glass bottles and filtered the same day, using 0.7 µm glass fiber filters (GF/F Whatman). After homogenization, aliquots of 500 mL were prepared and completed to 1 L with Milli-Q water in order to avoid potential matrix effects occurring during the SPE loading step. The initial pH and dissolved organic carbon (DOC) were 8.5 and 1.89 mg C/L, respectively. Some samples were acidified, either to pH 6.5 with 50 µL of formic acid or to pH 2–3 with 400 µL of formic acid according to recommendations for specific SPE cartridges.

Ten liters of treated effluent were collected in June 2019 from the wastewater treatment plant of *Seine Amont* (Valenton, France) in 10 L amber glass bottles and filtered on 0.7 µm glass fiber filters (GF/F Whatman). The initial pH and DOC were 7.9 and 7.1 mg C/L, respectively. One liter subsamples were acidified to pH 6.5 with 50 µL of formic acid.

3.3. SPE Cartridges

3.3.1. Selection of Cartridges

Strata-X cartridges (500 mg, 6 mL) (Phenomenex, Le Pecq, France) are ionic-based polymeric sorbents, chosen to retain compounds according to their pKa: Strata X-AW (for pKa < 5), Strata X-CW (for pKa > 8), Strata X-A (for pKa < 2), Strata X-C (for pKa > 10.5). Oasis HLB (200 mg, 6 mL) (Waters, Milford, MA, USA) is a reversed-phase cartridge used to select a wide range of substances, including neutral, basic, acidic, and polar ones. Silica-based cartridges Strata C18 (200 mg, 6 mL) and Strata SDBL (500 mg, 6 mL) (Phenomenex, Le Pecq, France) are employed to retain neutral hydrophobic and non-polar molecules. ENV+ (500 mg, 6 mL) (Biotage, Glamorgen, UK) is a polymeric phase appropriate for the retention of polar analytes. C18/ENV+ (400 mg, 6 mL) (Biotage, Glamorgen, UK) is a layered cartridge composed of the previous ENV+ phase as the bottom layer and a C18 phase as the top layer to improve the range of analytes possibly extractable. Finally, a homemade Multilayer cartridge made of Oasis HLB (200 mg), Isolute ENV+ (150 mg), Strata X-AW (100 mg), and Strata X-CW (100 mg) was prepared according to the method developed by Kern et al. [27].

3.3.2. Extraction Protocol

Samples were spiked with the mix of internal standards at 100 ng/L, 24 h prior to extraction and stored at 4 °C. Experimental conditions for conditioning, loading, and elution on each cartridge followed recommendations of suppliers, or the method developed by Kern et al. for the Multilayer cartridge [27], and are described in Table 5. All extractions were carried out on Visiprep (Sigma-Aldrich, Augsburg, Germany) and Autotrace (AT280, Caliper) SPE systems. All cartridges were loaded with 1 L of sample at pH 6–7, except for Strata X-A (pH 2–3 recommended) and Oasis HLB, for which both pHs were tested. Before elution, the cartridges were dried for 30 min. After elution, the extracts were stored in the dark at 4 °C prior to their analysis. SPE extracts were evaporated under a stream of nitrogen, then reconstituted in 1 mL of Milli-Q water and MeOH (80/20, v/v), and filtered through 0.2 µm PTFE filters before a 10 µL injection on the analytical system.

Table 5. Description of the methods used for water sample extraction on the different SPE cartridges.

Cartridge	Sample	Conditioning	Washing	Eluting
X-A	1 L pH 2–3 Marne	10 mL MeOH 10 mL Milli-Q	10 mL MeOH	10 mL MeOH + 5% formic acid
X-AW	1 L pH 6–7 Marne, MQ, WWe	10 mL MeOH 10 mL Milli-Q	10 mL MeOH	5 mL MeOH + 5% ammonia 5 mL MeOH + 5% formic acid
X-C	1 L pH 6–7 Marne	10 mL MeOH 10 mL Milli-Q pH = 2	10 mL MeOH + 0.1 M HCl	5 mL MeOH + 0.1M HCl 5 mL MeOH + 5% ammonia
X-CW	1 L pH 6–7 Marne, MQ, WWe	10 mL MeOH 10 mL Milli-Q	10 mL MeOH	5 mL MeOH + 5% formic acid 5 mL MeOH + 5% ammonia
HLB	1 L pH 2–3 Marne 1 L pH 6–7 Marne, MQ, WWe	10 mL MeOH 10 mL AcEt 5 mL DCM 10 mL Milli-Q	No washing	5 mL MeOH 5 mL AcEt 5 mL DCM
ENV+	1 L pH 6–7 Marne, MQ, WWe	10 mL MeOH 10 mL Milli-Q	5 mL Milli-Q /MeOH (95/5, v/v)	5 mL MeOH 5 mL acetone + 5% ammonia
C18	1 L pH 6–7 Marne, MQ, WWe	10 mL MeOH 10 mL Milli-Q	No washing	5 mL MeOH 5 mL acetone + 5% ammonia
C18/ENV+	1 L pH 6–7 Marne	10 mL MeOH 10 mL Milli-Q	No washing	5mL MeOH 5mL acetone + 5% ammonia
SDBL	1 L pH 6–7 Marne	10 mL MeOH 10 mL Milli-Q pH = 4	No washing	5 mL MeOH 5 mL acetone + 5% ammonia
Multilayer	1 L pH 6–7 Marne, MQ, WWe	10 mL MeOH 10 mL Milli-Q	No washing	6 mL AcEt/MeOH (50/50, v/v) + 1.43% ammonia 3 mL AcEt/MeOH (50/50, v/v) + 1.7% formic acid

3.4. UV-Visible and 3D Fluorescence Analyses

Spectroscopic measurements were performed on samples before and after filtration on the SPE cartridges (loading phase). UV-Visible analyses were conducted on a spectrophotometer (UviLine 9400, Secomam, Aqualabo, Champigny-sur-Marne, France), and 3D-fluorescence spectra were obtained by a spectrofluorometer (FP-8300, Jasco, Pfungstadt, Germany). Those apparatuses were both equipped with a Xenon lamp and a 1 cm-quartz cell. For the fluorescence spectra, EEM were generated by scanning excitation wavelengths from 240 to 450 nm (every 5 nm), and the emission wavelengths were detected between 250 to 600 nm (every 2 nm), at a scan speed of 1000 nm/min and response at 0.1 s. To avoid any effect of the intern filter, samples with an absorbance at 254 nm higher than 0.1 were diluted.

3D-fluoresence indexes where chose according to previous studies [40,43]. Fluorescence regional integration (FRI) was performed, regions I to V were calculated and compared.

3.5. LC-HRMS Analyses

LC-HRMS analyses of the Marne River samples were conducted on SYNAPT HDMS QTOF (Waters) coupled with a Nano ACQUITY UPLC system (Waters). The column was an ACQUITY UPLC Peptide BEH C18, 130 Å (100 μm × 100 mm, 1.7 μm). The mobile phase was Milli-Q water with 0.1% formic acid (A) and ACN (B) following the gradient described in Supplementary Materials Table S1, for a total run time of 40 min. Analyses were performed in positive mode (ESI+) with screening between 50 and 1000 m/z. LC-HRMS analyses of Milli-Q and WWe samples were performed with a Vion–UPLC-IMS-QTOF (Waters) equipped with an ACQUITY UPLC BEH C18 (2.1 × 100 mm, 1.7 μm) column and the corresponding pre-column. The mobile phase was Milli-Q water + 0.1% formic acid (A) and ACN + 0.1% formic acid (B) following the gradient described in Supplementary Materials Table S1, for a total run time of 34 min. Analyses were performed in positive mode (ESI+) with screening between 50 and 1000 m/z. To ensure data quality, several steps were implemented.

Quality insurance procedure: Before each analysis, a quality reference standard, consisting of 9 components (Acetaminophen, Caffeine, Leucine enkephalin, Reserpine, Sulfadimethoxine, Sulfaguanidine, Terfenadine, Val-tyr-val, Verapamil), was injected five times to check the system performance by calculating the mass error deviation, the average peak width and, when available, the CCS (collision cross section) error for those compounds. If the mass error deviation was higher than 2 ppm or the peak width was longer than 3.0 sec, a system calibration was conducted. The same procedure was repeated until the expected conditions were met. Each sampling sequence began with three blank injections to wash the column and five pool injections to stabilize the column. The pool sample consists in a mix of equal volume of each sample and is used as a quality control. Each sample was then injected in randomized triplicate to minimize the effect of instrumental deviation. For data treatment, only the features detected in every replicate of a given sample were considered. Every 10 injections, a pool was re-injected to monitor the system. The clustering of these pool samples in the PCA was checked to assess the reproducibility of the system.

Raw HRMS data were exported directly after acquisition and converted in *.mzML format via MSConvert (Version 3) [50,51] for further treatment in R software (Version 3.6.2). Pre-treatment of raw data was performed with the *Patroon* package [52], using OpenMS for peak picking and alignment. Features that were not present in all replicates of a given sample were discarded. PCA was performed with the *FactoMineR* [53] and *factoextra* [54] packages, and Upset diagrams were plotted with the *UpSetR* package [55].

The extraction efficiency of an SPE cartridge was determined based on the area of detected internal standards in the samples compared to the area of a standard mix injected. The number of detected features and the total area of the signal (i.e., sum of the areas of each feature) were compared for each cartridge. The average m/z and average retention of a sample were calculated from the m/z values and retention times of each feature detected in the sample. Retention was expressed as the percentage of ACN needed to elute a given feature at the corresponding retention time. Values of average m/z and average retention were also weighed by the area of each feature.

4. Conclusions

The goal of this work was to compare the efficiency of several commercially available SPE cartridges and of a homemade Multilayer cartridge (as developed by Kern et al. [27]) for non-target screening purposes. Various parameters were monitored, such as the recovery of a limited number of analytical standards, the number and properties of detected features in HRMS, or the global retention of DOM (through 3D fluorescence measurements) on each cartridge. The Multilayer cartridge was the most effective at retaining internal standards and a large range of HRMS features, and it was not as affected by the adsorption of DOM as other cartridges (i.e., adsorption competition effects and/or matrix effects). It seemed to take advantage of the four phases involved in its composition (HLB, ENV+, XCW, and XAW) in terms of diversity of compounds (i.e., polarity and molecular size).

Although the other cartridges had some specificity missed by this Multilayer cartridge (e.g., an important number of specific features was only retained on the ENV+ cartridge), its efficiency in extracting a great number of compounds was clearly demonstrated. This observation could be further confirmed by using different analytical tools (e.g., gas chromatography) or different ionization modes and sources to increase the range of molecules detected. Moreover, a further investigation involving the tentative identification of some retained features could be performed. The optimization of extraction conditions (e.g., masses and combinations of sorbents, types, and volumes of eluting solvents) could also be investigated to retain an even greater number of features. It would finally be useful to better characterize the chemical space (log Kow, pKa, chemical functions, etc.) covered by each cartridge. A larger number of internal standards covering a broader range of chemical properties could thus be studied. This could help determine the relevance of each type of cartridge towards specific purposes (e.g., the detection of a larger number of polar or ionic compounds or some specific families of molecules).

Supplementary Materials: The following are available online. Figure S1: EEM matrices of Marne river (A) before extraction, (B) after extraction on X-A, (C) after extraction on HLB, (D) after extraction on C18/ENV+, (E) after extraction on SDBL, (F) after extraction on the Multilayer cartridge, Table S1: Description of the gradient used for non-target analysis.

Author Contributions: Conceptualization, A.B., J.L.R. and R.M.; methodology, A.B., C.S.P., E.C. and J.L.R.; software, J.L.R. and N.H.; validation, A.B., E.C., J.L.R. and N.H.; formal analysis, E.C. and N.H.; investigation, C.S.P. and N.H.; resources, A.B., E.C., J.L.R. and R.M; data curation, E.C. and N.H.; writing—original draft preparation, E.C. and N.H.; writing—review and editing, A.B., E.C., J.L.R., N.H. and R.M.; visualization, N.H.; supervision, A.B., J.L.R. and R.M.; project administration, A.B.; funding acquisition, A.B., E.C. and J.L.R. All authors have read and agreed to the published version of the manuscript.

Funding: This study has been conducted within the projects Cosmet'eau (OFB/AESN),Screenatm'eau (OSU-EFLUVE) and WaterOmics (ANR-17-CE34-0009). The authors gratefully acknowledge the French Biodiversity Agency (OFB, formerly ONEMA), Agence de l'Eau Seine-Normandie (AESN), the French National Agency for Research (ANR) and the Observatoire des Sciences de l'Univers—Enveloppes FLUides de la Ville à l'Exobiologie, OSU-EFLUVE) for financial support.

Data Availability Statement: Raw HRMS data of Marne river samples are available at https://doi.org/10.5281/zenodo.5589621, accessed on 18 November 2021.

Acknowledgments: The authors would like to thank Damien Habert (CRRET) and Emmanuelle Mebold (PRAMMICS) for their technical support on HRMS instruments.

Conflicts of Interest: The authors declare no conflict of interest.

Sample Availability: Samples of the compounds are available from the authors.

References

1. European Parliament and of the Council, EU Directive 2000/60/CE, Water Framework Directive on Water, Last Revised August 2018. 2000. Available online: https://eur-lex.europa.eu/legal-content/FR/TXT/?uri=celex%3A32000L0060 (accessed on 18 November 2021).
2. Hollender, J.; Schymanski, E.L.; Singer, H.P.; Ferguson, P.L. Nontarget Screening with High Resolution Mass Spectrometry in the Environment: Ready to Go? *Environ. Sci. Technol.* **2017**, *51*, 11505–11512. [CrossRef]
3. González-Gaya, B.; Lopez-Herguedas, N.; Bilbao, D.; Mijangos, L.; Iker, A.M.; Etxebarria, N.; Irazola, M.; Prieto, A.; Olivares, M.; Zuloaga, O. Suspect and Non-Target Screening: The Last Frontier in Environmental Analysis. *Anal. Methods* **2021**, *13*, 1876–1904. [CrossRef]
4. Rodriguez-Aller, M.; Gurny, R.; Veuthey, J.-L.; Guillarme, D. Coupling Ultra High-Pressure Liquid Chromatography with Mass Spectrometry: Constraints and Possible Applications. *J. Chromatogr. A* **2013**, *1292*, 2–18. [CrossRef]
5. Schymanski, E.L.; Singer, H.P.; Slobodnik, J.; Ipolyi, I.M.; Oswald, P.; Krauss, M.; Schulze, T.; Haglund, P.; Letzel, T.; Grosse, S.; et al. Non-Target Screening with High-Resolution Mass Spectrometry: Critical Review Using a Collaborative Trial on Water Analysis. *Anal. Bioanal. Chem.* **2015**, *407*, 6237–6255. [CrossRef]
6. Hernández, F.; Pozo, Ó.J.; Sancho, J.V.; López, F.J.; Marín, J.M.; Ibáñez, M. Strategies for Quantification and Confirmation of Multi-Class Polar Pesticides and Transformation Products in Water by LC–MS2 Using Triple Quadrupole and Hybrid Quadrupole Time-of-Flight Analyzers. *TrAC Trends Anal. Chem.* **2005**, *24*, 596–612. [CrossRef]
7. Cortéjade, A.; Kiss, A.; Cren, C.; Vulliet, E.; Buleté, A. Development of an Analytical Method for the Targeted Screening and Multi-Residue Quantification of Environmental Contaminants in Urine by Liquid Chromatography Coupled to High Resolution Mass Spectrometry for Evaluation of Human Exposures. *Talanta* **2016**, *146*, 694–706. [CrossRef]
8. Ibáñez, M.; Sancho, J.V.; Hernández, F.; McMillan, D.; Rao, R. Rapid Non-Target Screening of Organic Pollutants in Water by Ultraperformance Liquid Chromatography Coupled to Time-of-Light Mass Spectrometry. *TrAC Trends Anal. Chem.* **2008**, *27*, 481–489. [CrossRef]
9. Diaz, R.; Ibáñez, M.; Sancho, J.V.; Hernández, F. Qualitative Validation of a Liquid Chromatography–Quadrupole-Time of Flight Mass Spectrometry Screening Method for Organic Pollutants in Waters. *J. Chromatogr. A* **2013**, *1276*, 47–57. [CrossRef]
10. Deeb, A.A.; Stephan, S.; Schmitz, O.J.; Schmidt, T.C. Suspect Screening of Micropollutants and Their Transformation Products in Advanced Wastewater Treatment. *Sci. Total Environ.* **2017**, *601–602*, 1247–1253. [CrossRef]
11. Schollée, J.E.; Bourgin, M.; von Gunten, U.; McArdell, C.S.; Hollender, J. Non-Target Screening to Trace Ozonation Transformation Products in a Wastewater Treatment Train Including Different Post-Treatments. *Water Res.* **2018**, *142*, 267–278. [CrossRef]
12. Singer, H.P.; Wössner, A.E.; McArdell, C.S.; Fenner, K. Rapid Screening for Exposure to "Non-Target" Pharmaceuticals from Wastewater Effluents by Combining HRMS-Based Suspect Screening and Exposure Modeling. *Environ. Sci. Technol.* **2016**, *50*, 6698–6707. [CrossRef]
13. Krauss, M.; Singer, H.; Hollender, J. LC–High Resolution MS in Environmental Analysis: From Target Screening to the Identification of Unknowns. *Anal. Bioanal. Chem.* **2010**, *397*, 943–951. [CrossRef]
14. Schulz, W.; Lucke, T.; Balsaa, P.; Hinnenkamp, V.; Brüggen, S.; Dünnbier, U.; Liebmann, D.; Fink, A.; Götz, S.; Geiss, S.; et al. *Non-Target Screening in Water Analysis—Guideline for the Application of LC-ESI-HRMS for Screening*; Water Chemistry Society, Division of the Gesellschaft Deutscher Chemiker: Mülheim an der Ruhr, Germany, 2019. Available online: http://www.wasserchemische-gesellschaft.de (accessed on 11 April 2020).
15. Hohrenk, L.L.; Itzel, F.; Baetz, N.; Tuerk, J.; Vosough, M.; Schmidt, T.C. Comparison of Software Tools for Liquid Chromatography–High-Resolution Mass Spectrometry Data Processing in Nontarget Screening of Environmental Samples. *Anal. Chem.* **2020**, *92*, 1898–1907. [CrossRef]
16. Pluskal, T.; Uehara, T.; Yanagida, M. Highly Accurate Chemical Formula Prediction Tool Utilizing High-Resolution Mass Spectra, MS/MS Fragmentation, Heuristic Rules, and Isotope Pattern Matching. *Anal. Chem.* **2012**, *84*, 4396–4403. [CrossRef]
17. Hufsky, F.; Böcker, S. Mining Molecular Structure Databases: Identification of Small Molecules Based on Fragmentation Mass Spectrometry Data. *Mass Spectrom. Rev.* **2017**, *36*, 624–633. [CrossRef]
18. Hug, C.; Ulrich, N.; Schulze, T.; Brack, W.; Krauss, M. Identification of Novel Micropollutants in Wastewater by a Combination of Suspect and Nontarget Screening. *Environ. Pollut.* **2014**, *184*, 25–32. [CrossRef]
19. Alygizakis, N.A.; Gago-Ferrero, P.; Hollender, J.; Thomaidis, N.S. Untargeted Time-Pattern Analysis of LC-HRMS Data to Detect Spills and Compounds with High Fluctuation in Influent Wastewater. *J. Hazard. Mater.* **2019**, *361*, 19–29. [CrossRef]
20. Krauss, M.; Hug, C.; Bloch, R.; Schulze, T.; Brack, W. Prioritising Site-Specific Micropollutants in Surface Water from LC-HRMS Non-Target Screening Data Using a Rarity Score. *Environ. Sci. Eur.* **2019**, *31*, 45. [CrossRef]
21. Müller, A.; Schulz, W.; Ruck, W.K.L.; Weber, W.H. A New Approach to Data Evaluation in the Non-Target Screening of Organic Trace Substances in Water Analysis. *Chemosphere* **2011**, *85*, 1211–1219. [CrossRef]
22. Smith, R.M. Before the Injection—Modern Methods of Sample Preparation for Separation Techniques. *J. Chromatogr. A* **2003**, *1000*, 3–27. [CrossRef]
23. Chon, K.; Chon, K.; Cho, J. Characterization of Size Fractionated Dissolved Organic Matter from River Water and Wastewater Effluent Using Preparative High Performance Size Exclusion Chromatography. *Org. Geochem.* **2017**, *103*, 105–112. [CrossRef]
24. Fontanals, N.; Marcé, R.M.; Borrull, F. New Materials in Sorptive Extraction Techniques for Polar Compounds. *J. Chromatogr. A* **2007**, *1152*, 14–31. [CrossRef]

25. Nürenberg, G.; Schulz, M.; Kunkel, U.; Ternes, T.A. Development and Validation of a Generic Nontarget Method Based on Liquid Chromatography – High Resolution Mass Spectrometry Analysis for the Evaluation of Different Wastewater Treatment Options. *J. Chromatogr. A* **2015**, *1426*, 77–90. [CrossRef]
26. Moschet, C.; Piazzoli, A.; Singer, H.; Hollender, J. Alleviating the Reference Standard Dilemma Using a Systematic Exact Mass Suspect Screening Approach with Liquid Chromatography-High Resolution Mass Spectrometry. *Anal. Chem.* **2013**, *85*, 10312–10320. [CrossRef]
27. Kern, S.; Fenner, K.; Singer, H.P.; Schwarzenbach, R.P.; Hollender, J. Identification of Transformation Products of Organic Contaminants in Natural Waters by Computer-Aided Prediction and High-Resolution Mass Spectrometry. *Environ. Sci. Technol.* **2009**, *43*, 7039–7046. [CrossRef]
28. Bastian, S.; Youngjoon, J.; Sarit, K.; Amy, H.L.; Pradeep, D.; Jake, O.; Maria Jose, G.R.; Sara, G.G.; Jochen, M.F.; Kevin, T.V.; et al. An Assessment of Quality Assurance/Quality Control Efforts in High Resolution Mass Spectrometry Non-Target Workflows for Analysis of Environmental Samples. *TrAC Trends Anal. Chem.* **2020**, *133*, 116063. [CrossRef]
29. Drotleff, B.; Lämmerhofer, M. Guidelines for Selection of Internal Standard-Based Normalization Strategies in Untargeted Lipidomic Profiling by LC-HR-MS/MS. *Anal. Chem.* **2019**, *91*, 9836–9843. [CrossRef]
30. Köppe, T.; Jewell, K.S.; Dietrich, C.; Wick, A.; Ternes, T.A. Application of a Non-Target Workflow for the Identification of Specific Contaminants Using the Example of the Nidda River Basin. *Water Res.* **2020**, *178*, 115703. [CrossRef]
31. Peter, K.T.; Wu, C.; Tian, Z.; Kolodziej, E.P. Application of Nontarget High Resolution Mass Spectrometry Data to Quantitative Source Apportionment. *Environ. Sci. Technol.* **2019**, *53*, 12257–12268. [CrossRef]
32. Albergamo, V.; Schollée, J.E.; Schymanski, E.L.; Helmus, R.; Timmer, H.; Hollender, J.; de Voogt, P. Nontarget Screening Reveals Time Trends of Polar Micropollutants in a Riverbank Filtration System. *Environ. Sci. Technol.* **2019**. [CrossRef]
33. Samanipour, S.; Baz-Lomba, J.A.; Reid, M.J.; Ciceri, E.; Rowland, S.; Nilsson, P.; Thomas, K.V. Assessing Sample Extraction Efficiencies for the Analysis of Complex Unresolved Mixtures of Organic Pollutants: A Comprehensive Non-Target Approach. *Anal. Chim. Acta* **2018**, *1025*, 92–98. [CrossRef] [PubMed]
34. Hogenboom, A.C.; van Leerdam, J.A.; de Voogt, P. Accurate Mass Screening and Identification of Emerging Contaminants in Environmental Samples by Liquid Chromatography–Hybrid Linear Ion Trap Orbitrap Mass Spectrometry. *J. Chromatogr. A* **2009**, *1216*, 510–519. [CrossRef]
35. Schymanski, E.L.; Singer, H.P.; Longrée, P.; Loos, M.; Ruff, M.; Stravs, M.A.; Ripollés Vidal, C.; Hollender, J. Strategies to Characterize Polar Organic Contamination in Wastewater: Exploring the Capability of High Resolution Mass Spectrometry. *Environ. Sci. Technol.* **2014**, *48*, 1811–1818. [CrossRef]
36. Ruff, M.; Mueller, M.S.; Loos, M.; Singer, H.P. Quantitative Target and Systematic Non-Target Analysis of Polar Organic Micro-Pollutants along the River Rhine Using High-Resolution Massspectrometry e Identification of Unknown Sources and Compounds. *Water Res.* **2015**, *87*, 145–154. [CrossRef]
37. Gago-Ferrero, P.; Schymanski, E.L.; Bletsou, A.A.; Aalizadeh, R.; Hollender, J.; Thomaidis, N.S. Extended Suspect and Non-Target Strategies to Characterize Emerging Polar Organic Contaminants in Raw Wastewater with LC-HRMS/MS. *Environ. Sci. Technol.* **2015**, *49*, 12333–12341. [CrossRef] [PubMed]
38. Nurmi, J.; Pellinen, J.; Rantalainen, A.-L. Critical Evaluation of Screening Techniques for Emerging Environmental Contaminants Based on Accurate Mass Measurements with Time-of-Flight Mass Spectrometry. *J. Mass Spectrom.* **2012**, *47*, 303–312. [CrossRef]
39. Díaz, R.; Ibáñez, M.; Sancho, J.V.; Hernández, F. Target and Non-Target Screening Strategies for Organic Contaminants, Residues and Illicit Substances in Food, Environmental and Human Biological Samples by UHPLC-QTOF-MS. *Anal. Methods* **2012**, *4*, 196–209. [CrossRef]
40. Guillossou, R.; Le Roux, J.; Goffin, A.; Mailler, R.; Varrault, G.; Vulliet, E.; Morlay, C.; Nauleau, F.; Guérin, S.; Rocher, V.; et al. Fluorescence Excitation/Emission Matrices as a Tool to Monitor the Removal of Organic Micropollutants from Wastewater Effluents by Adsorption onto Activated Carbon. *Water Res.* **2021**, *190*, 116749. [CrossRef] [PubMed]
41. Duan, J.; Li, W.; Sun, P.; Lai, Q.; Mulcahy, D.; Guo, S. Rapid Determination of Nine Haloacetic Acids in Wastewater Effluents Using Ultra-Performance Liquid Chromatography-Tandem Mass Spectrometry. *Anal. Lett.* **2013**, *46*, 569–588. [CrossRef]
42. Wickramasekara, S.; Hernández-Ruiz, S.; Abrell, L.; Arnold, R.; Chorover, J. Natural Dissolved Organic Matter Affects Electrospray Ionization during Analysis of Emerging Contaminants by Mass Spectrometry. *Anal. Chim. Acta* **2012**, *717*, 77–84. [CrossRef] [PubMed]
43. Sgroi, M.; Roccaro, P.; Korshin, G.V.; Greco, V.; Sciuto, S.; Anumol, T.; Snyder, S.A.; Vagliasindi, F.G.A. Use of Fluorescence EEM to Monitor the Removal of Emerging Contaminants in Full Scale Wastewater Treatment Plants. *J. Hazard. Mater.* **2017**, *323*, 367–376. [CrossRef]
44. Coble, P.G. Characterization of Marine and Terrestrial DOM in Seawater Using Excitation-Emission Matrix Spectroscopy. *Mar. Chem.* **1996**, *51*, 325–346. [CrossRef]
45. Parlanti, E.; Wörz, K.; Geoffroy, L.; Lamotte, M. Dissolved Organic Matter Fluorescence Spectroscopy as a Tool to Estimate Biological Activity in a Coastal Zone Submitted to Anthropogenic Inputs. *Org. Geochem.* **2000**, *31*, 1765–1781. [CrossRef]
46. Chen, J.; LeBoeuf, E.J.; Dai, S.; Gu, B. Fluorescence Spectroscopic Studies of Natural Organic Matter Fractions. *Chemosphere* **2003**, *50*, 639–647. [CrossRef]
47. Zhang, T.; Lu, J.; Ma, J.; Qiang, Z. Fluorescence Spectroscopic Characterization of DOM Fractions Isolated from a Filtered River Water after Ozonation and Catalytic Ozonation. *Chemosphere* **2008**, *71*, 911–921. [CrossRef]

48. Jeanneau, L.; Faure, P.; Jardé, E. Influence of Natural Organic Matter on the Solid-Phase Extraction of Organic Micropollutants: Application to the Water-Extract from Highly Contaminated River Sediment. *J. Chromatogr. A* **2007**, *1173*, 1–9. [CrossRef]
49. Lacey, C.; McMahon, G.; Bones, J.; Barron, L.; Morrissey, A.; Tobin, J.M. An LC–MS Method for the Determination of Pharmaceutical Compounds in Wastewater Treatment Plant Influent and Effluent Samples. *Talanta* **2008**, *75*, 1089–1097. [CrossRef] [PubMed]
50. Kessner, D.; Chambers, M.; Burke, R.; Agus, D.; Mallick, P. ProteoWizard: Open Source Software for Rapid Proteomics Tools Development. *Bioinformatics* **2008**, *24*, 2534–2536. [CrossRef]
51. Chambers, M.C.; Maclean, B.; Burke, R.; Amodei, D.; Ruderman, D.L.; Neumann, S.; Gatto, L.; Fischer, B.; Pratt, B.; Egertson, J.; et al. A Cross-Platform Toolkit for Mass Spectrometry and Proteomics. *Nat. Biotechnol.* **2012**, *30*, 918–920. [CrossRef]
52. PatRoon, Version 1. 2019. Available online: https://github.com/rickhelmus/patRoon (accessed on 18 November 2021).
53. FactoMineR, Version 2.4. 2020. Available online: https://cran.r-project.org/web/packages/FactoMineR/index.html (accessed on 18 November 2021).
54. Factoextra, Versions 1.0.7. 2020. Available online: https://cran.r-project.org/web/packages/factoextra/index.html (accessed on 18 November 2021).
55. Conway, J.; Gehlenborg, N. pSetR: A More Scalable Alternative to Venn and Euler Diagrams for Visualizing Intersecting Sets, Version 1.4.0. 2019. Available online: https://cran.r-project.org/web/packages/UpSetR/index.html (accessed on 18 November 2021).

Article

Ultraviolet Photodissociation for Non-Target Screening-Based Identification of Organic Micro-Pollutants in Water Samples

Christian Panse [1,2], Seema Sharma [3], Romain Huguet [3], Dennis Vughs [4], Jonas Grossmann [1,2] and Andrea Mizzi Brunner [4,*]

1 Functional Genomics Center Zurich, University of Zurich/ETH Zurich, Winterthurerstrasse 190, CH-8057 Zürich, Switzerland; cp@fgcz.ethz.ch (C.P.); jg@fgcz.ethz.ch (J.G.)
2 SIB Swiss Institute of Bioinformatics, Quartier Sorge-Batiment Amphipole, 1015 Lausanne, Switzerland
3 Thermo Fisher Scientific, San Jose, CA 95134, USA; seema.sharma@thermofisher.com (S.S.); romain.huguet@thermofisher.com (R.H.)
4 KWR Water Research Institute, P.O. Box 1072, 3430 BB Nieuwegein, The Netherlands; dennis.vughs@kwrwater.nl
* Correspondence: andrea.brunner@kwrwater.nl

Academic Editor: Thomas Letzel
Received: 26 August 2020; Accepted: 10 September 2020; Published: 12 September 2020

Abstract: Non-target screening (NTS) based on the combination of liquid chromatography coupled to high-resolution mass spectrometry has become the key method to identify organic micro-pollutants (OMPs) in water samples. However, a large number of compounds remains unidentified with current NTS approaches due to poor quality fragmentation spectra generated by suboptimal fragmentation methods. Here, the potential of the alternative fragmentation technique ultraviolet photodissociation (UVPD) to improve identification of OMPs in water samples was investigated. A diverse set of water-relevant OMPs was selected based on k-means clustering and unsupervised artificial neural networks. The selected OMPs were analyzed using an Orbitrap Fusion Lumos equipped with UVPD. Therewith, information-rich MS2 fragmentation spectra of compounds that fragment poorly with higher-energy collisional dissociation (HCD) could be attained. Development of an R-based data analysis workflow and user interface facilitated the characterization and comparison of HCD and UVPD fragmentation patterns. UVPD and HCD generated both unique and common fragments, demonstrating that some fragmentation pathways are specific to the respective fragmentation method, while others seem more generic. Application of UVPD fragmentation to the analysis of surface water enabled OMP identification using existing HCD spectral libraries. However, high-throughput applications still require optimization of informatics workflows and spectral libraries tailored to UVPD.

Keywords: mass spectrometry; non-target screening; ultraviolet photodissociation; higher-energy collisional dissociation; organic micropollutants; water quality; small molecule fragmentation; cheminformatics; data analysis

1. Introduction

1.1. Challenges of Monitoring Drinking Water Quality

Reliable identification of organic micro-pollutants (OMPs) in drinking water and its sources is essential to risk assessment and prediction of the behavior of a substance in the environment and during water treatment. Non-target screening (NTS) based on the combination of liquid chromatography coupled to high-resolution mass spectrometry (LC-HRMS/MS) has become the key method to identify

OMPs in water samples, as it has the potential to detect all ionizable compounds that are amenable to the selected chromatographic separation, within a defined mass range [1].

The unambiguous identification of an OMP from NTS data relies on the accurate mass and isotopic pattern from the full scan MS1 spectrum to determine the elemental formula of the compound. The addition of MS2 fragmentation spectral data then allows to determine the molecular structure, given that the fragmentation event causes reproducible bond cleavages that result in diagnostic and interpretable fragment ions representative of the structure of the molecule. The MS2-based structural identification typically relies on matching of the experimental spectrum with entries in spectral libraries or *in silico* predicted fragmentation spectra. For compounds that show poor fragmentation spectra generated by higher-energy collisional dissociation (HCD) fragmentation, the fragmentation technique routinely applied in Orbitrap based NTS, confident structural elucidation often remains elusive. Alternative fragmentation techniques that cause different bond cleavages may remedy structural elucidation in these cases.

1.2. Interpreting Fragmentation Spectra from Ultraviolet Photodissociation

Ultraviolet photodissociation (UVPD) is a fragmentation technique achieved with a UV laser. Its main application to date is protein characterization with proteomics and intact protein MS. However, it also allows structural elucidation of small molecules that cannot be identified by HCD alone [2]. For instance, UVPD was shown to facilitate characterization of various lipid classes [3], to generate unique fragments or enhance detection of kinetically unfavorable fragments of flavonoids, phenylpropanoids and chalconoids [4,5].

In the Orbitrap Fusion Lumos mass spectrometer, a Q3 series passively Q-switched Nd:YAG laser (CryLaS GmbH) that outputs the 5th harmonic at 213 nm is interfaced to the rear of the ion trap, in the low pressure cell of which the UV photoactivation occurs [6]. The laser pulse energy is a 1.5 ± 0.2 μJ/pulse at 2.5 kHz repetition rate. With 450 ± 200 μm, including the divergence at the center of the ion trap, the beam diameter is slightly larger than the simulated ion cloud diameter at normal AGC targets and no focusing optics are required. With the incorporation of this compact and robust solid state laser into a commercial instrument, UVPD could be routinely implemented in NTS workflows.

To date, however, UVPD fragmentation has not been applied to the NTS-based monitoring of small molecules. This is due to the fact that apart from the compounds mentioned above, little is known about UVPD fragmentation pathways of small molecules, the kinetics of fragment formation, and the influence of reaction time on fragmentation patterns. Moreover, as it is a novel fragmentation technique in the small molecule field, spectral library entries with UVPD spectra are still lacking, and the usability of HCD spectra for spectral matching of UVPD spectra remains to be explored. Furthermore, the aptness of *in silico* prediction algorithms for the prediction of UVPD spectra has not yet been demonstrated. Combinatorial prediction algorithms that do not apply any rules of fragmentation, but use a bond dissociation approach, could potentially be used for UVPD data.

1.3. UVPD Fragmentation for Water Quality Monitoring

Here, the potential of the fragmentation technique UVPD to improve the structural identification of small molecules, in particular OMPs in water samples, was evaluated using the Orbitrap Fusion Lumos Tribrid [6]. After application of a cheminformatics strategy to select water relevant OMPs that cover a wide chemical space and development of a data analysis workflow in R, HCD and UVPD fragmentation patterns of selected OMPs could be characterized and compared. The two fragmentation techniques generated both unique and common fragments, demonstrating that some fragmentation pathways are specific to the respective fragmentation method whilst others seem more generic. Application to environmental water samples showed that HCD spectral libraries can be used for UVPD based OMP identification, in particular when high collision energy (CE) spectra are available. However, to increase successful feature annotation of NTS data with UVPD fragmentation, UVPD spectra need to be added to spectral libraries.

2. Results and Discussion

2.1. Proof of Principle: Manual Spectral Interpretation of Single Compounds

To investigate the potential of UVPD for OMP identification, the three compounds triadimenol, gemfibrozil and sucralose were selected as model compounds. These compounds are both relevant for the water sector and known to not fragment well using standard HCD fragmentation settings, i.e., CEs ranging from 20 to 50 (arbitrary units). As little was known about UVPD fragmentation pathways of these OMPs, we applied the combinatorial prediction algorithm of MetFrag [7], which does not rely on fragmentation rules, but uses a bond dissociation approach to predict potential fragments and matches these to the experimentally observed.

UVPD provided unique fragmentation information for structural elucidation of triadimenol, a fungicide that can be found in drinking water sources (Figure 1). HCD fragmentation at 35CE, the CE range typically used in NTS experiments, resulted in a predominant fragment at 70 *m/z*, a minor fragment at 99 *m/z* and a low intensity fragment (see 10× zoom-in) at 141 *m/z*. These fragments could be assigned to the *in silico* predicted fragments $[C_2H_2N_3 + H] + H^+$, $[C_6H_{12}O\text{-}H]^+$ and $[C_7H_5ClO] + H^+$. In contrast, UVPD fragmentation led to more and different fragment ions. The peaks detected with HCD were also detected in the UVPD fragmentation spectra when using shorter reaction times (25 to 100 ms), but decreased with increasing UVPD reaction times. Concurrently, peaks at 112, 168, and 261 *m/z* increased with increasing reaction times. These could be matched to the *in silico* predicted fragments $[C_4H_5N_3O] + H^+$, $[C_8H_{14}N_3O] + H^+$, and $[C_{14}H_{18}N_3O_2] + H^+$. A fragment at 227 *m/z* was detected with UVPD at short reaction times, i.e., 25 to 50ms only and could be matched to $[C_{12}H_{16}ClO_2]^+$. These promising results showed that UVPD could lead to informative spectral information of an OMP that did not fragment well with HCD, and that the *in silico* prediction using MetFrag could successfully be applied for UVPD spectral annotation.

Next, UVPD fragmentation of OMPs that ionize in negative ionization mode were investigated, starting with the pharmaceutical gemfibrozil. Two peaks could be annotated in the HCD spectrum with 35CE and UVPD spectrum with 25ms reaction time, namely $[C_8H_9O]^-$ at 121 *m/z*, and $[C_7H_{13}O_2\text{-}H]\text{-}H^-$ at 127 *m/z* (see Figure S1a). In the UVPD spectra with longer reaction times, only the 121 *m/z* peak could be matched. The predominant UVPD peak at 112 *m/z* increased with reaction times. This peak was also present in the HCD spectrum. However, it could not be matched to an *in silico* predicted fragment mass, nor could any of the other UVPD peaks. The base peak in all UVPD spectra was the precursor ion, the absolute intensity of which decreased with increasing reaction times. As observed previously with UVPD of lower charged negative DNA ions, this could be due to an electron detachment-induced charge reduction [2]. Electron detachment dissociation usually involves two or more negatively charged species. For single ion negative UVPD, the mechanism for electron detachment may be more favorable than fragment generation. However, these data were based on 193 nm UVPD, and whether 213 nm UVPD would have the same effects remains unknown.

As a second OMP analyzed in negative ionization mode, the artificial sweetener sucralose was fragmented. With the standard HCD fragmentation with 35CE, six peaks could be annotated (see SI Figure 1b). Four of the annotations, i.e., $[C_2H_4O_2]\text{-}H^-$ at 59 *m/z*, $[C_3H_5O_2\text{-}H]\text{-}H^-$ at 71 *m/z*, $[C_3H_5O_2]^-$ at 73 *m/z* and $[C_3H_5O_3\text{-}H]\text{-}H^-$ at 87 *m/z*, were low mass range fragments only detected using HCD fragmentation. The other two, $[C_4H_7O_3\text{-}H]\text{-}H^-$ at 101 *m/z* and $[C_6H_{10}O_4\text{-}2H]\text{-}H^-$ at 143 *m/z*, were present in both HCD and UVPD spectra with 50, 100, 200 ms and—in the latter—400 ms reaction time. At 25 ms, the species, $[C_6H_{10}O_4\text{-}H]\text{-}H^-$ at 144 *m/z*, was present instead. At 400 ms reaction time, the 143 *m/z* peak was the only one that could be annotated in an overall noisy low intensity fragmentation spectrum. Corresponding to what was observed for triadimenol, long UVPD reaction times negatively affected spectral quality as well as the duty cycle in the case of sucralose.

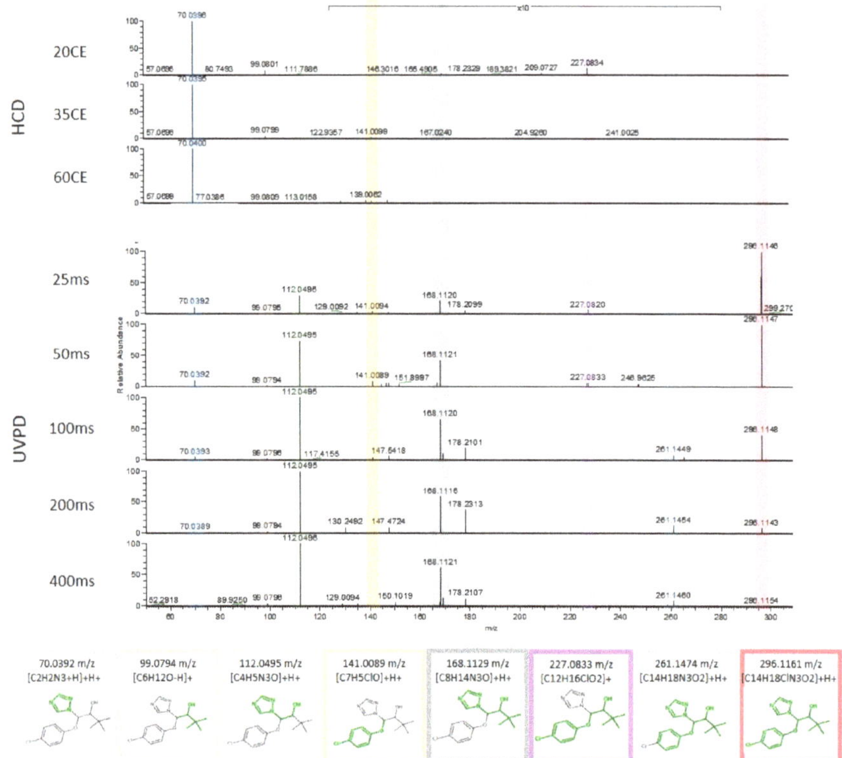

Figure 1. Comparison of higher-energy collisional dissociation (HCD) (top) and ultraviolet photodissociation (UVPD) (bottom) fragmentation spectra of the fungicide triadimenol acquired with 20, 35 and 60 collision energy (CE) and 25, 50, 100, 200 and 400 ms reaction time. Annotated fragments are highlighted in color. The *m/z* range 130–280 is 10× enlarged.

In the spectra of all shorter reaction times, i.e., 25 to 200 ms, the fragments $[C_6H_{10}O_5\text{-}2H]\text{-}H^-$ at 159 *m*/*z* and $[C_6H_9Cl_2O_3]\text{-}H^-$ at 197 *m*/*z* could be annotated; at lower reaction times also $[C_6H_9Cl_2O_3]^-$ at 198 *m/z* and $[C_6H_9Cl_2O_3\text{-}H]\text{-}H^-$ at 196 *m/z*. At 50 ms, another annotated fragment, $[C_6H_{10}ClO_5\text{-}H]\text{-}H^-$ at 195 *m*/*z*, was present. This reaction time resulted, thus, in the most informative spectra, in particular in combination with the HCD spectrum of low mass range annotated fragments.

The UVPD fragmentation data of the three model compounds of which one ionized in positive and two in negative ionization mode suggested that UVPD could facilitate structural elucidation of some OMPs for which HCD spectra did not contain enough information. While fragmentation of the positively charged triadimenol led to more fragments and higher fragment intensities with UVPD compared to HCD, the negatively charged compound gemfibrozil fragmented poorly with both HCD and UVPD, and UVPD fragmentation of sucralose resulted in complementary fragments to HCD.

2.2. Selection of Reference Standards Based on Clustering

To further investigate the applicability of UVPD for OMP identification, a representative selection of compounds regarding their distribution in the chemical space and water relevance was made. First, a k-means clustering was performed using Pubchem extended fingerprints, resulting in 20 defined clusters (Figure 2a). The molecular discrimination of these clusters was confirmed using the unsupervised artificial neural network self-organizing maps (SOM, Figure 2b). In the SOM, the selected compounds were colored according to their k-means cluster number. As compounds of the same color are in close vicinity in the SOM, this shows that the different clusters successfully separated these compounds.

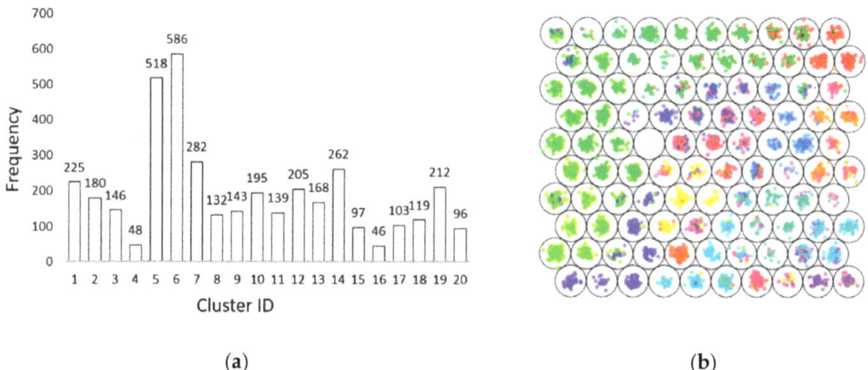

(a) (b)

Figure 2. Selection of organic micro-pollutants (OMPs) by two complementary approaches: (**a**) k-means clustering of Pubchem extended fingerprints; (**b**) self-organizing map. Compounds are colored according to the k-means cluster number.

Depending on in-house reference standard availability, one to four compounds were selected per cluster for fragmentation experiments, apart from cluster 19, for which no standard was available. In addition, a selection of disinfection by-products known to be relevant for drinking water treatment was added to the set of compounds, to be fragmented by HCD and UVPD.

2.3. An R-Based Data Analysis Workflow and Shiny Application Interface to Explore (Novel) Fragmentation Techniques

To enable high throughput data analysis of the LC-HRMS data including UVPD fragmentation spectra, an R-based workflow was developed that takes the extracted ion chromatogram (XIC) of a given compounds based on its simplified molecular-input line-entry specification (SMILES), determines the peak apex and extracts the corresponding retention time (RT). The three MS2 spectra with highest intensity neighboring the apex were used to match experimental spectra with *in silico* predicted fragmentation spectra of the given SMILES.

For a user friendly output and to support exploratory data analysis [8], a Shiny application based interface was created to further examine the data (https://CRAN.R-project.org/package=shiny). The Shiny application is provided with the 'uvpd' R package (https://github.com/cpanse/uvpd/) and can be accessed at http://fgcz-ms-shiny.uzh.ch:8080/p2722-uvpd/. Its user interface is split into an input and output part. On the left, the input panel provides a selection for the input data, compounds, ionization mode and cut-off values for the relative and absolute mass errors of the precursor mass, precursor signal removal in the MS2 spectra and cluster ID. This selection then determines the respective output in the several tab panels on the right. These tabs provide data visualization and tables of the selected compound. Table 1 describes the output tabs and whether the selected filtering is applied to a given tab.

Table 1. Description of Shiny application output panels and applied filtering parameters.

Tab Panel	**Description**	**Selected Filter Option Applied to the Tab**				
		Compound	**Remove Precursor Items**	**(+/−) Ion Type**	**Ppm Error Cut-Off**	**Absolute Error Cut-Off**
stacked fragments	1) Bivariate scatterplots of scores 1, 2 and 3 per fragmentation mode. 2) Two stacked bar charts of the logarithmically transformed fragment ion intensities of the matched fragment ions and types, respectively, per fragmentation mode. 3) Bivariate scatterplots of the total ion count (TIC) of the MS2 spectrum and the corresponding master intensity for the three most abundant master intensities of each raw file per fragmentation mode. 4) Boxplots of the absolute error distribution (in Dalton) per fragmentation mode.	X	X	X	X	X
summary	1) Statistics of the overall data and the applied filter setting. 2) Frequency value per fragmentation mode. 3) Histograms of ppm and absolute error distribution over the entire data set and selected compound, including a maximum-likelihood fitting, assuming an underlying normal distribution.	X	X	X	X	X
ms2	1) Table of detected fragment ions and ion types. 2) Fragmentation spectra per fragmentation mode.	X	X	X	X	X
data	All quantitative and qualitative data.	X	X	X	X	X
scores	1) Scores 1, 2 and 3. 2) Plots of the scores.		X	X	X	X
frequencies	Downloadable frequency table, per compound and fragmentation type		X	automatic	X	X
predicted ion	*In silico* predicted, i.e., theoretical fragment ions predicted with 'metfRag: frag.generateFragments'	X				
help	Help page					

2.4. Higher-Throughout Comparison and Interpretation of UVPD and HCD Fragmentation Spectra

For a thorough comparison of UVPD and HCD fragmentation spectra, 46 selected water-relevant OMPs covering a wide chemical space were analyzed with LC-HRMS using UVPD with 25–800 ms reaction time, and HCD with 20, 35 and 60 CE. Eight of the compounds could not be detected with electrospray ionization (ESI), one eluted too early with reverse-phase (RP) LC for peak detection, one had an intensity below the cut-off threshold and two were not picked by the data analysis workflow due to a Na-adduct and Cl-salt, respectively (see Table S1). The remaining 34 compounds belonged to 11 different clusters, with one to four compounds per cluster. Their fragmentation behavior varied substantially and, based on fragmentation, four different groups of compounds could be distinguished (Table S1); such with poor fragmentation with both UVPD and HCD (Figure 3a), a preference for HCD (Figure 3b) or UVPD (Figure 3 c), or good fragmentation with both UVPD and HCD (Figure 3d). These groups, however, did not seem related to the cluster number.

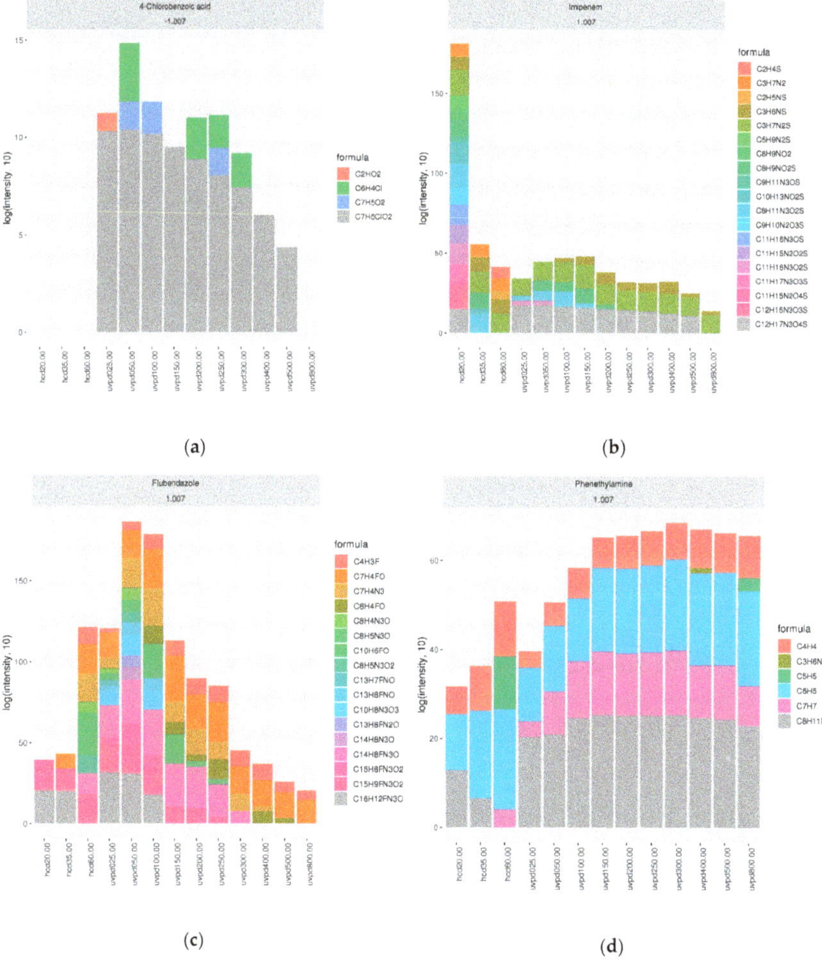

Figure 3. Fragmentation spectra annotation of the different ion types per fragmentation condition. Stacked bar plots from Shiny application output showing the summed intensities of the annotated fragments of (**a**) 4-chlorobenzoic acid; (**b**) imipenem; (**c**) flubendazole; (**d**) phenetylamine.

The first group of compounds did not generate information-rich spectra, as illustrated in Figure 3a, which shows the summed intensities of the annotated fragments from HCD and UVPD spectra of 4-chlorobenzoic acid; poor fragmentation was observed for ibuprofen, 3-nitrophenol and 4-nitrophenol with both fragmentation techniques under all conditions, for 4-chlorobenzoic acid, 4-nitrophthalic acid and 5-nitroisophtalic acid in particular with HCD. In the case of 2-methyl-4-nitrophenol and gemfibrozil, spectra were still poor, but slightly better with HCD.

A second group of compounds showed a preference for HCD compared to UVPD (Figure 3b) either with increasing CEs, for instance 3-nitroindole, with low CEs, for instance imipenem at 20CE, with a specific CE, for instance 2-methyl-4-chlorophenoxyacetic acid (MCPA) at 35CE, or with all CEs, for instance N-Desmethyl Clarithromycin. In contrast, for another group of compounds, no fragmentation was observed with HCD, but good fragmentation with a range of UVPD reaction times (Figure 3c), for instance for benzocaine and to a lesser degree 4-nitroanthranilic acid. In the case of fenofibric acid, flubendazole and triadimenol good information rich spectra were generated with a range of UVPD reaction times, but only with HCD at 60 CE.

A fourth group of compounds exhibited good fragmentation with both fragmentation techniques (Figure 3d). Some of these had an optimum at a specific fragmentation condition, for instance 2-amino-3-nitrobenzoic acid with UVPD at 50 ms, 3,5-Dinitrosalicylic acid with HCD at 35CE and UVPD at 50 ms, aflatoxin B2, dinoterb, epoxiconazole and JWH-250 with HCD at 60CE and 2-hydroxy-4-nitro-benzoic acid at 20 CE. Others fragmented well with a range of conditions, such as 2,4 Dinitrophenol, which showed more informative spectra with higher CEs, and 2-methoxy-4-nitrophenol and phenethylamine with higher HCD CEs and longer UVPD reaction times. Good fragmentation was observed for both (ranges of) UVPD and HCD conditions in the case of 2-methoxy-4,6-dinitrophenol, 4-hydroxy-3-nitrobenzenesulfonic acid, 4-nitrobenzenesulfonic acid and 5-nitrovanillin. In the case of nitrofurazone, more informative spectra were generated with UVPD.

Cluster numbers could not be related to these four broad groups of fragmentation behavior. For instance, while benzocaine and 4-Nitroanthranilic acid both fragmented well with UVPD and poorly with HCD, the other two compounds from cluster 13, piperacillin and 5-Nitroisophthalic acid, showed good fragmentation for both UVPD and HCD and poor fragmentation, respectively. Regarding cluster 11, aflatoxin b2 and fenofibric acid both exhibited information rich spectra at multiple different UVPD reaction times. However, gemfibrozil, a compound of the same cluster, did not fragment well with both UVPD and HCD. This lack of similar fragmentation behavior within a cluster could indicate that fragmentation behavior depends only on a few of the descriptors used for clustering.

In particular, UV absorbing compounds such as aromatic compounds, and compounds with double bonds are expected to fragment well with UVPD. Compound class information for each of the compounds that fragmented well with UVPD, including the lipids [3], flavonoids, phenylpropanoids and chalconoids [4,5] published previously could be utilized to predict UVPD fragmentation. Furthermore, if certain compounds of classes with good UV absorbance did not fragment well, further clustering within that compound class could be utilized to improve our understanding and ultimately the prediction of UVPD fragmentation behavior.

In the UVPD spectra, fragment ion intensities decreased with increasing UVPD reaction times when normalized to precursor intensity, as illustrated in Figure S2. Overall, UVPD fragmentation was beneficial for multiple compounds, often leading to a number of annotated fragments that were unique to the fragmentation technique. This complementarity of UVPD makes it an attractive addition to HCD that can be implemented in data-dependent decision trees during NTS data acquisition. Optimal UVPD reaction times depended on the compound, analogous to HCD where the optimal CE varied amongst compounds. Interestingly, in the HCD experiments, oftentimes, CEs higher (60 CE) and lower (20 CE) than the 35 CE routinely used in NTS experiments were needed to generate informative fragmentation spectra. This should be considered in future studies to increase the confidence of OMP identification, in particular when UVPD is not available.

2.5. NTS of a Meuse River Surface Water Sample

To investigate the applicability of currently available spectral libraries and NTS workflows to UVPD data, a surface water sample from the river Meuse was acquired with HCD and UVPD fragmentation and analyzed using the NTS data analysis software Compound Discoverer (Thermo Fisher Scientific, San Jose, USA). This software enables suspect screening based on spectral matching with the spectral library mzCloud, which consists of collision-induced dissociation (CID) and HCD fragmentation spectra. The mzCloud score of a tentatively identified compound is a measure for the confidence of identification. It is based on the number of fragments that match the experimental and library spectra, with a score of 100 indicating a perfect match. Comparison of mzCloud scores of spectra acquired with HCD and UVPD showed that the overall score distribution was similar (no significant difference, see Appendix A), visualized in the combined box and violin plots in Figure 4a. However, individual compounds differed strongly in their scores. For instance, known water relevant OMPs on average showed a mzCloud score with UVPD fragmentation that was 15 points lower than the HCD score; atrazine scored 75.1 with UVPD versus 95.4 with HCD, caffeine 75.5 versus 91, carbamazepine 69.1 versus 96.6 and terbuthylazine 87.4 versus 98.6. UVPD spectra were matched with HCD library spectra of high CEs, i.e., 70 CE, 40 CE, 80 CE and 90 CE for the four different OMPs. In contrast, metoprolol showed a similar mzCloud score with UVPD, i.e., 78.9, when matched with an HCD 30 CE spectrum, compared to an HCD score of 77.6.

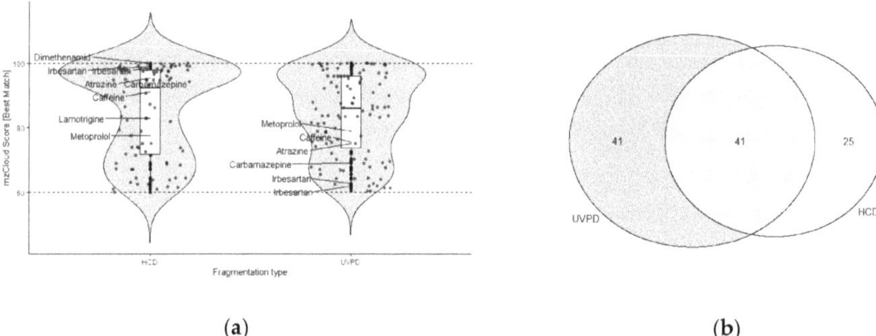

(**a**) (**b**)

Figure 4. Comparison of HCD and UVPD NTS MS2 data of river Meuse water (**a**) mzCloud Score distribution. The scores of common water relevant compounds are labelled by compound name; (**b**) overlap in features annotations with an mzCloud score above 60.

Half of the compound annotations with mzCloud in the NTS data with UVPD were not assigned in the HCD data (Figure 4b). The HCD data was manually checked for features with accurate mass and retention time matching these unique compounds. Information on whether these features were detected in the HCD data and their annotation is available in Table S2. Twenty-five features were detected in the HCD data, but had no mzCloud hit, and six were annotated with a different compound based on the mzCloud matching. Ten features were not detected. This is most likely due to differences in peak picking during data analysis. Manual inspection of the UVPD assignments showed that in most cases when there was no assignment in the HCD data, the annotated UVPD spectrum consisted of only a precursor signal and—if any—low intensity ions close to the noise cut-off (see SI Table S2 Compounds detected in UVPD NTS experiments). In these cases, the match was usually against a low energy CID or HCD library spectrum (CE10 to 20) that also only contained the precursor. Consequently, these can likely be false positive assignments. The high scores for matches based on the precursor signal alone are problematic. In future studies, more appropriate scoring algorithms should be considered.

In contrast, the assignments where multiple fragments were matched, i.e., 1-methylbenzotriazole with 15, 3-hydroxyfluorene with 11, acetyl norfentanyl with three, mandipropamid with seven and metolachlor with 17 fragments, were all based on high energy HCD library spectra (60 to 130 CE) except for 3-hydroxfluorene (20 CE). This was in correspondence with the assignments of known water relevant OMPs, and indicates that UVPD-induced fragmentation pathways in these molecules resemble those of higher energy HCD. As routinely lower HCD CEs are applied, i.e., 20–50 CE, this can explain why these assignments were only made in the case of UVPD fragmentation, emphasizing the benefit of this alternative fragmentation technique and/or higher CEs for NTS based identification of OMPs. Moreover, UVPD annotations could be used to exclude false positive annotations of sparse HCD MS2 spectra (precursor only matches) and vice versa. While HCD spectral libraries proved to be of (limited) use for UVPD spectral annotation, for the routine implementation of UVPD data in NTS workflows, spectral libraries need to be extended with UVPD spectra.

3. Materials and Methods

3.1. Selection of Reference Standards Through Clustering

Selection criteria for OMPs to be fragmented with UVPD and HCD included their relevance for the water sector and a good coverage of the chemical space, as the compound structures were expected to affect fragmentation. To select water relevant OMPs, 4000 compounds were randomly selected from the NORMAN Substance Database, which is compiled of multiple suspect lists relevant for environmental monitoring (SusDat, https://www.norman-network.com/nds/susdat/). To select compounds with diverse chemical structures, these compounds were clustered [9]. To this end, the Simplified molecular-input line-entry systems (SMILES) of each compound were parsed and configured for atom typing and isotoping using the R package rcdk [10]. Next, for each compound, the extended fingerprint, a binary vector of 1024 dimensions, was extracted. A k-means clustering was conducted of the computed Tanimoto Distance matrix between all pairs of fingerprints [11]. The optimal number of clusters was determined by the elbow method [12]. To investigate molecular discrimination by the clusters, we trained a self-organizing map (SOM) [13] as a complementary approach. The SOM grid was initialized with 10×10 nodes. Each fingerprint selected in the training phase was colored by the corresponding k-means cluster ID for visualization. The entire *in silico* data analysis was performed using R version 3.5.1 to 4.0.1 running on Linux, Windows and MacOSX systems [14]. All code snippets are available as an R package through https://github.com/cpanse/uvpd/.

3.2. LC-HRMS Analysis with UVPD Fragmentation

Selected reference compounds listed in SI Table S2 were prepared in ultra-pure water with a final concentration of 10 µg/L. The surface water (SW) sample was collected from the river Meuse, the Netherlands, 16.666× concentrated using Oasis-HLB SPE columns-based extraction and diluted 50× for the LC-HRMS analysis. The internal standards (IS) atrazine-d5 (CDN isotopes, Pointe-Claire, Quebec, Canada), benzotriazole-d4 and bentazon-d6 (LGC Standards, Wesen, Germany) were added to the SW sample to a final concentration of 1 µg/L. Samples were filtered using 0.2 µm Phenex^{TM}-RC 15 mm Syringe Filters (Phenomenex, Torrance, USA) prior to analysis. Blank samples were prepared correspondingly, through spike-in of IS to ultra-pure water followed by filtration. In total, 100 µL of sample were injected into the LC-HRMS.

Compounds were analyzed using reverse phase (RP) LC-HRMS/MS with a Vanquish Horizon UHPLC system (Thermo Fisher Scientific, San Jose, CA, USA) coupled to an Orbitrap Fusion Lumos equipped with ultraviolet photodissociation (UVPD) and the acquisition software AcquireX (Thermo Fisher Scientific, San Jose, CA, USA). An XBridge BEH C18 XP column (150 mm × 2.1 mm I.D., particle size 2.5 µm, Waters, Etten-Leur, The Netherlands) was used in combination with a 2.0 mm × 2.1 mm I.D. Phenomenex SecurityGuard Ultra column (Phenomenex, Torrance, CA, USA), at a temperature of 25 °C. The LC gradient started with 5% acetonitrile, 95% water and 0.05% formic acid (*v*/*v*/*v*), increased to 100% acetonitrile, 0.05% formic acid in 25 min and then remained constant for 4 min. The flow rate was 0.25 mL/min.

For the reference standards, fragmentation spectra were acquired using targeted methods with mass triggers. The fixed collision energies (CEs) 20, 35 and 60 were used for HCD fragmentation, and UVPD reaction times ranging from 25 to 800 ms for UVPD fragmentation. The full scan mass range was 100–800 *m*/*z* with 120k resolution at FWHM for the MS1 scans, and 50–500 *m*/*z* with 15k resolution at FWHM for the MS2 scans (due to a corrupted data file, the disinfection by-products data with 100 ms and 25 0 ms UVPD reaction time is lacking in the data set).

For the SW sample, NTS analyses were performed with data dependent acquisition (ddA), using the AcquireX deepscan functionality that ensures MS2 scans are acquired for most features, Top Speed and 35 CE for the HCD and 100 and 150 ms reaction time for the UVPD experiments. The full scan mass range was set at 80–1300 *m/z* with 120k resolution, the MS2 at 50–500 *m/z* with 15k resolution.

3.3. Manual Annotation of Fragmentation Spectra

Thermo Fisher Scientific raw files were viewed with Thermo Xcalibur Browser (Thermo Fisher Scientific, San Jose, CA, USA). MS2 peak lists of HCD and UVPD fragmentation spectra were exported and used for fragment annotation with the MetFrag web tool (https://msbi.ipb-halle.de/MetFragBeta/).

3.4. An R-Based LC-HRMS Data Analysis Workflow to Explore Novel Fragmentation Techniques

Fragment ions of the selected reference standard compounds were predicted with tree depth 1 and 2, using the R package MetFrag [7] in a preprocessing step. Charge configurations were derived for the predicted singly charged fragments $[M]^+$, $[M + H]^+$ and $[M + 2H]^+$ and $[M]^-$, $[M - H]^-$ and $[M - 2H]^-$ for the positive and negative ionization mode, respectively. The predicted fragment ions were stored and made available as a dataset in the R package UVPD.

Thermo Fisher Scientific raw files were processed with the R package rawDiag [15]. The in profile mode recorded data were centroided using the centroid method of the R package protViz (https://CRAN.R-project.org/package=protViz, [16]). For all compounds measured in all fragmentation modes, the retention time (RT), the area under the curve (AUC) of the APEX extracted ion chromatography (XIC) of the protonated and deprotonated precursor species, i.e., $[M + H]^+$ and $[M - H]^-$ of the selected reference standard compounds, the master intensity and the total ion count (TIC) were determined (see Figure S3). The *m/z* of the $[M + H]^+$ and $[M - H]^-$ were calculated based on the compound SMILES. The top three highest intensity spectra of each reference compound per fragmentation mode were used to assess the performance of the different fragmentation modes. The peaks of the centroided fragmentation spectra were annotated with the previously predicted fragment ions if the match was within a given mass window. A default cut-off value for fragment matching was set to 1 Da, further refinement can be made in the Shiny application (1–100 ppm, 10e–4 to 0.5 Da). The default values in the Shiny application are relative and absolute cut-off of 10 ppm and 0.02 Da, respectively. These are also the tolerances used throughout the manuscript.

The quantitative, e.g., MS1 derived XIC and master intensity, and MS2 derived TIC and fragment intensities, and qualitative fragmentation data were joined by the raw file name and the scan number. To compare fragment ion annotation qualitatively and quantitatively across all compounds and fragmentation modes, we implemented three different scores:

$$Score\ 1 = \frac{n_{exp\ frags\ matched}}{n_{theor\ frags}} \tag{1}$$

$$Score\ 2 = \frac{n_{exp\ frags\ matched}}{n_{exp\ frags}} \tag{2}$$

$$Score\ 3 = \frac{int_{exp\ frags\ matched}}{int_{\mathrm{MS1}\ precursor}} \tag{3}$$

where *exp frags* are the experimentally detected fragments, *exp frags* matched the experimentally detected fragment ions that could be matched to the *in silico* predicted, and *theor frags* the *in silico* predicted theoretically possible fragment ions. *n* indicates the number of fragments and *int* their intensity.

All data and results are visualized and can be interactively accessed in the R shiny application provided with the uvpd R package and through http://fgcz-ms-shiny.uzh.ch:8080/p2722-uvpd/. The entire workflow is shown in Figure S2.

3.5. NTS Data Analysis

NTS data were processed with Compound Discoverer 3.1 (Thermo Fisher Scientific, San Jose, CA, USA) for peak picking, componentization and suspect screening using the spectral library mzCloud. The output feature list, i.e., a table with accurate mass/retention time pairs (features) and their intensity, information on whether an MS2 spectrum was acquired for a given feature and the mzCloud spectral matching scores were imported into R Studio for further data analysis and visualization. R version 3.6.3. and R-Studio version 1.1.463 were used for the data analysis [14,17].

4. Conclusions

Combining the novel fragmentation technique UVPD and cheminformatics tools, we showed the potential of UVPD for structural elucidation of water-relevant OMPs in NTS data. Based on the two complementary methods k-means clustering and SOMs, a set of OMPs could be selected that was representative for the water cycle and a wide chemical space. An R-based LC-HRMS data analysis workflow and interactive interface for data visualization was developed to investigate UVPD fragmentation of these OMPs in a high-throughput manner.

Information-rich UVPD fragmentation spectra were achieved for 62% of the examined OMPs, in 15% of the cases also for OMPs that fragmented poorly with HCD. For 26% of the OMPs, neither fragmentation technique generated informative spectra; the remaining 12% HCD spectra were information-rich. UVPD and HCD generated both unique as well as overlapping fragments, demonstrating that some fragmentation pathways are specific to the respective fragmentation methods, while others seem to be more generic. These unique fragments provided additional information for structural identification complementary to HCD spectra. Based on these results, implementation of UVPD as a second fragmentation option in data dependent decision trees during NTS data acquisition is an attractive strategy to improve the confidence in OMP identification.

Analysis of NTS UVPD data with existing NTS software and the spectral library mzCloud enables annotation of features in the UVPD data using HCD library spectra of high CEs. For the routine implementation of UVPD fragmentation in NTS workflows, however, databases need to be extended with UVPD spectra, which would allow the full potential of this novel fragmentation technique to be exploited.

Supplementary Materials: The following are available online, Figure S1: Negative ionization mode model OMPs (a) gemfibrozil and (b) sucralose; Figure S2: Fragment intensity decreases with increasing UVPD reaction times. Total ion counts (TIC) for a given fragmentation mode versus master intensity; Figure S3: Data analysis and visualization workflow, Table S1: List of reference standards; Table S2: Compounds detected in UVPD NTS experiments.

Author Contributions: Conceptualization, A.M.B.; methodology, A.M.B. and C.P.; software, C.P.; validation, A.M.B. and C.P.; formal analysis, A.M.B. and C.P.; investigation, A.M.B., S.S., R.H., D.V. and C.P.; resources, A.M.B., S.S., R.H. and C.P.; data curation, C.P.; writing—original draft preparation, A.M.B. and C.P.; writing—review and editing, A.M.B., S.S., R.H., J.G., D.V. and C.P.; visualization, A.M.B. and C.P.; project administration, A.M.B.; funding acquisition, A.M.B. and C.P. All authors have read and agreed to the published version of the manuscript.

Funding: This research was funded by the Joint Research Program of the Dutch and Belgian drinking water companies.

Acknowledgments: The authors would like to thank Robert Bijlsma, University Jaume I, for the kind contribution of reference standards, Tessa Pronk for input in the clustering analyses, Thomas ter Laak and the Dutch and Belgian drinking water companies, in particular Eelco Pieke for critical reading, and the anonymous reviewers, especially reviewer 1, for constructive comments.

Conflicts of Interest: The authors declare no conflict of interest. The funders had no role in the design of the study; in the collection, analyses or interpretation of data; in the writing of the manuscript, or in the decision to publish the results.

Availability: Used code snippets available at https://github.com/cpanse/uvpd/; Shiny app available at http://fgcz-ms-shiny.uzh.ch:8080/p2722-uvpd/; LC-HRMS raw data of set of reference standards available at https://doi.org/10.5281/zenodo.4001653.

Appendix A

There was no significant difference between mzCloud scores of NTS data with HCD and UVPD fragmentation:

t.test(data$mzCloud.Best.Match ~ data$frag)

Welch Two Sample t-test

data: data$mzCloud.Best.Match by data$frag
$t = 0.44183$, df = 177.56, p-value = 0.6591
alternative hypothesis: true difference in means is not equal to 0
95 percent confidence interval:
−2.969521 4.682821
sample estimates:
mean in group HCD mean in group UVPD
84.99091 84.13426

References

1. Hollender, J.; Schymanski, E.L.; Singer, H.P.; Ferguson, P.L. Nontarget Screening with High Resolution Mass Spectrometry in the Environment: Ready to Go? *Environ. Sci. Technol.* **2017**, *51*, 11505–11512. [CrossRef] [PubMed]
2. Brodbelt, J.S. Photodissociation mass spectrometry: New tools for characterization of biological molecules. *Chem. Soc. Rev.* **2014**, *43*, 2757–2783. [CrossRef] [PubMed]
3. Morrison, L.J.; Parker, W.R.; Holden, D.D.; Henderson, J.C.; Boll, J.M.; Trent, M.S.; Brodbelt, J.S. UVliPiD: A UVPD-Based Hierarchical Approach for De Novo Characterization of Lipid A Structures. *Anal. Chem.* **2016**, *88*, 1812–1820. [CrossRef] [PubMed]
4. Huguet, R.; Stratton, T.; Weisbrod, C.; Berhow, M. The "ETD-like" Fragmentation of Small Molecules. In Proceedings of the ASMS, San Antonio, TX, USA, 5–9 June 2016.
5. Huguet, R.; Stratton, T.; Sharma, S.; Mullen, C.; Canterbury, J.; Zabrouskov, V. Utilizing UVPD Fragmentation for Plant Molecules: Phenylpropanoids. In Proceedings of the ASMS, Indianapolis, Indiana, 4–8 June 2017.
6. Mullen, C.; Weisbord, C.; Zhuk, E.; Huguet, R.; Schwartz, J. Implementation of 213 nm Ultra Violet Photodissociation (UVPD) on a Modified Orbitrap Fusion Lumos. In Proceedings of the ASMS, Indianapolis, Indiana, 4–8 June 2017.
7. Ruttkies, C.; Schymanski, E.L.; Wolf, S.; Hollender, J.; Neumann, S. MetFrag relaunched: Incorporating strategies beyond in silico fragmentation. *J. Cheminform.* **2016**, *8*, 3. [CrossRef] [PubMed]
8. Tukey, J.W. *Exploratory Data Analysis*; Addison-Wesley: Boston, MA, USA, 1977.
9. Karmaus, A.L.; Filer, D.L.; Martin, M.T.; Houck, K.A. Evaluation of food-relevant chemicals in the ToxCast high-throughput screening program. *Food Chem. Toxicol.* **2016**, *92*, 188–196. [CrossRef] [PubMed]
10. Guha, R. Chemical Informatics Functionality in R. *J. Stat. Softw.* **2007**, *18*, 16. [CrossRef]
11. Hartigan, J.A.; Wong, M.A. Algorithm AS 136: A K-Means Clustering Algorithm. *J. R. Stat. Soc. Ser. C (Appl. Stat.)* **1979**, *28*, 100–108. [CrossRef]
12. Charrad, M.; Ghazzali, N.; Boiteau, V.; Niknafs, A. NbClust: An R Package for Determining the Relevant Number of Clusters in a Data Set. *J. Stat. Softw.* **2014**, *61*, 1–36. [CrossRef]
13. Kohonen, T. Essentials of the self-organizing map. *Neural Netw.* **2013**, *37*. [CrossRef] [PubMed]
14. R Core Team. *R: A Language and Environment for Statistical Computing*; R Foundation for Statistical Computing: Vienna, Austria, 2017.

15. Trachsel, C.; Panse, C.; Kockmann, T.; Wolski, W.E.; Grossmann, J.; Schlapbach, R. rawDiag: An R Package Supporting Rational LC–MS Method Optimization for Bottom-up Proteomics. *J. Proteome Res.* **2018**, *17*, 2908–2914. [CrossRef] [PubMed]
16. Panse, C.; Grossmann, J. protViz: Visualizing and Analyzing Mass Spectrometry Related Data in Proteomics. R Package Version 0.6. 2020. Available online: https://CRAN.R-project.org/package=protViz (accessed on 12 September 2020).
17. RStudio Team. RStudio: Integrated Development for R. 2015. Available online: https://rstudio.com/products/rstudio/.

Sample Availability: Samples of the compounds are not available from the authors.

Article

Untargeted/Targeted 2D Gas Chromatography/Mass Spectrometry Detection of the Total Volatile Tea Metabolome

Joshua Morimoto [1], Marta Cialiè Rosso [2], Nicole Kfoury [1], Carlo Bicchi [2], Chiara Cordero [2] and Albert Robbat, Jr. [1,*]

[1] Department of Chemistry, Tufts University, Medford, MA 02155, USA; joshua.morimoto@tufts.edu (J.M.); nicole.kfoury@tufts.edu (N.K.)

[2] Dipartimento di Scienza e Tecnologia del Farmaco, Università degli Studi di Torino, 10125 Turin, Italy; marta.cialierosso@unito.it (M.C.R.); carlo.bicchi@unito.it (C.B.); chiara.cordero@unito.it (C.C.)

* Correspondence: albert.robbat@tufts.edu; Tel.: +1-617-627-3474

Received: 29 July 2019; Accepted: 15 October 2019; Published: 18 October 2019

Abstract: Identifying all analytes in a natural product is a daunting challenge, even if fractionated by volatility. In this study, comprehensive two-dimensional gas chromatography/mass spectrometry (GC×GC-MS) was used to investigate relative distribution of volatiles in green, pu-erh tea from leaves collected at two different elevations (1162 m and 1651 m). A total of 317 high and 280 low elevation compounds were detected, many of them known to have sensory and health beneficial properties. The samples were evaluated by two different software. The first, GC Image, used feature-based detection algorithms to identify spectral patterns and peak-regions, leading to tentative identification of 107 compounds. The software produced a composite map illustrating differences in the samples. The second, Ion Analytics, employed spectral deconvolution algorithms to detect target compounds, then subtracted their spectra from the total ion current chromatogram to reveal untargeted compounds. Compound identities were more easily assigned, since chromatogram complexities were reduced. Of the 317 compounds, for example, 34% were positively identified and 42% were tentatively identified, leaving 24% as unknowns. This study demonstrated the targeted/untargeted approach taken simplifies the analysis time for large data sets, leading to a better understanding of the chemistry behind biological phenomena.

Keywords: tea; metabolomics; GC/MS; software; database; MS subtraction; spectral deconvolution; 2DGC; volatilomics

1. Introduction

Obtaining the total metabolome of a natural or biological sample is a significant challenge. Even when analyzed by both gas and liquid chromatography (GC, LC), analyzing complex mixtures results in the detection of hundreds, if not thousands, of compounds. For example, we found that climate-induced tea plant (*Camellia sinensis* (L.) Kuntze) interactions cause significant differences in both the presence and concentration of secondary metabolites [1–3]. Abiotic pressures alone influence the relative distribution of more than 750 volatile compounds, with ~450 detected in any given sample. Differences in elevation, seasonal temperature, and rainfall from 2013 to 2016 caused two-thirds of the metabolites to either increase or decrease in concentration, one-third of them by more than 100%. We also measured nonvolatile concentration differences, including the catechins (flavan-3-ol derivatives with high antioxidant power). Only five days after the onset of the East Asian monsoon rains began, catechin concentrations decreased while total polyphenols and antioxidant potential increased. In these studies, we consider the tea "environmental," since we microwaved the leaves in the field to stop enzymatic oxidation; they were not processed to produce commercial tea.

We previously relied on automated sequential, 2D gas chromatography/mass spectrometry (GC-GC/MS) to build the environmental tea database and on GC/MS to quantify the analyte distribution in the samples. In this study, we used comprehensive 2D GC/MS (GC×GC/MS) to accomplish both tasks, building from the existing database. When comparing each technique, library-building by GC-GC/MS relied upon fifty 1-min sample transfers (heart-cuts) from the first (polar) to the second (nonpolar) column to obtain clean spectra. The total time to obtain 50 data files was 4.5 days. In contrast, GC×GC/MS, produced one data file in one hour. Although runtimes significantly differed, data analysis of the respective files for library-building purposes took about the same amount of time.

The terms targeted, untargeted, and feature detection are often applied to GC/MS and LC/MS data. In targeted analysis, a predetermined list of compounds is selected for the analysis, whereas untargeted involves the evaluation of all detectable compounds in the sample [4,5]. The following criteria affirm compound identity [6–8]. Positive identification requires confirmation by at least two independent measurements, such as sample and reference compound mass spectra and linear retention indices (LRI). Tentative identification is based on an acceptable match between sample and commercial library, database, or literature spectra. Because natural products and biological metabolomes contain hundreds to thousands of compounds [9], it is impractical to purchase reference standards to confirm all sample identities. Therefore, assignments are typically tentative for most compounds.

Given the numerous and diverse compounds in natural products and in metabolomic datasets, feature-based peak detection algorithms have become popular. *m/z* peak-retention time pairs are used to assess peak region commonalities and differences in large data sets to reduce data complexity. Because feature-based peak detection is overly sensitive to algorithmic parameters, coelution and instrument noise, it produces many more features than actual compounds [10–12], with less than 2% yielding identifiable analytes [13]. Although peak detection tools can differentiate samples [11,14,15], identifying why they differ is difficult, since features themselves provide no biological context [16,17]. Despite these drawbacks, feature-based detection software is popular among the metabolomics community [15].

In this study, we used GC×GC/MS to reveal compositional differences in farmer-processed green, pu-erh teas (herein called pu-erh), collected from the same farm, plants, elevations, and time-period as environmental tea. 2D data were analyzed by GC Image and Ion Analytics. The first was adopted to detect 2D-peaks and to delineate peak-regions covering the full chromatographic space. Similarities and differences between high and low elevation samples were highlighted by both peak-region feature distribution and visual feature rendering [18]. Additionally, we identified compounds where possible. The second provided an untargeted/targeted workflow, based on spectral deconvolution and spectral subtraction of analytes found in the sample, to produce clean mass spectra and quantifiable constituent differences. New algorithms facilitated the annotation process.

2. Results

2.1. Molecular Feature Detection

Figure 1 shows a comparative visualization, based on a visual features approach [18], highlighting compositional similarities and differences in the high (reference sample) and low (analyzed sample) elevation pu-erh teas. This pair-wise comparison is done on a composite chromatogram obtained by summing each sample's 2D chromatograms ($n = 2$) after transformation and re-alignment. In total, 1450 peak-regions were delineated; 107 with spectra clean enough to make tentative compound assignments, see Supplementary Table S1. Each pixel in Figure 1 corresponds to detector acquisition points (scans) that show response differences. The algorithm subtracts the pixel value registered in the reference image from the corresponding value in the analyzed image, with the difference divided by the larger of the pixel values. The brighter the pixel, the larger the difference is relative to the analyzed pixel value; the darker the pixel, the larger the difference is relative to the reference pixel value. Medium gray indicates the difference is small relative to the value. The hue of a pixel indicates whether the analyzed image (green) or reference image (magenta) has the higher value. The saturation (color vs. grey) of a pixel indicates the magnitude of the difference ratio between the analyzed and reference images, with grey indicating equal pixel values and bold colors large differences.

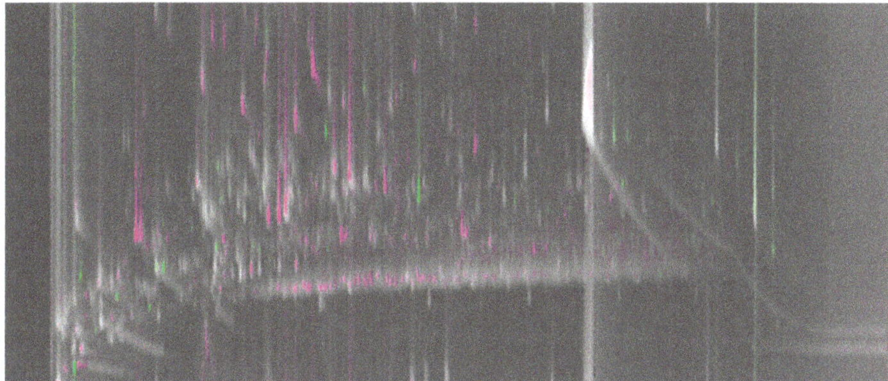

Figure 1. A comparative visualization between composite GC×GC/MS chromatograms of the high (reference image) and low (analyzed image) elevation teas. The magenta and green peak regions show the areas in the chromatogram where relative distribution of common analytes is higher in the high and low elevation teas, respectively. Light grey peak regions correspond to analytes with similar percentage response in the two chromatograms.

2.2. Untargeted/Targeted Analysis

Shown in Figure 2 are the (a) total and (b and c) reconstructed ion current (TIC and RIC) chromatograms for high elevation tea. Using the environmental database as target compounds, spectral deconvolution produced chromatogram (b), which reveals the 187 compounds in pu-erh tea that survived farmer processing. After subtracting the target compound spectra from the TIC chromatogram, untargeted analysis revealed another 130 compounds in the sample (c). Table 1 lists the breakdown of the number of compounds by their identification levels, showing 107, 132, and 78 positively and tentatively identified compounds, and unknowns in the high elevation pu-erh tea, respectively. Although 25% of the volatile constituents are unknowns, should statistical analysis reveal their importance, we know where in the separation to collect the compound by GC-GC for further analysis. In addition to retention and spectral information, the database includes sample type, the sensory characteristics of the compound in the sample, and known health benefits.

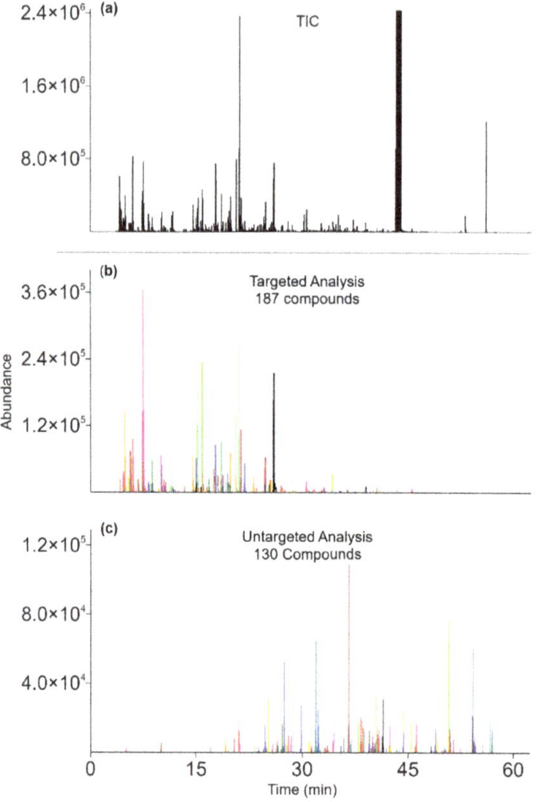

Figure 2. The TIC (**a**) and RIC chromatograms of targeted (**b**, in environmental tea) and untargeted (**c**, due to farmer processing) compounds in high elevation pu-erh tea.

Table 1. Targeted and untargeted compounds found by Ion Analytics.

	High Elevation		Low Elevation	
Identity Level	**Targeted**	**Untargeted**	**Targeted**	**Untargeted**
Positive	92	15	82	13
Tentative	78	54	69	43
* Unknown	17	61	15	58
Total	187	130	166	114

* Unknowns are assigned a numerical identifier in the database.

Compared to the 450 compounds we typically detect by stir bar sorptive extraction [19], head space solid phase microextraction (HS-SPME) produced a lower extraction yield, as expected, with 317 and 280 compounds in the high and low elevation samples, see Table 1. The high elevation sample contained 42 unique compounds; whereas only five were found in the low elevation teas. Also listed in the table are the number of targeted and untargeted compounds, see Supplementary Table S2 for identities. Once the total, detectable profile was established, GC Image (by pattern recognition) confirmed 272 of the 322 (84%) assignments, see Supplementary Table S1.

3. Discussion

Supplementary Figure S1 illustrates the process we use to analyze complex natural products such as botanicals, essential oils, herbs, and spices caused by environmental and/or manufacturing perturbations. In this study, we were interested in learning how sensory-active and health beneficial compounds in pu-erh tea differ in plants grown at high (1651 m) and low (1162 m) elevations. Since the leaves are processed in the same manner, differences in chemistry are primarily due to differences in temperature (Δ3 °C). Target compounds are leaf metabolites that survived farmer processing. Untargeted compounds are those produced by the fire-heated kettle process when making pu-erh tea. Spectral deconvolution of the target (environmental database) compounds, ~60% (187/317 and 166/280) of the constituents in each sample, followed by subtraction of their spectra simplified the analysis.

Figure 3 shows the TIC peak of three coeluting target compounds (acetic acid, 2-methylfuran, and hexane) in high elevation tea. Although the spectrum at each peak scan is different (Figure 3b), Ion Analytics correctly identified each analyte by deconvolving their spectral signature. Shown in the bottom panel are the RIC peaks for acetic acid (green), 2-methylfuran (pink), and hexane (blue), whose identities were confirmed using reference standards. The RIC peaks provide the means to measure the relative distribution of each analyte in the sample.

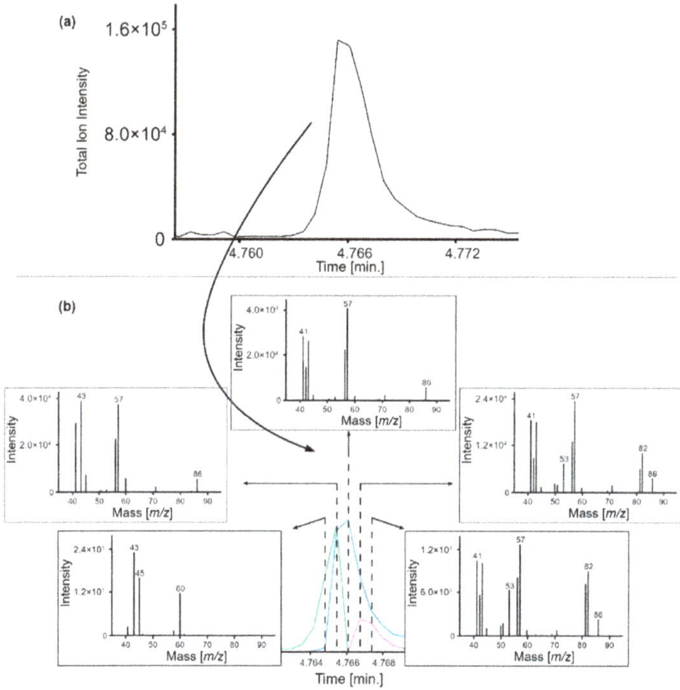

Figure 3. (**a**) Targeted analysis example of high elevation tea. (**b**) Spectral deconvolution of acetic acid (green), hexane (blue), and 2-methylfuran (pink) ions and relative abundances.

Figure 4 is an example of an untargeted analysis where the spectrum (c) for the target compound, hexanal, is subtracted from the TIC (a), see example spectra (b), exposing the spectrum of an unknown. Since spectrum (d) is constant at each peak scan, the MS subtraction algorithm reveals another compound. The retention time and mass spectrum match mesityl oxide (e). Subtracting the mesityl oxide reference spectrum from the remaining signal yields baseline noise (f and g), positively confirming the peak assignments and that no other compounds elute in the corresponding peak-region. Also shown in (g) are the RIC peaks for both hexanal (blue) and mesityl oxide (green).

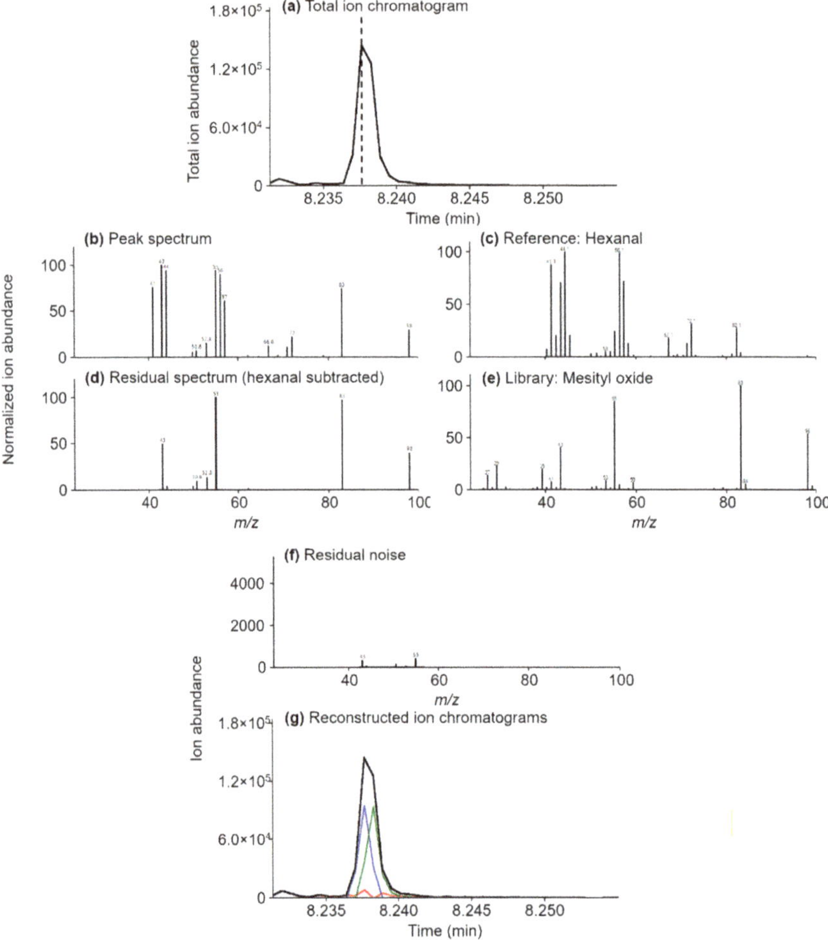

Figure 4. Untargeted analysis example, high elevation tea. Spectral deconvolution and MS subtraction of hexanal, the target compound, ions and relative abundances (**c**) from the TIC (**a**) peak spectra (**b**) yields the residual spectrum (**d**) for mesityl oxide (**e**). Subtraction of the ions and relative abundances of both compounds results in baseline noise (**f**,**g**). The RIC chromatograms for hexanal (blue) and mesityl oxide (green) are also in (**g**). Experimental spectra were acquired in the range of 40–280 *m*/*z*, and therefore, ions below 40 *m*/*z* are missing from spectra (**b**–**d**,**f**).

This approach reduces the complexity of each subsequent analysis by using the results of preceding samples as target compounds. By annotating the database, analytes can be tracked independent of sample type and from sample-to-sample, year-to-year. Both software provide data to analyze statistically. How one starts depends on the goal of the investigation. How one ends depends on whether speciation provides meaningful input into the system under investigation, see below.

Effects of Processing and Elevation on Tea

The most flavorful teas in Yunnan, China, are those grown at the highest elevations [20,21]. Earlier, we showed that high compared to low elevation "environmental" teas contained more and higher concentrations of metabolites that exhibit sweet, floral notes [1]. We assumed, therefore, that the farmer-processed pu-erh tea would as well. The results of this study support this finding. The high elevation tea retained more of the sensory-pleasing environmental metabolites than the low elevation tea, which possessed more of the earthy, harsher tasting aromatics. The comparative chromatogram, depicted in Figure 1, shows more magenta peaks overall, indicating a decrease in volatiles in the low elevation pu-erh compared to the high. Further examination with Ion Analytics showed that of the 275 common analytes, 134 were higher and 19 were lower in abundance in the high elevation tea at the 95% confidence interval of the average relative percent difference, namely, 23% ± 46% ($n = 2$ samples), with 119 and 60 higher than 100% and 200%, respectively. We found substantial concentration differences in the oxygenated monoterpenes. The high elevation tea contained more of the *trans*- (14000%) and *cis*- (3400%) furanoid linalool oxides, compounds characterized as sweet and floral, than the low elevation tea. Similarly, the *cis*- (700%) and *trans*- (750%) pyranoid linalool oxides, which exhibit citrusy, woody notes, were also higher in concentration.

Table 2 lists the 42 compounds unique to high elevation pu-erh and the five compounds unique to low elevation pu-erh, along with their sensory and/or health beneficial properties. Although high elevation spring teas are higher in quality, the occurrence of the summer monsoon rains offsets the elevational (temperature) effect, resulting in less desirable tea, as evidenced by farmers receiving 50% less for the summer compared to spring teas [22]. These findings illustrate why it is important to measure the total volatile profile, especially when evaluating the health benefits of green tea in clinical trials [23–26].

Table 2. Unique compounds in high and low elevation pu-erh tea and their sensory active and/or health beneficial properties.

High Elevation Compounds	Aroma *	Health Benefits
furfural	woody, almond, baked bread	—
18	—	—
(2E)-hexenal	green, banana, aldehydic	antimicrobial [27]
2-furanmethanol	sweet, caramel, burnt	—
(2E)-hexenol	leafy, fruity, unripe banana	—
2-heptanol	fruity, oily, fatty	—
2,5-dimethylpyrazine	cocoa, roasted nuts	—
2(5H)-furanone	buttery	—
heptanol	musty, leafy, herbal, peony	cardioprotective [28]
(3E)-hexenoic acid	fruity, honey, acidic	—
101	—	—
(3Z)-hexenyl acetate	green, banana, apple	—
heptanoic acid	rancid, sour, sweat	—
2-methoxyphenol	phenolic, smoke, spice	—
maltol	caramel, cotton candy, fruity	antianxiety [29], antioxidant [30]
114	—	—
511	—	—
(3Z)-hexenyl butyrate	green apple, fruity, wine	—
(2E)-hexenyl butyrate	green, apricot, ripe banana	—
hexyl butyrate	fruity, apple, waxy	—
512	—	—
514	—	—
nerol	neroli, citrus, magnolia	antibacterial [31], antifungal [32], antinociceptive/anti-inflammatory [33]
(3Z)-hexenyl valerate	apple, kiwi, unripe banana, tropical	—
(3Z)-hexenyl isovalerate	green apple, tropical, pineapple	—
phenylethyl acetate	rose, fruity	—
pentyl hexanoate	pineapple, apple, pear	—
1-nitro-2-phenyl ethane	floral, spice	cardioprotective [34]
γ-nonalactone	coconut, creamy, waxy, buttery	—

Table 2. *Cont.*

High Elevation Compounds	Aroma *	Health Benefits
(3Z)-hexenyl hexenoate	waxy, pear, winey, grassy, pineapple	—
hexyl hexanoate	fresh cut grass, vegetable	—
(2E)-hexenyl caproate	cognac, herbal, waxy	—
(Z)-jasmone	floral, woody, herbal, spicy	antibacterial [35], anticancer [36]
(E,E)-α-farnesene	citrus, lavender, bergamot,	—
2,4-di-tert-butylphenol	phenolic	antioxidant [37]
δ-cadinene	thyme, woody	—
(Z)-calamenene	herb, spice	antimalarial [38], antitumor [39]
dodecanoic acid	fatty, coconut, bay oil	cardioprotective [40], antibacterial/anti-inflammatory [41]
caryophyllene oxide	woody, spicy	anticancer/analgesic/anti-inflammatory [42]
τ-muurolol	herbal, spicy, honey	antibacterial [43], antioxidant [44]
α-cadinol	herbal, woody	antibacterial/antioxidant [43], anticancer [45], anti-inflammatory [45]
bancroftinone	—	—
Low Elevation Compounds	**Aroma ***	**Health Benefit**
ethyl acetate	weedy, green	—
isoamyl alcohol	alcoholic, banana	antifungal [46]
(2E)-pentenal	pungent, green apple, orange, tomato	—
m-ethyltoluene	—	—
118	—	—

* Aroma information was obtained from The Good Scents Company (http://thegoodscentscompany.com/).

4. Materials and Methods

4.1. Sample Collection

Tea samples (var. *assamica*) were collected from Nannuo Mountain, Menghai County in Yunnan Province, China, in 2014. Low elevation teas, grown at 1162 m, were collected from Xiang Yang Village from June 8—10. High elevation teas, grown at 1651 m, were collected from Ya Kou Old Village from June 10—12. Elevation differences correspond to a 3 °C cooler temperature at high elevation (https://www.spc.noaa.gov/exper/soundings/help/lapse.html). Farmers at the study site processed the samples as green, pu-erh tea [2,22].

4.2. Sample Preparation and HS-SPME

Added to a 20 mL glass vial, thermostatted at 50 °C, were 1.5 g of dried plant material and 2 mL of ultrapure water. Volatiles were sampled in the headspace of the vial by solid phase microextraction (SPME) for 50 min. The fibers were coated with 50/30 μm thick divinylbenzene/carboxen on polydimethyl siloxane (PDMS), which was 2 cm long (Supelco, Bellefonte, PA, USA). Fibers were preconditioned according to the manufacturer before the addition of internal standards, α- and β-thujone, which were sampled by exposing the SPME to 5 μL of a stock solution at 1000 mg/L for 20 min at 50 °C. The internal standards were used to normalize the GC×GC/MS peak responses. Two replicates were prepared per sample.

4.3. GC×GC/MS Instrumentation

Tea samples were analyzed using a Gerstel (Mülheim an der Ruhr, Germany) MPS-2 multipurpose sampler and Agilent (Santa Clara, CA, USA) models 6890 GC and 5975C MS. The MS, operated in electron ionization mode at 70 eV, scanned from 40 to 280 *m/z* at 30 Hz. The GC×GC was equipped with a two-stage KT 2004 loop thermal modulator (Zoex Corporation, Houston, TX, USA), which was cooled by liquid nitrogen. The hot jet pulse time was 250 ms. The modulation period was 4 s. A mass flow controller linearly reduced the cold-jet total flow from 40% to 8% by the end of the run. A deactivated fused silica capillary loop (1 m × 0.1 mm d_c) transferred sample portions from the first to the second column. The first column was an SE52 column (95% PDMS, 5% phenyl, 30 m × 0.25 mm × 0.25 μm), which was coupled to an OV1701 column (86% PDMS, 7% phenyl, 7% cyanopropyl, 1 m × 0.1 mm ×

0.10 μm) as the second column. Both columns were purchased from Mega (Legnano, Milan, Italy). The temperature program was 50 °C (1 min) to 210 °C at 3 °C/min, then at 280 °C (10 min) at 10 °C/min. The SPME fiber was thermally desorbed into the split/splitless injector for 5 min using a split ratio of 1:5. Helium served as the carrier gas, operating at constant flow (1.3 mL/min) with an initial head pressure of 298 kPa.

4.4. GC-GC/MS Instrumentation

Heart-cutting GC-GC/MS instrumentation and parameters for tea library-building have been described in detail in our previous works [1–3,19]. Briefly, PDMS-coated stir bars (Gerstel, Mülheim an der Ruhr, Germany) were used to extract volatiles from aqueous tea infusions. A thermal desorption unit (TDU, Gerstel) was used to provide splitless transfer of the sample into the programmable temperature vaporization inlet (CIS, Gerstel), held at −100 °C. The TDU was heated from 40 °C (0.70 min) to 275 °C (3 min) at 600 °C/min under 50 mL/min helium gas flow. After 0.1 min, the CIS was heated to 275 °C at 12 °C/min and held for 5 min. The first GC (Agilent 6890, Santa Clara, CA, USA) housed column 1 (C1, 30 m × 250 μm × 0.25 μm Rtx-Wax, Restek, Bellefonte, PA, USA) and was equipped with a flame ionization detector. The temperature of C1 was programmed from 40 °C (1 min) to 240 °C at 5 °C/min. C1 was connected to the CIS with a TDU on one end and to a five-port crosspiece (Gerstel) on the other. The second oven (Agilent 6890) housed column 2 (C2, 30 m × 250 μm × 0.25 μm Rxi-5MS, Restek), which was connected to the crosspiece through a cryogenic freeze trap (CTS1, Gerstel) on one end and to the MS (Agilent 5975) on the other. The oven temperature was held at 40 °C for 1 min, and then increased to 280 °C at a rate of 5 °C/min. Both columns operated at 1.2 mL/min constant helium flow. The MS operating conditions were: 70 eV electron ionization source, 230 °C ion source, 150 °C quadrupole, and 40 to 250 *m/z* scan range. A multipurpose autosampler (MPS, Gerstel) was used for automated sample injection, and a multicolumn switching device (MCS, Gerstel) supplied countercurrent flow to the crosspiece. A heart-cut was made each minute for a total of 40 heart-cuts per sample.

4.5. Data Analysis

GC Image v 2.8 (LLC, Lincoln, NE, USA) was used for untargeted/targeted fingerprinting based on peak-region features and for comparative visualization of composite chromatogram pairs for high and low elevation tea samples. Two replicate preparations of each sample were used in this approach. The software identified compounds when the LRI match was within ± 10 units of published values and the NIST spectral match for that compound was >950. After the analysis, positively matched peaks were used to re-align chromatograms [47,48] and combined into a single, composite chromatogram. 2D-peak detection was performed on the composite chromatogram with peak outlines recorded as peak-region objects in the chromatographic plane. Then, all metadata belonging to positively matched peaks and peak-region objects (chemical names, retention times, *m/z* fragmentation pattern, retention index, and additional information about matching results) were combined into a template of features.

We used the Ion Analytics software (Gerstel, Mülheim an der Ruhr, Germany and Andover, MA, USA) as follows. First, we selected the environmental tea database to target known compounds in unprocessed tea by spectral deconvolution of the chromatogram. For each target compound, the normalized ion intensity, $I_i(t)$ (relative to the user-defined main ion, usually the base ion, specified by $i = 1$), at scan (t) is

$$I_i(t) = \frac{A_i(t)}{R_i A_1}, \tag{1}$$

where $A_i(t)$ is the i-th qualifier ion intensity at scan (t) and R_i is the expected relative ion abundance ratio obtained from reference standards or NIST, Wiley, Adams, etc. All qualifier ions are normalized to the main ion ($I_1 = 1$). The spectral match, ΔI, is calculated as:

$$\Delta I = \frac{\sum_{i=1}^{N-1} \sum_{j=i+1}^{N} Abs\left(I_i - I_j\right)}{\sum_{i=1}^{N-1} i}, \tag{2}$$

where ΔI is the average normalized ion intensity deviation of each of the N specified qualifier ions. As ΔI approaches zero, the quality of the match increases. Between three to five ions, including the main ion, were used for each compound. Another constraint is that the scan-to-scan variance, ΔE, must be ≤ 7, and is calculated as:

$$\Delta E = \Delta I \cdot \log A_1, \tag{3}$$

Additive ion signal due to coelution is eliminated by comparing all ion ratios against each other. The relative error is computed at each scan. If the ion ratio exceeds the expected ratio, and if all other ions are in agreement, the residual signal is subtracted from the matrix-affected ion. An acceptable match fits the criterion:

$$\Delta I \leq K + \frac{\Delta_0}{A_1}, \tag{4}$$

where K is the adjustable percentage difference and Δ_0 is the cumulative error from background signal and/or instrument noise. Target compound identification occurs when ΔE or ΔI is ≤ 7 in at least three consecutive peak scans. Additionally, the qualifier ion ratio deviation must be $\leq 20\%$ at each scan across the peak to ensure consistency across the peak. The Q-value must be ≥ 90. The Q-value measures compound hit quality on a scale of integers 1 to 100, where a higher value indicates a higher quality match between sample and library spectra. The Q-value is determined as the maximum of either $100 - D$ or 1, where D is calculated as:

$$D = \frac{100 \sum_{i=1}^{N} \left| r_i^e - r_i^0 \right| \cdot \left(\log\left(100 \cdot r_i^e + 1\right)\right)^2}{21.3 \sum_{i=0}^{N} r_i^e}, \tag{5}$$

where r_i^e is the expected qualifier ion ratio for the i-th qualifier ion, r_i^0 is the observed qualifier ion ratio for the i-th qualifier ion, and N is the number of qualifier ions. The Q-ratio must also be within $\leq$ 20% of the relative abundance. The Q-ratio is the peak area ratio of the extracted i-th and main ions, calculated as:

$$D = \frac{100 \sum_{i=1}^{N} \left| r_i^e - r_i^0 \right| \cdot \left(\log\left(100 \cdot r_i^e + 1\right)\right)^2}{21.3 \sum_{i=0}^{N} r_i^e}, \tag{6}$$

where S_i is the peak area extracted for the i-th common ion over the hit and S_1 is the peak area for the main ion over the hit. Since GC×GC produces multiple modulated peaks per compound, the deconvolution algorithm searches for up to five peaks per compound by default, which can be increased when high concentration analytes are found in the sample.

Then, the mass spectrum of each target compound was subtracted from corresponding peak scans to reveal untargeted compounds. Peak assignments were made by matching sample and library spectra and retention indices, which were then confirmed using available reference standards. If assignments were not possible, numerical identifiers were used to assign peak names. Finally, once all peak assignments were made, the mass spectrum for each compound was subtracted from the TIC chromatogram to reveal missed peaks. This was established when residual signals approximated the baseline (background noise) of the chromatogram.

5. Conclusions

Our overall objective is to demonstrate two complimentary data analysis approaches to differentiating complex samples using tea as the model system. First, we used molecular feature detection to show which regions of the chromatograms differed, then untargeted/targeted analysis to identify all compounds in the sample, both in regions of the chromatogram that differed significantly and where it did not. The latter showed the quantifiable differences at the molecular level in high and low elevation tea. From a sensory perspective, our findings point to compounds consistent with local perspectives of quality in high and low elevation green tea. In this paper, we used comprehensive GC×GC/MS, rather than GC/MS or GC-GC/MS, to produce data on the same time scale as GC/MS with the resolution of GC-GC/MS, without taking days to produce data on one sample. The technology and approach taken differentiates complex samples at the molecular level, critically important to the study of systems biology.

Supplementary Materials: The following are available online, Figure S1: Ion Analytics and GC Image workflow schematic for the analysis of targeted and untargeted compounds, Table S1: Features detected by GC Image and confirmatory analysis, Table S2: Compounds identified by Ion Analytics. The unprocessed GC×GC /MS data can be found at: http://dx.doi.org/10.25833/jxyj-8w58.

Author Contributions: Conceptualization, A.R., C.C. and C.B.; methodology, A.R., C.C., and C.B.; software, A.R. and N.K.; validation, J.M., M.C.R. and C.C.; formal analysis, J.M. and M.C.R.; investigation, M.C.R. and N.K.; resources, A.R., C.C., and C.B.; data curation, J.M. and M.C.R.; writing—original draft preparation, J.M. and A.R.; writing—review and editing, J.M., A.R., N.K., M.C.R., and C.C.; visualization, J.M., A.R., M.C.R., and C.C.; supervision, A.R., C.C., and C.B.; project administration, A.R.; funding acquisition, A.R.

Funding: This work did not receive any specific funding from public, commercial, or not-for-profit agencies.

Acknowledgments: We thank Stephen Reichenbach and Qingping Tao for providing GC Image support.

Conflicts of Interest: A.R. is the founder of Ion Analytics and developer of the software.

References

1. Kfoury, N.; Morimoto, J.; Kern, A.; Scott, E.R.; Orians, C.M.; Ahmed, S.; Griffin, T.; Cash, S.B.; Stepp, J.R.; Xue, D.Y.; et al. Striking changes in tea metabolites due to elevational effects. *Food Chem.* **2018**, *264*, 334–341. [CrossRef] [PubMed]
2. Kowalsick, A.; Kfoury, N.; Robbat, A., Jr.; Ahmed, S.; Orians, C.; Griffin, T.; Cash, S.B.; Stepp, J.R. Metabolite profiling of Camellia sinensis by automated sequential, multidimensional gas chromatography/mass spectrometry reveals strong monsoon effects on tea constituents. *J. Chromatogr. A* **2014**, *1370*, 230–239. [CrossRef] [PubMed]
3. Robbat, A.; Kfoury, N.; Baydakov, E.; Gankin, Y. Optimizing targeted/untargeted metabolomics by automating gas chromatography/mass spectrometry workflows. *J. Chromatogr. A* **2017**, *1505*, 96–105. [CrossRef]
4. Cordero, C.; Liberto, E.; Bicchi, C.; Rubiolo, P.; Reichenbach, S.E.; Tian, X.; Tao, Q.P. Targeted and Non-Targeted Approaches for Complex Natural Sample Profiling by GCxGC-qMS. *J. Chromatogr. Sci.* **2010**, *48*, 251–261. [CrossRef]
5. Skogerson, K.; Wohlgemuth, G.; Barupal, D.K.; Fiehn, O. The volatile compound BinBase mass spectral database. *BMC Bioinf.* **2011**, *12*, 321. [CrossRef]
6. Sumner, L.W.; Amberg, A.; Barrett, D.; Beale, M.H.; Beger, R.; Daykin, C.A.; Fan, T.W.M.; Fiehn, O.; Goodacre, R.; Griffin, J.L.; et al. Proposed minimum reporting standards for chemical analysis. *Metabolomics* **2007**, *3*, 211–221. [CrossRef]
7. Creek, D.J.; Dunn, W.B.; Fiehn, O.; Griffin, J.L.; Hall, R.D.; Lei, Z.; Mistrik, R.; Neumann, S.; Schymanski, E.L.; Sumner, L.W.; et al. Metabolite identification: are you sure? And how do your peers gauge your confidence? *Metabolomics* **2014**, *10*, 350–353. [CrossRef]
8. Sumner, L.W.; Lei, Z.; Nikolau, B.J.; Saito, K.; Roessner, U.; Trengove, R. Proposed quantitative and alphanumeric metabolite identification metrics. *Metabolomics* **2014**, *10*, 1047–1049. [CrossRef]
9. Alseekh, S.; Fernie, A.R. Metabolomics 20 years on: What have we learned and what hurdles remain? *Plant J.* **2018**, *94*, 933–942. [CrossRef]

10. Mahieu, N.G.; Patti, G.J. Systems-Level Annotation of a Metabolomics Data Set Reduces 25,000 Features to Fewer than 1000 Unique Metabolites. *Anal. Chem.* **2017**, *89*, 10397–10406. [CrossRef]
11. Tautenhahn, R.; Bottcher, C.; Neumann, S. Highly sensitive feature detection for high resolution LC/MS. *BMC Bioinf.* **2008**, *9*, 504. [CrossRef] [PubMed]
12. Stolt, R.; Torgrip, R.J.O.; Lindberg, J.; Csenki, L.; Kolmert, J.; Schuppe-Koistinen, I.; Jacobsson, S.P. Second-order peak detection for multicomponent high-resolution LC/MS data. *Anal. Chem.* **2006**, *78*, 975–983. [CrossRef] [PubMed]
13. da Silva, R.R.; Dorrestein, P.C.; Quinn, R.A. Illuminating the dark matter in metabolomics. *P. Natl. Acad. Sci. USA* **2015**, *112*, 12549–12550. [CrossRef] [PubMed]
14. Smith, C.A.; Want, E.J.; O'Maille, G.; Abagyan, R.; Siuzdak, G. XCMS: Processing mass spectrometry data for metabolite profiling using Nonlinear peak alignment, matching, and identification. *Anal. Chem.* **2006**, *78*, 779–787. [CrossRef]
15. Weber, R.J.M.; Lawson, T.N.; Salek, R.M.; Ebbels, T.M.D.; Glen, R.C.; Goodacre, R.; Griffin, J.L.; Haug, K.; Koulman, A.; Moreno, P.; et al. Computational tools and workflows in metabolomics: An international survey highlights the opportunity for harmonisation through Galaxy. *Metabolomics* **2016**, *13*, 12. [CrossRef]
16. Papadimitropoulos, M.-E.P.; Vasilopoulou, C.G.; Maga-Nteve, C.; Klapa, M.I. Untargeted GC-MS Metabolomics. In *Metabolic Profiling: Methods and Protocols*; Theodoridis, G.A., Gika, H.G., Wilson, I.D., Eds.; Springer: New York, NY, USA, 2018; pp. 133–147.
17. Cho, K.; Mahieu, N.G.; Johnson, S.L.; Patti, G.J. After the feature presentation: technologies bridging untargeted metabolomics and biology. *Curr. Opin. Biotechnol.* **2014**, *28*, 143–148. [CrossRef]
18. Reichenbach, S.E.; Tian, X.; Cordero, C.; Tao, Q. Features for non-targeted cross-sample analysis with comprehensive two-dimensional chromatography. *J. Chromatogr. A* **2012**, *1226*, 140–148. [CrossRef]
19. Kfoury, N.; Baydakov, E.; Gankin, Y.; Robbat, A. Differentiation of key biomarkers in tea infusions using a target/nontarget gas chromatography/mass spectrometry workflow. *Food Res. Int.* **2018**, *113*, 414–423. [CrossRef]
20. Han, W.Y.; Huang, J.G.; Li, X.; Li, Z.X.; Ahammed, G.J.; Yan, P.; Stepp, J.R. Altitudinal effects on the quality of green tea in east China: a climate change perspective. *Eur. Food Res. Technol.* **2017**, *243*, 323–330. [CrossRef]
21. Chen, G.-H.; Yang, C.-Y.; Lee, S.-J.; Wu, C.-C.; Tzen, J.T.C. Catechin content and the degree of its galloylation in oolong tea are inversely correlated with cultivation altitude. *J. Food Drug Anal.* **2014**, *22*, 303–309. [CrossRef]
22. Ahmed, S.; Stepp, J.R.; Orians, C.; Griffin, T.; Matyas, C.; Robbat, A.; Cash, S.; Xue, D.Y.; Long, C.L.; Unachukwu, U.; et al. Effects of Extreme Climate Events on Tea (Camellia sinensis) Functional Quality Validate Indigenous Farmer Knowledge and Sensory Preferences in Tropical China. *PLoS ONE* **2014**, *9*, e109126. [CrossRef] [PubMed]
23. Chen, I.J.; Liu, C.-Y.; Chiu, J.-P.; Hsu, C.-H. Therapeutic effect of high-dose green tea extract on weight reduction: A randomized, double-blind, placebo-controlled clinical trial. *Clin. Nutr.* **2016**, *35*, 592–599. [CrossRef] [PubMed]
24. Tsao, A.S.; Liu, D.; Martin, J.; Tang, X.-M.; Lee, J.J.; El-Naggar, A.K.; Wistuba, I.; Culotta, K.S.; Mao, L.; Gillenwater, A.; et al. Phase II Randomized, Placebo-Controlled Trial of Green Tea Extract in Patients with High-Risk Oral Premalignant Lesions. *Cancer Prev. Res.* **2009**, *2*, 931–941. [CrossRef] [PubMed]
25. Gee, J.R.; Saltzstein, D.R.; Kim, K.; Kolesar, J.; Huang, W.; Havighurst, T.C.; Wollmer, B.W.; Stublaski, J.; Downs, T.; Mukhtar, H.; et al. A Phase II Randomized, Double-blind, Presurgical Trial of Polyphenon E in Bladder Cancer Patients to Evaluate Pharmacodynamics and Bladder Tissue Biomarkers. *Cancer Prev. Res.* **2017**, *10*, 298–307. [CrossRef]
26. Samavat, H.; Ursin, G.; Emory, T.H.; Lee, E.; Wang, R.; Torkelson, C.J.; Dostal, A.M.; Swenson, K.; Le, C.T.; Yang, C.S.; et al. A Randomized Controlled Trial of Green Tea Extract Supplementation and Mammographic Density in Postmenopausal Women at Increased Risk of Breast Cancer. *Cancer Prev. Res.* **2017**, *10*, 710–718. [CrossRef]
27. Battinelli, L.; Daniele, C.; Cristiani, M.; Bisignano, G.; Saija, A.; Mazzanti, G. In vitro antifungal and anti-elastase activity of some aliphatic aldehydes from *Olea europaea* L. fruit. *Phytomedicine* **2006**, *13*, 558–563. [CrossRef]
28. Tse, G.; Yeo, J.M.; Tse, V.; Kwan, J.; Sun, B. Gap junction inhibition by heptanol increases ventricular arrhythmogenicity by reducing conduction velocity without affecting repolarization properties or myocardial refractoriness in Langendorff-perfused mouse hearts. *Mol. Med. Rep.* **2016**, *14*, 4069–4074. [CrossRef]

29. Kumar, D.; Kumar, S. Isolation and Characterization of Bioactive Phenolic Compounds from Abies Pindrow Aerial Parts. *Pharm. Chem. J.* **2017**, *51*, 205–210. [CrossRef]
30. Shiratsuchi, H.; Chang, S.; Wei, A.; El-Ghorab, A.H.; Shibamoto, T. Biological activities of low-molecular weight compounds found in foods and plants. *J. Food Drug Anal.* **2012**, *20*, 359–365.
31. Iscan, G. Antibacterial and Anticandidal Activities of Common Essential Oil Constituents. *Rec. Nat. Prod.* **2017**, *11*, 374–388.
32. Miron, D.; Battisti, F.; Silva, F.K.; Lana, A.D.; Pippi, B.; Casanova, B.; Gnoatto, S.; Fuentefria, A.; Mayorga, P.; Schapoval, E.E.S. Antifungal activity and mechanism of action of monoterpenes against dermatophytes and yeasts. *Rev. Bras. Farm.* **2014**, *24*, 660–667. [CrossRef]
33. González-Ramírez, A.E.; González-Trujano, M.E.; Orozco-Suárez, S.A.; Alvarado-Vásquez, N.; López-Muñoz, F.J. Nerol alleviates pathologic markers in the oxazolone-induced colitis model. *Eur. J. Pharmacol.* **2016**, *776*, 81–89. [CrossRef] [PubMed]
34. de Lima, A.B.; Santana, M.B.; Cardoso, A.S.; da Silva, J.K.R.; Maia, J.G.S.; Carvalho, J.C.T.; Sousa, P.J.C. Antinociceptive activity of 1-nitro-2-phenylethane, the main component of Aniba canelilla essential oil. *Phytomedicine* **2009**, *16*, 555–559. [CrossRef] [PubMed]
35. Muroi, H.; Kubo, I. Combination Effects of Antibacterial Compounds in Green Tea Flavor Against *Streptococcus-mutans*. *J. Agric. Food Chem.* **1993**, *41*, 1102–1105. [CrossRef]
36. Tong, Q.S.; Jiang, G.S.; Zheng, L.D.; Tang, S.T.; Cai, J.B.; Liu, Y.; Zeng, F.Q.; Dong, J.H. Natural jasmonates of different structures suppress the growth of human neuroblastoma cell line SH-SY5Y and its mechanisms. *Acta Pharmacol. Sin.* **2008**, *29*, 861–869. [CrossRef]
37. Choi, S.J.; Kim, J.K.; Kim, H.K.; Harris, K.; Kim, C.-J.; Park, G.G.; Park, C.-S.; Shin, D.-H. 2,4-Di-tert-butylphenol from Sweet Potato Protects Against Oxidative Stress in PC12 Cells and in Mice. *J. Med. Food* **2013**, *16*, 977–983. [CrossRef]
38. Afoulous, S.; Ferhout, H.; Raoelison, E.G.; Valentin, A.; Moukarzel, B.; Couderc, F.; Bouajila, J. Helichrysum gymnocephalum Essential Oil: Chemical Composition and Cytotoxic, Antimalarial and Antioxidant Activities, Attribution of the Activity Origin by Correlations. *Molecules* **2011**, *16*, 8273–8291. [CrossRef]
39. Takei, M.; Umeyama, A.; Arihara, S. T-cadinol and calamenene induce dendritic cells from human monocytes and drive Th1 polarization. *Eur. J. Pharmacol.* **2006**, *537*, 190–199. [CrossRef]
40. Alves Naiane Ferraz, B.; Queiroz Thyago, M.; Almeida Travassos, R.; Magnani, M.; Andrade Braga, V. Acute Treatment with Lauric Acid Reduces Blood Pressure and Oxidative Stress in Spontaneously Hypertensive Rats. *Basic Clin. Pharmacol.* **2016**, *120*, 348–353. [CrossRef]
41. Wang, J.; Lu, J.; Xie, X.; Xiong, J.; Huang, N.; Wei, H.; Jiang, S.; Peng, J. Blend of organic acids and medium chain fatty acids prevents the inflammatory response and intestinal barrier dysfunction in mice challenged with enterohemorrhagic Escherichia coli O157:H7. *Int. Immunopharmacol.* **2018**, *58*, 64–71. [CrossRef]
42. Nguyen, L.T.; Myslivečková, Z.; Szotáková, B.; Špičáková, A.; Lněničková, K.; Ambrož, M.; Kubíček, V.; Krasulová, K.; Anzenbacher, P.; Skálová, L. The inhibitory effects of β-caryophyllene, β-caryophyllene oxide and α-humulene on the activities of the main drug-metabolizing enzymes in rat and human liver in vitro. *Chem. Biol. Interact.* **2017**, *278*, 123–128. [CrossRef] [PubMed]
43. Guerrini, A.; Sacchetti, G.; Grandini, A.; Spagnoletti, A.; Asanza, M.; Scalvenzi, L. Cytotoxic Effect and TLC Bioautography-Guided Approach to Detect Health Properties of Amazonian Hedyosmum sprucei Essential Oil. *Evid. Based Complement. Altern. Med.* **2016**, *8*. [CrossRef]
44. Rossi, D.; Guerrini, A.; Maietti, S.; Bruni, R.; Paganetto, G.; Poli, F.; Scalvenzi, L.; Radice, M.; Saro, K.; Sacchetti, G. Chemical fingerprinting and bioactivity of Amazonian Ecuador *Croton lechleri* Mull. Arg. (Euphorbiaceae) stem bark essential oil: A new functional food ingredient? *Food Chem.* **2011**, *126*, 837–848. [CrossRef]
45. Tung, Y.T.; Yen, P.L.; Lin, C.Y.; Chang, S.T. Anti-inflammatory activities of essential oils and their constituents from different provenances of indigenous cinnamon (Cinnamomum osmophloeum) leaves. *Pharm. Biol.* **2010**, *48*, 1130–1136. [CrossRef] [PubMed]
46. Pimenta, R.S.; Silva, J.F.M.D.; Buyer, J.S.; Jansiewicz, W.J. Endophytic Fungi from Plums (Prunus domestica) and Their Antifungal Activity against Monilinia fructicola. *J. Food Prot.* **2012**, *75*, 1883–1889. [CrossRef] [PubMed]

47. Reichenbach, S.E.; Tian, X.; Boateng, A.A.; Mullen, C.A.; Cordero, C.; Tao, Q.P. Reliable Peak Selection for Multisample Analysis with Comprehensive Two-Dimensional Chromatography. *Anal. Chem.* **2013**, *85*, 4974–4981. [CrossRef]
48. Rempe, D.W.; Reichenbach, S.E.; Tao, Q.P.; Cordero, C.; Rathbun, W.E.; Zini, C.A. Effectiveness of Global, Low-Degree Polynomial Transformations for GCxGC Data Alignment. *Anal. Chem.* **2016**, *88*, 10028–10035. [CrossRef]

Sample Availability: Samples of the compounds are not available from the authors.

Article

A New UPLC-qTOF Approach for Elucidating Furan and 2-Methylfuran Metabolites in Human Urine Samples after Coffee Consumption

Simone Stegmüller [1], Nadine Beißmann [1], Jonathan Isaak Kremer [1], Denise Mehl [2], Christian Baumann [2] and Elke Richling [1,*]

[1] Technische Universität Kaiserslautern, Department of Chemistry, Division of Food Chemistry and Toxicology, Erwin-Schrödinger-Str. 52, 67663 Kaiserslautern, Germany; stegmueller@chemie.uni-kl.de (S.S.); beissmann@chemie.uni-kl.de (N.B.); kremer@chemie.uni-kl.de (J.I.K.)

[2] AB SCIEX Germany GmbH, 64293 Darmstadt, Germany; denise.mehl@sciex.com (D.M.); Christian.baumann@sciex.com (C.B.)

* Correspondence: richling@chemie.uni-kl.de; Tel./Fax.: +0049-631-205-4061 (ext. 3085)

Academic Editor: Thomas Letzel
Received: 6 October 2020; Accepted: 30 October 2020; Published: 3 November 2020

Abstract: We have investigated urine samples after coffee consumption using targeted and untargeted approaches to identify furan and 2-methylfuran metabolites in urine samples by UPLC-qToF. The aim was to establish a fast, robust, and time-saving method involving ultra-performance liquid chromatography-quantitative time-of-flight tandem mass spectrometry (UPLC-qToF-MS/MS). The developed method detected previously reported metabolites, such as Lys-BDA, and others that had not been previously identified, or only detected in animal or in vitro studies. The developed UPLC-qToF method detected previously reported metabolites, such as lysine-*cis*-2-butene-1,4-dial (Lys-BDA) adducts, and others that had not been previously identified, or only detected in animal and in vitro studies. In sum, the UPLC-qToF approach provides additional information that may be valuable in future human or animal intervention studies.

Keywords: furan; 2-methylfuran; UPLC-qToF; untargeted analysis; urinary metabolites

1. Introduction

Furan and 2-methylfuran (2-MF) are colorless liquids that are very slightly soluble in water. Furan has been classified by the International Agency for Research on Cancer (IARC) as a possible carcinogen (class 2B) [1,2].

Furan is mainly found in coffee and jarred food. Therefore, in Europe adults and infants reportedly have the highest furan intakes, and 95% of the adults' daily intake is reportedly through coffee, as the mean level of furan in whole roasted coffee beans is 4.58 mg/kg [3]. The furan content of coffee beverages depends on various factors, such as roasting temperature and duration, preparation method, and use of instant or ground coffee. Coffee beverages prepared by an espresso machine have higher furan levels than coffee prepared, from the same ground coffee, by a drip coffee maker (open system) [4,5]. The 2-MF content has been found to be ca. six-fold higher than the furan content and is the main furan derivative identified in coffee [6].

Furan and 2-MF are formed in food via various pathways and precursors, such as ascorbic acid, amino acids, carbohydrates, unsaturated fatty acids, and carotenoids. Furan formation also occurs during the thermal decomposition of serine and cysteine [7–9]. Furthermore, 2-MF can be formed from amino acid and α,β-unsaturated aldehydes [9].

There is no experimental evidence of furan genotoxicity in peripheral blood or bone marrow cells but chromosomal aberrations linked to its exposure have been detected in rat splenocytes, indicating that it may damage DNA in the liver within 28 days [10]. Moreover, in a two-year study administration by gavage reportedly increased incidences of mononuclear cell leukemia, cholangiocarcinoma, and hepatocellular neoplasms in livers of male and female rats [11]. However, hepatotoxicity and carcinogenicity depend on doses and time [12], and for hepatocarcinogenicity furan must be activated to *cis*-2-butene-1,4-dial (BDA) by cytochrome P450 2E1 [13].

Bioactivation also seems to be necessary for 2-MF activity [14]. In hepatic and pulmonary microsomal systems acetylacrolein (AcA; (2Z)-oxopent-2-enal) has been identified as a possible reactive intermediate of 2-MF [15]. Apoptotic hepatocytes, abnormally pigmented Kupffer cells, and inflammatory infiltration have been observed in rats after 2-MF intake [16], and a No Observed Adverse Effect Level (NOAEL) of 0.4 mg/kg bodyweight (BW) per day has been calculated for it [6].

Due to the high reactivity of the main metabolites, BDA and AcA, of furan and 2-MF, they can react in the human body with diverse biomolecules, such as amino acids, glutathione, and nucleosides. Thus, the aims of this study were to identify adducts of BDA and AcA in the urine of participants who had previously consumed a coffee brew and establish excretion kinetics for identified metabolites. Since the examination and evaluation of the collected urine samples with a targeted approach using conventional high-performance liquid chromatography-tandem mass spectrometry (HPLC-MS/MS) analysis would be very time-consuming and costly, a new approach using ultra-performance LC-quantitative time-of-flight MS/MS was applied. This approach requires less laborious sample preparation, and provides exact masses of analytes, with the associated possibility of database searches, recording all MS/MS spectra over the set mass range, as well as a significant reduction in measuring time. The urine samples tested had already been examined in a previous study with a targeted approach [17,18]. The results obtained in the study presented here are intended to complement these data by identifying previously unknown metabolites and show whether the new approach can provide at least as good results as the targeted approach.

2. Results and Discussion

The main objective of the study was to determine the effectiveness of the untargeted UPLC-qToF-MS/MS approach for identifying known and unknown metabolites in human urine without complex sample preparation and the possibility of using it to complement or even replace targeted UPLC-MS/MS analyses. A major advantage of the method is that a large amount of data is collected for each sample and can be retrospectively used to address various questions.

The specific aims of this study were to identify furan and 2-MF in metabolites in samples of urine collected from participants 36 h after they had consumed a bolus of coffee brew. Details of the study design have been previously published [17]. Due to limitations in time to use the qToF apparatus, only urine samples from five of the 10 participants were used in the UPLC-qToF-MS/MS measurements. The results were compared with results from previous targeted investigations. An overview of the identified metabolites of furan and 2-methylfuran are given in Tables 1 and 2.

Table 1. Exact mass, found mass, and retention time of each furan and 2-methylfuran metabolite identified by UPLC-qToF-MS/MS, and matrices in which they have been previously found (with supporting references). Meanings of the abbreviated metabolites' names are given in the text or present the meanings in a footnote.

Metabolite	Chemical Formula	Exact Mass (Da)	Found at *m/z*	Retention Time (min)	Matrices where Previously Found and References
GSH-BDA	$C_{14}H_{17}N_3O_6S$	355.083809	356.1000 $[M + H]^+$	7.63	human urine [19–21], urine, bile, and hepatocytes of rats [22]
Lys-BDA	$C_{10}H_{16}N_2O_3$	212.11609	213.1232 $[M + H]^+$	2.27	human urine [20,23]
AcLys-BDA	$C_{12}H_{18}N_2O_4$	254.12666	255.1345 $[M + H]^+$	4.36	human urine [20,23]; rat urine [19,23]
AcCys-BDA-Lys	$C_{15}H_{23}N_3O_5S$	357.13584	356.1167 $[M - H]^-$	7.58	human urine and rat urine [23]; rat urine [23,24]; rat hepatocytes [22]
AcCys-BDA-AcLys	$C_{17}H_{25}N_3O_6S$	399.14641	400.1471 $[M + H]^+$	1.24	urine, bile, and hepatocytes of rats [19,22–24]
AcCys (SO)-BDA-AcLys	$C_{17}H_{25}N_3O_7S$	415.14132	416.1484 $[M + H]^+$	4.46	rat urine [19,23,24]
methanethiol-BDA-Glu	$C_{10}H_{13}NO_4S$	243.05653	244.0793 $[M + H]^+$	1.05	rat urine [19,24]
GSH-BDA-Lys	$C_{20}H_{31}N_5O_8S$	501.18933	502.1740 $[M + H]^+$	7.57	rat urine [25]; rat hepatocytes [19,22]
GSH-BDA-Gln	$C_{19}H_{27}N_5O_9S$	501.15295	500.1478 $[M - H]^-$	7.59	rat urine [25]; rat hepatocytes [19,22]
GSH-BDA-AcLys	$C_{22}H_{33}N_5O_9S$	543.19990	544.2036 $[M + H]^+$	7.59	rat hepatocytes [19,22]
GSH-BDA-Glu	$C_{19}H_{26}N_4O_{10}S$	502.13696	503.1497 $[M + H]^+$	1.47	rat bile [19,22,26]
Cys-BDA-GSH	$C_{17}H_{24}N_4O_8S_2$	476.10356	475.0941 $[M - H]^-$	7.68	rat bile [19,26]
CysGyl-BDA-GSH	$C_{19}H_{27}N_5O_9S_2$	533.12502	532.1480 $[M - H]^-$	7.60	rat bile [19,26]
BDA-Lys-BDA	$C_{14}H_{18}N_2O_4$	278.12666	277.1187 $[M - H]^-$	7.57	not previously reported, but postulated
Cys-BDA	$C_7H_9NO_3S$	187.03031	188.0384 $[M + H]^+$	3.32	not yet reported in the literature, only theoretical

Table 2. Exact mass, found mass, and retention time of each furan and 2-methylfuran metabolite identified by UPLC-qToF-MS/MS, and matrices in which they have been previously found (with supporting references). Meanings of the abbreviated metabolites' names are given in the text or present the meanings in a footnote.

Metabolite	Chemical Formula	Exact Mass (Da)	Found at *m/z*	Retention Time (min)	Previously Detected in (References)
AcCys-BDA	$C_9H_{11}NO_5S$	245.25234	246.0431 $[M + H]^+$	1.46	not previously reported, but postulated
Lys-AcA	$C_{11}H_{20}N_2O_3$	228.14739	227.1386 $[M - H]^-$	3.87	not previously reported, but postulated
AcLys-AcA	$C_{13}H_{20}N_2O_4$	268.30890	267.1258 $[M - H]^-$	7.65	not previously reported, but postulated
AcCys-AcA	$C_{10}H_{13}NO_5S$	259.05144	258.0441 $[M - H]^-$	7.58	not previously reported, but postulated

2.1. Identification of Furan-Derived Metabolites

In previous studies by our group, the metabolites lysine-BDA (Lys-BDA), acetyl lysine-BDA (AcLys-BDA), and glutathione-BDA (GSH-BDA) were identified and quantified in samples of urine collected from the participants at various times after coffee intake [20]. The resulting kinetics are shown in Figure 1.

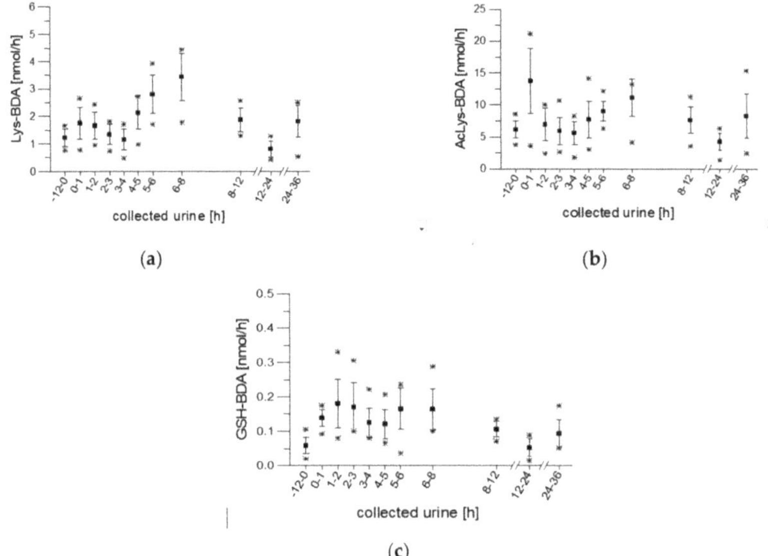

Figure 1. Excretion kinetics of Lys-BDA (**a**), AcLys-BDA (**b**), and GSH-BDA (**c**): HPLC-ESI-MS/MS measurements of contents in samples of urine from participants 1 to 10 at indicated times after coffee consumption (h). Data are previously published means with standard deviations and minimum/maximum values [20].

With the UPLC-qToF-MS/MS analysis, we detected Lys-BDA, NAcLys-BDA, and BDA-GSH in at least one urine sample from each participant at each selected time point, and observed similar changes with time in their concentrations (see Figure 2).

Using UPLC-qToF-MS/MS, we were able to determine exact masses and obtained MS^2-spectra for all three metabolites, which enabled us to confirm their structures. Furan-derived metabolites that have been detected in urine samples from rats after furan exposure were also detected in the urine samples of the participants using this approach. These included acetylcysteine butendial lysine (AcCys-BDA-Lys), acetylcysteine-SO-butendial acetyllysine (AcCys-(SO)-BDA-AcLys), methanethiol butendial glutamic acid (methanethiol-BDA-Glu), glutathione butendial lysine (GSH-BDA-Lys), and glutathione butendial glutamine (GSH-BDA-Gln). The metabolites AcCys-BDA-Lys and AcCys-(SO)-BDA-AcLys have already been detected in urine from smokers [23] but could not be quantified in the cited study as the signals were below the respective limits of quantification (LOQs). In the study presented here, we identified both metabolites by exact masses and MS^2 spectra, and obtained semi-quantitative kinetics, which correlated with the coffee intake, for the two metabolites in the urine samples of five participants (numbers three to six and eight; as shown in Figures 3 and 4).

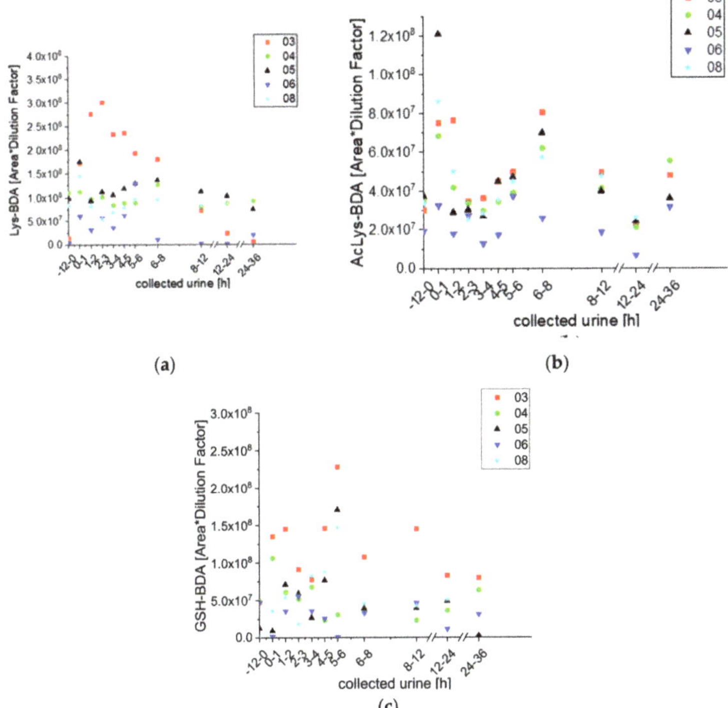

Figure 2. Excretion kinetics of Lys-BDA (**a**), AcLys-BDA (**b**), and GSH-BDA (**c**): UPLC-qToF-MS/MS measurements (as detailed in Materials and Methods) of contents in samples of urine from participants three, four, five, six, and eight at indicated times after coffee consumption (h). Excretion kinetics of Lys-BDA, AcLys-BDA, and GSH-BDA measured by UPLC-qToF (details see Materials and Methods section) over 36 h after coffee consumption. Shown are the individual kinetics of the participants three, four, five, six, and eight. The measurements were performed as triplicates and the data points are shown as mean values with standard deviation.

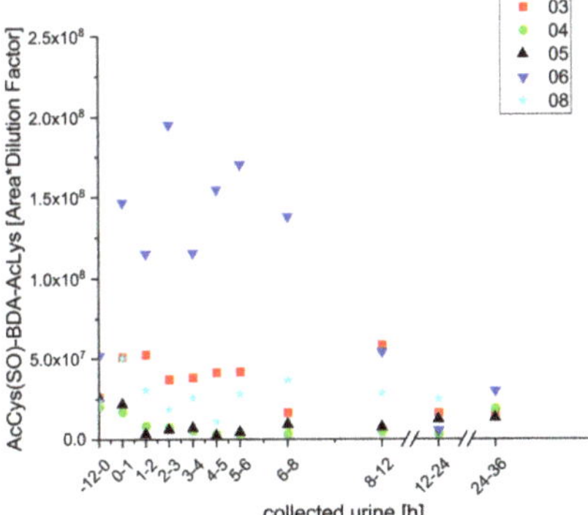

Figure 3. Excretion kinetics of AcCys-(SO)-BDA-AcLys (Rt = 4.46 min): UPLC-qToF-MS/MS measurements (as detailed in Section 3) of contents in samples of urine from participants three, four, five, six, and eight at indicated times after coffee consumption (h).

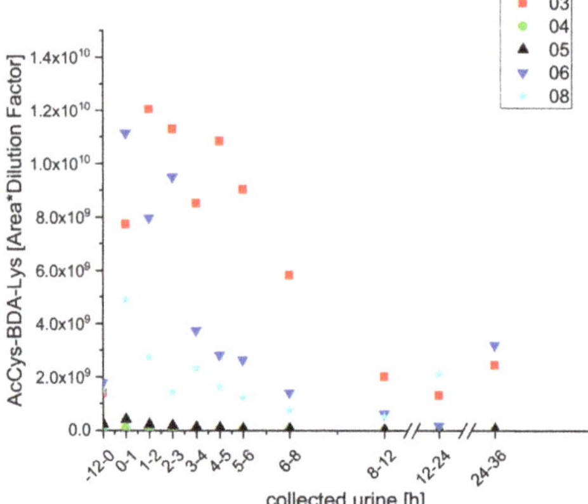

Figure 4. Excretion kinetics of AcCys-BDA-AcLys: UPLC-qToF-MS/MS measurements of contents in samples of urine from participants three, four, five, six, and eight at indicated times after coffee consumption (h).

Both AcCys-(SO)-BDA-AcLys and AcCys-BDA-AcLys were identified in samples from each of the five subjects included in the study presented here but the excretion kinetics differed, so excretion kinetic data for each individual are displayed rather than mean values for all five subjects.

The metabolite methanethiol-BDA-Glu has also been previously detected in rat urine [19] and we detected it in the urine of some of our participants but at lower levels than the other identified metabolites (too low to establish excretion kinetics).

However, we did detect two GSH-conjugates (GSH-BDA-Lys and GSH-BDA-Gln) that have been previously detected in hepatocytes and rat urine [19,25] and obtained semi-quantitative kinetics for them (Figures 5 and 6).

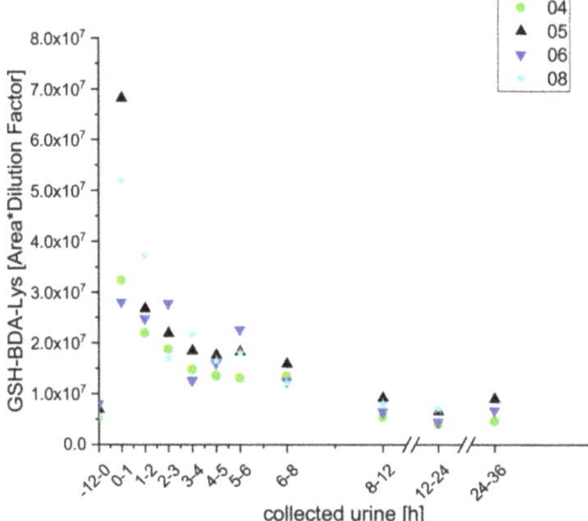

Figure 5. Excretion kinetics of GSH-BDA-Lys: UPLC-qToF-MS/MS measurements of contents in samples of urine from participants four, five, six, and eight at indicated times after coffee consumption (h).

Using UPLC-qToF-MS/MS, we also identified furan metabolites in the urine of the participants that had previously only been detected in vitro or in the bile of rats [19]. These included the GSH adducts glutathione butendial acetyl lysine (GSH-BDA-AcLys), glutathione butendial glutamic acid (GSH-BDA-Glu) cysteine butendial glutathione (Cys-BDA-GSH), and cysteine glycine butendial glutathione (Cys-Gly-BDA-GSH). All of these were identified, by exact masses and MS^2 spectra, for the first time in urine samples of human subjects, and coffee consumption-dependent kinetics were observed for two of them (GSH-BDA-AcLys and CysGly-BDA-GSH), as shown in Figures 7 and 8.

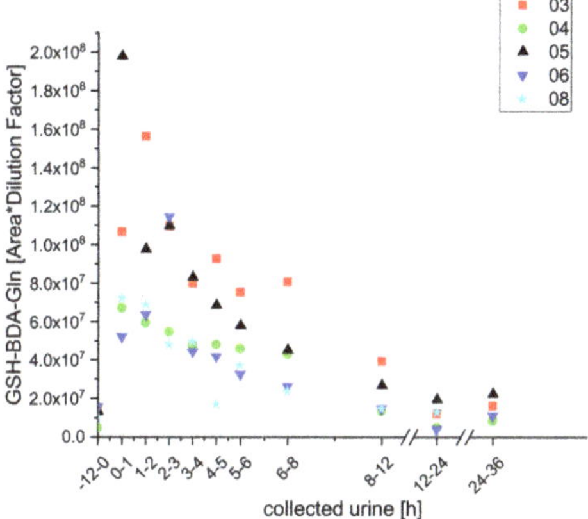

Figure 6. Excretion kinetics of GSH-BDA-Gln: UPLC-qToF-MS/MS measurements of contents in samples of urine from participants three, four, five, six, and eight at indicated times after coffee consumption (h). Excretion kinetics of the GSH-conjugate GSH-BDS-Lys from all five subjects were similar: in all cases, levels in the urine samples increased to a maximum within the first hour then continuously decreased until they reached the initial baseline level after 24 h. In contrast, there were clear between-individual differences in excretion kinetics of GSH-BDA-Gln. Levels of this metabolite peaked in urine from subjects four and eight within the first hour, but only after 3 h in urine from subjects five and six. Moreover, it was not detected at all in urine from participant three.

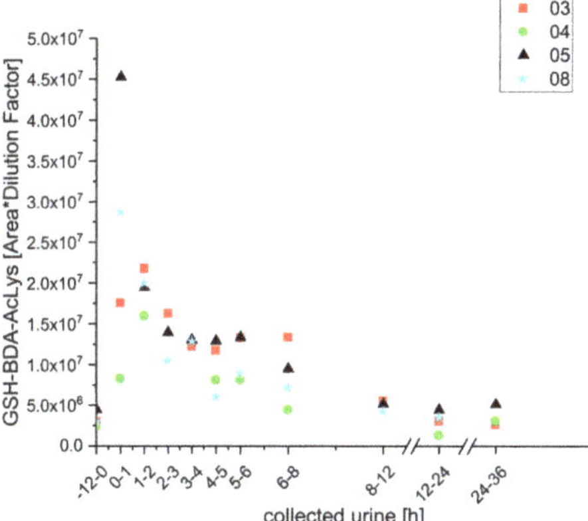

Figure 7. Excretion kinetics of GSH-BDA-AcLys: UPLC-qToF-MS/MS measurements of contents in samples of urine from participants three, four, five, and eight at indicated times after coffee consumption (h).

For GSH-BDA-Ac-Lys, we obtained similar initial values (in urine samples collected during the 12 h before coffee consumption) for all subjects. Moreover, its levels peaked in samples from all participants within the two hours following coffee consumption and declined to the baseline level within 24 h.

However, there was large variation in the triplicate determinations for samples from participant 6, due to low concentrations of the metabolite, and high uncertainties, so we excluded these data from further analysis.

Furthermore, we identified metabolites in our urine samples, which have not been previously described in the literature at all: a biadduct of BDA with lysine (BDA-Lys-BDA) and a BDA cysteine adduct (Cys-BDA). Levels of the BDA biadduct were clearly related to the coffee intake (Figure 9).

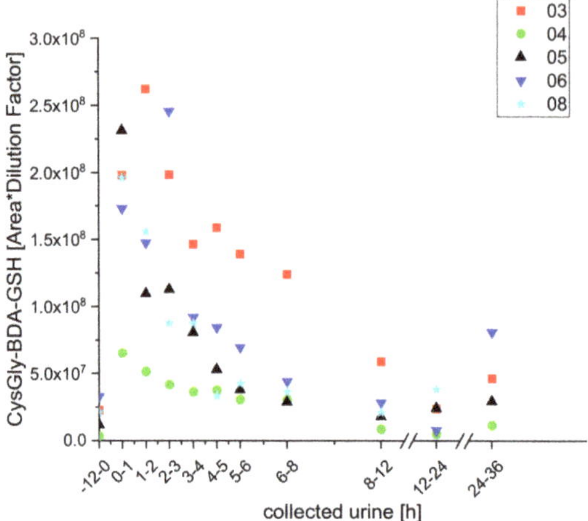

Figure 8. Excretion kinetics of CysGly-BDA-GSH: UPLC-qToF-MS/MS measurements of contents in samples of urine from participants three, four, five, six, and eight at indicated times after coffee consumption (h).

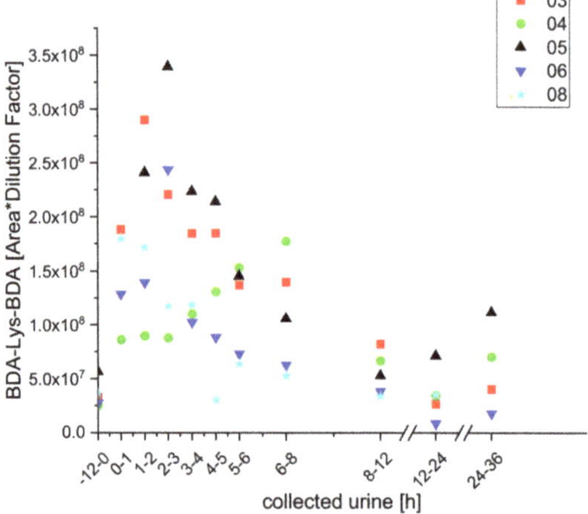

Figure 9. Excretion kinetics of BDA-Lys-BDA: UPLC-qToF-MS/MS measurements of contents in samples of urine from participants three, four, five, six, and eight at indicated times after coffee consumption (h).

Based on our findings regarding metabolites detected by the UPLC-qToF-MS/MS approach in human urine samples, we propose the metabolic pathways for furan taken in by the ingestion of a coffee brew shown in Figure 10.

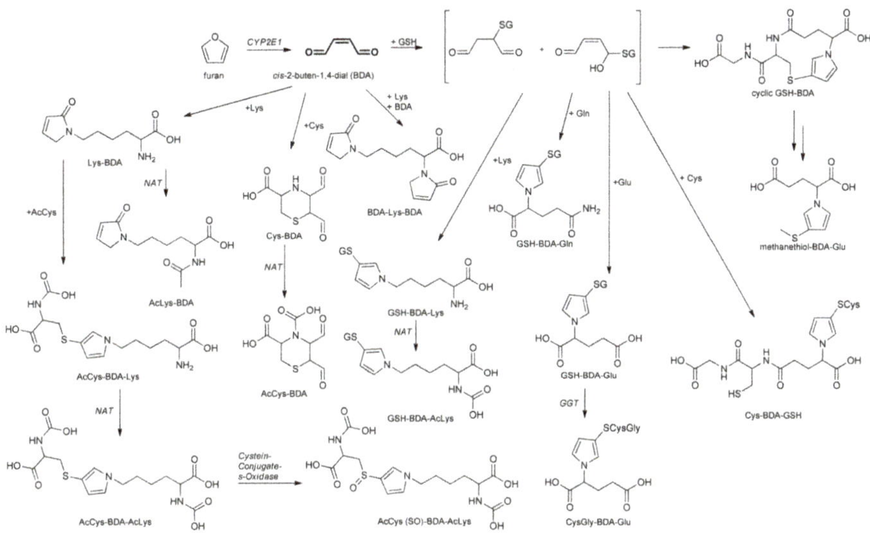

Figure 10. Proposed metabolic pathways for furan based on metabolites identified by UPLC-qToF-MS/MS in samples of urine from human participants after consumption of a bolus of coffee brew and previously identified pathways [19].

2.2. Identification of 2-Methylfuran Derived Metabolites

In analogy to the furan metabolites, we detected two 2-methylfuran conjugates with lysine (Lys-AcA) and acetyl lysine (AcLys-AcA) in samples of urine from the test subjects after coffee brew intake (Figures 11 and 12).

To our surprise, we also identified a conjugate of AcA and acetylcysteine, acetylcysteine acetylacrolein (AcCys-AcA), which has only been previously postulated as a theoretical metabolite, and observed kinetic changes in its levels that were strongly dependent on coffee consumption (Figure 13).

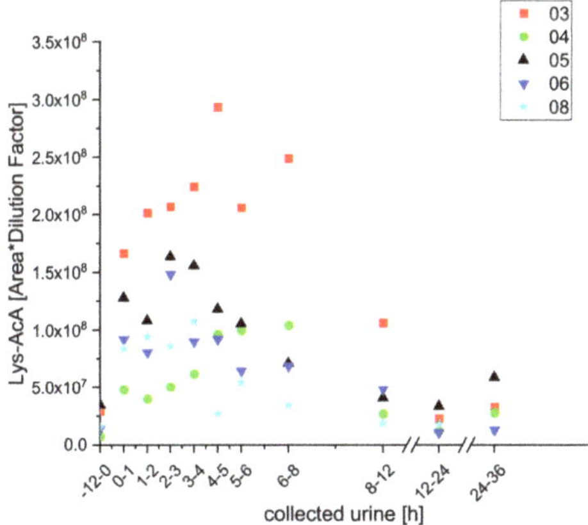

Figure 11. Excretion kinetics of Lys-AcA: UPLC-qToF-MS/MS measurements of contents in samples of urine from participants three, four, five, six, and eight at indicated times after coffee consumption (h).

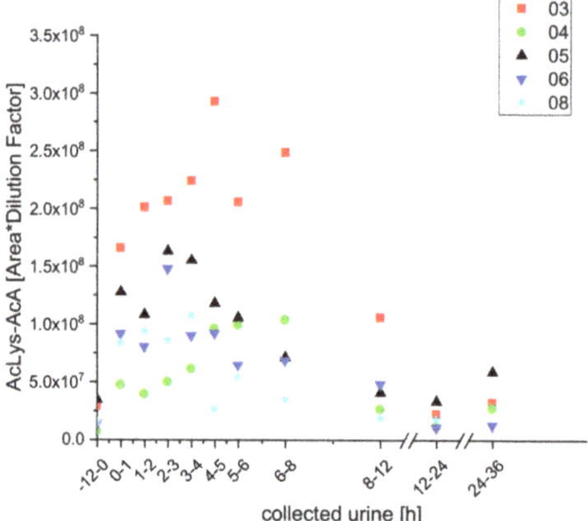

Figure 12. Excretion kinetics of AcLys-AcA: UPLC-qToF-MS/MS measurements of contents in samples of urine from participants three, four, five, six, and eight at indicated times after coffee consumption (h).

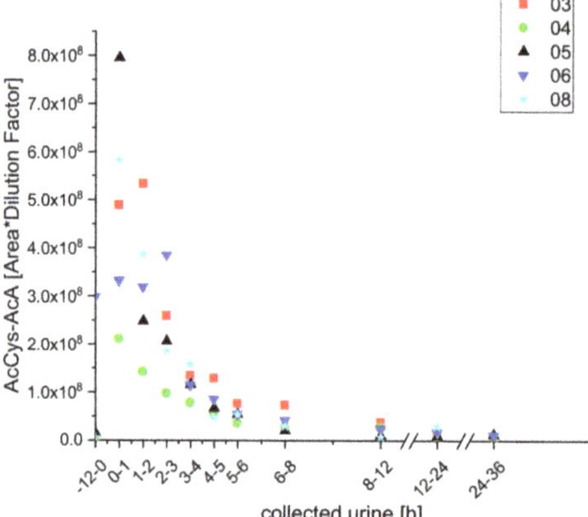

Figure 13. Excretion kinetics of AcCy-AcA: UPLC-qToF-MS/MS measurements of contents in samples of urine from participants three, four, five, six, and eight at indicated times after coffee consumption (h).

3. Materials and Methods

3.1. Chemicals and Reagents

Acetonitrile and formic acid (both LC-MS grade) were purchased from Honeywell (Charlotte, North Carolina, USA), and Biosolve (Valkenswaard, The Netherlands), respectively. Double distilled water produced with a BÜCHI (Essen, Germany) unit was used for the solvents and to dilute the urine samples.

3.2. Human Urine Samples

The analyzed urine specimens were obtained in a previously described human intervention study [17]. The study, which was approved by the ethical committee of the Rhineland-Palatinate medical commission (Approval No. 837.427.15(10195)) was performed in accordance with ethical standards of the Declaration of Helsinki (1964) [27]. For the exploratory investigation presented here, samples from five of the 10 subjects were examined. Portions of urine samples collected at time points indicated in the figures were membrane-filtered, transferred to autosampler vials, and subjected to UPLC-qToF-MS/MS analyses.

3.3. *Liquid Chromatography-High Resolution Mass Spectrometry (LC-HRMS)*

The urine samples were analyzed using a ExionLC^{TM} AD System (Darmstadt, Germany) coupled with a TripleTOF® 6600 system from Sciex. The UPLC conditions were chosen to detect the broadest possible range of metabolites. They were separated using a Synergi 4 u polar RP column (100 × 2.00 mm, Phenomenex, Torrance, CA, USA) with a column oven temperature of 20 °C and a mobile phase consisting of a mixture of 0.1% aqueous formic acid (A) and acetonitrile (B): 1% B for 0.1 min, linear increases to 30% B from 0.1 to 5.8 min, then to 80% B within 0.1 min, followed by a hold of 80% B for 2 min and re-equilibration to 1% B for a further 2 min (with a constant flow rate of 300 µl min^{-1}). Eluting analytes were quantified by the MS system operating in both positive and negative ionization modes. The conditions for positive mode were: curtain gas (CUR) 30 psi, ion source gas (GS) 1 and 2 pressure 60 and 70 psi, temperature 500 °C, ion spray voltage floating (ISVF) 5000 V. Conditions for negative mode were: CUR 40 psi, GS 1 and 2 pressure 60 and 70 psi, temperature 650 °C, ISVF 4000 V. First acquisitions of mass spectra were performed using positive and negative ionization modes with information-dependent acquisition (IDA) of product ions. In full-scan mode, the accumulation time was 100 ms for a mass range from 60 to 800 Da. Each IDA experiment had an accumulation time of 60 ms with the collision energy set to −35 or 35 eV with a spread of 15 eV in high-resolution mode. The IDA criteria were as follows: the 15 most intense ions (number of IDA experiments) with an intensity threshold of 200 cps and exclusion time of 5 s after one occurrence. In second acquisitions, spectra were acquired data-independently by the "sequential window acquisition of all theoretical fragment-ion spectra" (SWATH®) method with positive and negative ionization. Here the full scans also covered the 60 to 800 Da range with −10 and 10 eV collision energy and an acquisition time of 150 ms. Generally, the IDA data were used to identify the metabolites, but for those for which no MS^2 spectra were included in the IDA acquisition, the SWATH data were used.

The UPLC-qToF-System was controlled by Analyst® TF Software 1.7 (AB Sciex Pte. Ltd, Singapore) and the data were analyzed with Sciex OS 1.5.0.23389 (AB Sciex Pte. Ltd, Singapore/), MetabolitePilote^{TM} v. 2.0 (AB Sciex Pte. Ltd, Singapore), and MarkerView^{TM} v. 1.3.1 software (AB Sciex Pte. Ltd, Singapore).

3.4. *Data Evaluation and Illustration*

The data obtained from the IDA measurements were processed as follows.

To determine excretion kinetics of the detected metabolites, the peak areas obtained from the TOF-MS measurements were normalized with respect to the weights of the corresponding urine samples and plotted, in the form of area times dilution factor (y) values, against the time points when the samples were collected after coffee intake (x values, in hours).

The obtained kinetics of the single metabolites are semi-quantitative or relative evaluations because no standards were included in the measurements, which would have allowed determination of concentrations of the single metabolites.

Exact masses were assigned to the individual metabolites using the MS^2 spectra obtained, in conjunction with published data and spectral predictions. The SWATH data were used if no MS^2 spectrum of the metabolites was included in the IDA data. The recorded MS^2 spectra and the assignment of the fragments are shown in the Supplementary Materials.

The metabolites were searched for using a library in SciexOS based on the structures and masses known in the literature and on the other hand masses were extracted, which showed a significant difference between the first two points in time. Furthermore, for 2-MF derived adducts analogous reactions as for furan were assumed and explicitly searched for this.

4. Conclusions

The presented results show that the UPLC-qToF-MS/MS approach can identify metabolites in a complex matrix such as urine rapidly and robustly without extensive sample preparation. Using it, we identified both metabolites that have already been identified in human urine samples and others that have only been previously detected in animal studies or theoretically postulated. Moreover, we obtained information on the excretion kinetics of x identified metabolites following the single intake of a coffee brew, and kinetics observed for y metabolites matched those obtained in previous targeted analyses.

Although these data provide a good overview of furan and 2-MF metabolites excreted in urine after the intake of a coffee brew, the study has several limitations. Quantification of the metabolites, and thus balancing of the total excretion of the furan and 2-methylfuran taken in from a coffee brew, is almost impossible with the available data, and at least one standard compound would have to be added to the samples for calibrated quantification. In addition, the subjects' drinking behavior, and thus their urine volumes, varied substantially. Standardization of the urine's dilution factors, e.g., using creatinine concentrations, would be helpful for enhancing the approach.

In summary, the new UPLC-qToF-MS/MS approach has several advantages that the targeted analysis of metabolites in a complex matrix such as urine lacks. If the analytical goal is considered when acquiring the samples, it can potentially provide abundant additional information in future animal or human intervention studies. The opportunities it provides for statistical evaluation of the data and rapid identification of metabolites responsible for differences in kinetic profiles of samples may be particularly valuable.

Supplementary Materials: The following materials are available online: Figure S1: MS/MS spectrum of Lys-BDA; Figure S2: MS/MS spectrum of AcLys-BDA; Figure S3: MS/MS spectrum of GSH-BDA; Figure S4: MS/MS spectrum of AcCys(SO)-BDA-AcLys; Figure S5: MS/MS spectrum of AcCys-BDA-Lys; Figure S6: MS/MS spectrum of GSH-BDA-Lys; Figure S7: MS/MS spectrum of GSH-BDA-Gln; Figure S8: MS/MS spectrum of GSH-BDA-AcLys; Figure S9: MS/MS spectrum of CysGly-BDA-GSH; Figure S10: MS/MS spectrum of BDA-Lys-BDA; Figure S11: MS/MS spectrum of Lys-AcA; Figure S12: MS/MS spectrum of AcLys-AcA; Figure S13: MS/MS spectrum of AcCys-AcA; Figure S14: Urinary metabolites of 2-methylfuran.

Author Contributions: Conceptualization, E.R.; methodology, S.S., C.B., D.M., and J.I.K.; software, C.B., N.B., S.S., and D.M.; investigation, S.S.; data curation, N.B. and S.S.; project administration, E.R.; resources, E.R.; writing—original draft preparation, S.S., E.R., and N.B.; writing—review and editing, E.R., J.I.K., N.B., and S.S.; visualization, N.B.; formal analysis, S.S.; validation, S.S., J.I.K., and N.B.; supervision, E.R. All authors have read and agreed to the published version of the manuscript.

Funding: No external funding was received for this research.

Acknowledgments: We thank the volunteers for their participation, and Tamara Bakuradze for her support in the human intervention study.

Conflicts of Interest: The authors declare no conflict of interest.

References

1. IARC. *Dry Cleaning, Some Chlorinated Solvents and Other Industrial Chemicals*; International Agency for Research on Cancer: Lyon, France, 1995; ISBN 978-92-832-1563-9.
2. National Center for Biotechnology Information. 2-Methylfuran. Available online: https://pubchem.ncbi.nlm.nih.gov/compound/2-methylfuran (accessed on 23 March 2020).
3. European Food Safety Authority. Update of results on the monitoring of furan levels in food. *EFSA J.* **2010**, *8*, 1702. [CrossRef]
4. Altaki, M.; Santos, F.J.; Galceran, M. Occurrence of furan in coffee from Spanish market: Contribution of brewing and roasting. *Food Chem.* **2011**, *126*, 1527–1532. [CrossRef]
5. Fromberg, A.; Fagt, S.; Granby, K. Furan in heat processed food products including home cooked food products and ready-to-eat products. *EFSA Support. Publ.* **2009**, *6*. [CrossRef]
6. Knutsen, H.K.; Alexander, J.; Barregård, L.; Bignami, M.; Brüschweiler, B.; Ceccatelli, S.; Cottrill, B.; Dinovi, M.; Edler, L.; Grasl-Kraupp, B.; et al. Risks for public health related to the presence of furan and methylfurans in food. *EFS2* **2017**, *15*. [CrossRef]

7. Locas, C.P.; Yaylayan, V.A. Origin and mechanistic pathways of formation of the parent furan—A food toxicant. *J. Agric. Food Chem.* **2004**, *52*, 6830–6836. [CrossRef]
8. Limacher, A.; Kerler, J.; Davidek, T.; Schmalzried, F.; Blank, I. Formation of furan and methylfuran by maillard-type reactions in model systems and Food. *J. Agric. Food Chem.* **2008**, *56*, 3639–3647. [CrossRef]
9. Adams, A.; Bouckaert, C.; Van Lancker, F.; De Meulenaer, B.; De Kimpe, N. Amino acid catalysis of 2-alkylfuran formation from lipid oxidation-derived α,β-unsaturated aldehydes. *J. Agric. Food Chem.* **2011**, *59*, 11058–11062. [CrossRef]
10. Neuwirth, C.; Mosesso, P.; Pepe, G.; Fiore, M.; Malfatti, M.; Turteltaub, K.; Dekant, W.; Mally, A. Furan carcinogenicity: DNA binding and genotoxicity of furan in rats in vivo. *Mol. Nutr. Food Res.* **2012**, *56*, 1363–1374. [CrossRef]
11. National Toxicology Program. Toxicology and Carcinogenesis Studies of Furan (CAS No. 110-00-9) in F344 rats and B6C3F1 mice (gavage studies). *Natl. Toxicol. Program Tech. Rep. Ser.* **1993**, *402*, 1–286.
12. De Conti, A.; Kobets, T.; Escudero-Lourdes, C.; Montgomery, B.; Tryndyak, V.P.; Beland, F.A.; Doerge, D.R.; Pogribny, I.P. Dose- and time-dependent epigenetic changes in the livers of fisher 344 rats exposed to furan. *Toxicol. Sci.* **2014**, *139*, 371–380. [CrossRef]
13. Peterson, L.A.; Naruko, K.C.; Predecki, D.P. A Reactive metabolite of furan,cis-2-butene-1,4-dial, is mutagenic in the ames assay. *Chem. Res. Toxicol.* **2000**, *13*, 531–534. [CrossRef]
14. Palmen, N.; Evelo, C.T. Glutathione depletion in human erythrocytes and rat liver: A study on the interplay between bioactivation and inactivation functions of liver and blood. *Toxicol. Vitr.* **1996**, *10*, 273–281. [CrossRef]
15. Ravindranath, V.; Burka, L.; Boyd, M. Reactive metabolites from the bioactivation of toxic methylfurans. *Science* **1984**, *224*, 884–886. [CrossRef]
16. Gill, S.S.; Kavanagh, M.; Cherry, W.; Barker, M.; Weld, M.; Cooke, G.M. A 28-day Gavage Toxicity Study in Male Fischer 344 Rats with 2-methylfuran. *Toxicol. Pathol.* **2013**, *42*, 352–360. [CrossRef] [PubMed]
17. Kremer, J.I.; Gömpel, K.; Bakuradze, T.; Eisenbrand, G.; Richling, E. Urinary excretion of niacin metabolites in humans after coffee consumption. *Mol. Nutr. Food Res.* **2018**, *62*, e1700735. [CrossRef]
18. Kremer, J.I.; Pickard, S.; Stadlmair, L.F.; Glaß-Theis, A.; Buckel, L.; Bakuradze, T.; Eisenbrand, G.; Richling, E. Alkylpyrazines from coffee are extensively metabolized to pyrazine carboxylic acids in the human body. *Mol. Nutr. Food Res.* **2019**, *63*, e1801341. [CrossRef]
19. Moro, S.; Chipman, J.K.; Wegener, J.W.; Hamberger, C.; Dekant, W.; Mally, A. Furan in heat-treated foods: Formation, exposure, toxicity, and aspects of risk assessment. *Mol. Nutr. Food Res.* **2012**, *56*, 1197–1211. [CrossRef]
20. Kremer, J.I.; Karlstetter, D.; Klier, C.; Rohtermund, J.; Thomas, H.; Becker, H.; Lang, L.; Bakuradze, T.; Eisenbrand, G.; Richling, E. Detection and stability of BDA-lysine and -glutathione metabolites as urinary exposure biomarkers for furan in humans. *Naunyn-Schmiedebergs Arch. Pharmacol.* **2020**, *393* (Suppl. 1), S1–S97. [CrossRef]
21. Peterson, L.A.; Cummings, M.E.; Chan, J.Y.; Vu, C.C.; Matter, B.A. Identification of acis-2-butene-1,4-dial-derived glutathione conjugate in the urine of furan-treated rats. *Chem. Res. Toxicol.* **2006**, *19*, 1138–1141. [CrossRef]
22. Lu, D.; Sullivan, M.M.; Phillips, M.B.; Peterson, L.A. Degraded protein adducts ofcis-2-butene-1,4-dial are urinary and hepatocyte metabolites of furan. *Chem. Res. Toxicol.* **2009**, *22*, 997–1007. [CrossRef]
23. Grill, A.E.; Schmitt, T.; Gates, L.A.; Lu, D.; Bandyopadhyay, D.; Yuan, J.M.; Murphy, S.E.; Peterson, L.A. Abundant rodent furan-derived urinary metabolites are associated with tobacco smoke exposure in humans. *Chem. Res. Toxicol.* **2015**, *28*, 1508–1516. [CrossRef] [PubMed]
24. Kellert, M.; Wagner, S.; Lutz, U.; Lutz, W.K. Biomarkers of furan exposure by metabolic profiling of rat urine with liquid chromatography-tandem mass spectrometry and principal component analysis. *Chem. Res. Toxicol.* **2008**, *21*, 761–768. [CrossRef] [PubMed]
25. Rietjens, I.M.C.M.; Recker, T.; Günther, H.; Hanlon, P.; Honda, H.; Mally, A.; O'Hagan, S.; Scholz, G.; Seidel, A.; Swenberg, J.; et al. Exposure assessment of process-related contaminants in food by biomarker monitoring. *Arch. Toxicol.* **2018**, *92*, 15–40. [CrossRef] [PubMed]

26. Hamberger, C.; Kellert, M.; Schauer, U.M.; Dekant, W.; Mally, A. Hepatobiliary toxicity of furan: Identification of furan metabolites in bile of male F344/N rats. *Drug Metab. Dispos.* **2010**, *38*, 1698–1706. [CrossRef]
27. World Medical Association. WMA Declaration of Helsinki: Ethical Principles for Medical Research Involving Human Subjects. Available online: https://www.wma.net/policies-post/wma-declaration-of-helsinki-ethical-principles-for-medical-research-involving-human-subjects/ (accessed on 18 September 2020).

Article

Mass Spectrometry-Based Untargeted Metabolomics and α-Glucosidase Inhibitory Activity of Lingzhi (*Ganoderma lingzhi*) During the Developmental Stages

Dedi Satria [1,2], Sonam Tamrakar [1], Hiroto Suhara [3], Shuhei Kaneko [4,†] and Kuniyoshi Shimizu [1,*]

1 Division of Systematic Forest and Forest Products Sciences, Department of Agro-Environmental Sciences, Graduate School of Bioresource and Bioenvironmental Sciences, Kyushu University, Fukuoka 812-8581, Japan; dedi.satria@umsb.ac.id or dedi.satoria@gmail.com (D.S.); tamrakar.snm@gmail.com (S.T.)
2 Faculty of Health and Sciences, Muhammadiyah University of Sumatera Barat, Bukittinggi 26181, Indonesia
3 Miyazaki Prefectural Wood Utilization Research Center, Miyazaki 885-0037, Japan; suhara-hiroto@pref.miyazaki.lg.jp
4 Fukuoka Prefecture Forest Research & Extension Center, Fukuoka 818-8549, Japan; shu-k@kir.biglobe.ne.jp
* Correspondence: shimizu@agr.kyushu-u.ac.jp; Tel.: +81-92-802-4675
† (S.K. formerly worked in Fukuoka Prefecture Forest Research & Extension Center).

Academic Editor: Thomas Letzel
Received: 26 April 2019; Accepted: 27 May 2019; Published: 29 May 2019

Abstract: Lingzhi is a *Ganoderma* mushroom species which has a wide range of bioactivities. Analysis of the changes in metabolites during the developmental stages of lingzhi is important to understand the underlying mechanism of its biosynthesis, as well as its bioactivity. It may also provide valuable information for the cultivation efficiency of lingzhi. In this study, mass spectrometry based untargeted metabolomics was carried out to analyze the alteration of metabolites during developmental stages of lingzhi. Eight developmental stages were categorized on the basis of morphological changes; starting from mycelium stage to post-mature stage. GC/MS and LC/MS analyses along with multivariate analysis of lingzhi developmental stages were performed. Amino acids, organic acids, sugars, polyols, fatty acids, fatty alcohols, and some small polar metabolites were extracted as marker metabolites from GC/MS analysis, while, lanostane-type triterpenoids were observed in LC/MS analysis of lingzhi. The marker metabolites from untargeted analysis of lingzhi developmental stages were correlated with the α-glucosidase inhibitory activity. Two metabolites, compounds **34** and **35**, were identified as potential contributors of the α-glucosidase inhibitory activity. The current result shows that some metabolites are involved in the developmental process and α-glucosidase inhibitory activity of lingzhi.

Keywords: *Ganoderma lingzhi*; developmental stages; untargeted metabolomics; GC/MS; LC/IT-TOF-MS; α-glucosidase inhibitory activity

1. Introduction

Lingzhi (*Ganoderma lingzhi*) is a non-edible woody mushroom native to the eastern parts of Asia, including China, Japan, and Korea [1,2]. For over two millennia, *G. lingzhi* has been traditionally used for medicinal purposes [3]. Recent studies of *G. lingzhi* show that it has a wide range of bioactivities such as anti-androgenic, anti-cancer, anti-hypertension, anti-virus, anti-melanocyte, and anti-diabetes [4–6]. In particular, *G. lingzhi* triterpenoids are known to be responsible for the inhibition of α-glucosidase and aldose reductase, leading to their anti-diabetic properties [7–10].

Owing to the wide range of bioactivities, the commercialization of *G. lingzhi* products is increasing rapidly [11]. Commercial products derived from mycelia, spores or fruiting bodies of *G. lingzhi* are mostly sourced from artificial cultivation [12]. The artificial cultivation not only alleviates the difficulties in obtaining wild *G. lingzhi* in nature, but also helps to fulfill the ever increasing global demand [12]. However, it is important to note that the quality of *G. lingzhi* products varies considerably depending on different strains, cultivation conditions, seasonal variations, as well as the variations in the developmental stage of the harvested fruiting bodies [13].

Metabolomics approaches by means of mass spectrometry (MS) have been gaining interest not only for analysis of plants, but also for mushrooms [14,15]. Gas chromatography-mass spectrometry (GC/MS) is a platform which is frequently used in metabolomic studies for the analysis of primary metabolites, whereas liquid chromatography-mass spectrometry (LC/MS) is commonly used for the analysis of secondary metabolites [16]. To increase the coverage of metabolites, the integration of GC/MS and LC/MS platforms has been reported frequently [17,18]. For instance, it has been used to analyze the alteration of metabolites during developmental stages of plants or ripening phases of fruit [19,20]. This analysis is important to understand the underlying mechanism of the biosynthesis of metabolites, as well as their bioactivity.

Although, targeted analysis of some *G. lingzhi* triterpenoids during its various developmental stages has been reported before [13,21]; untargeted metabolomics studies analyzing the chemical changes of *G. lingzhi* at different developmental stages has not been reported to date. In this study, untargeted metabolomics was performed by using GC/MS and LC/IT-TOF-MS in combination with multivariate analysis, to investigate the changes of metabolites in *G. lingzhi* at different developmental stages. Moreover, we demonstrated the variability in alpha-glucosidase inhibitory activity throughout the developmental stages of *G. lingzhi*.

2. Results

2.1. Primary Metabolite Profiling of G. lingzhi at the Eight Developmental Stages by GC/MS

The morphological changes of the fruiting bodies of *G. lingzhi* during its development were categorized into eight stages, as shown in Figure 1: stage one (mycelia), stage two (primordia), stage three (bud-breaking, brown stipe), stage four (early cap formation), stage five (cap formation, white edge), stage six (immature stage, yellow edge), stage seven (mature stage, spore not dispersed), and stage eight (post-mature stage, after spore dispersed).

Figure 1. Experimental design of lingzhi (*Ganoderma lingzhi*) analysis harvested at eight developmental stages.

To investigate the changes of primary metabolites in *G. lingzhi* fruiting bodies at the eight developmental stages, GC/MS spectral data was combined with multivariate analysis. Unsupervised principal component analysis (PCA) was performed to check the clustering trends of features by reduction of dimensionality. A total of 9368 features which were extracted from data preprocessing by XCMS Online, were then analyzed by PCA. From the PCA score plot ($R2X_{(cum)} = 0.807$, $Q2_{(cum)} = 0.624$; Figure 2A), each of the samples tended to gather according to the developmental stage, into four clusters (stage one, stage two, stages three to six, and stages seven to eight). Stages one-two and stages seven-eight were clearly separated by PC1, explained by 34.5% of variance; while stages three-six were distinguished from other stages along PC2 by 21.0% of variance. Furthermore, a supervised partial least squares discriminant analysis (PLS-DA) model ($R2X_{(cum)} = 0.630$, $R2Y_{(cum)} = 0.361$, $Q2_{(cum)} = 0.222$;

Figure 2B) was used to reveal the responsible metabolites for discrimination of developmental stages. The selection of variables represents the primary metabolites in the PLS-DA score plot based on the VIP value (>1.0) and *p*-value (<0.05). The identification of metabolites was examined on the basis of retention times, retention indices of reported metabolites [22–28], and mass fragmentation patterns, with reference to the National Institute of Standards and Technology (NIST) library.

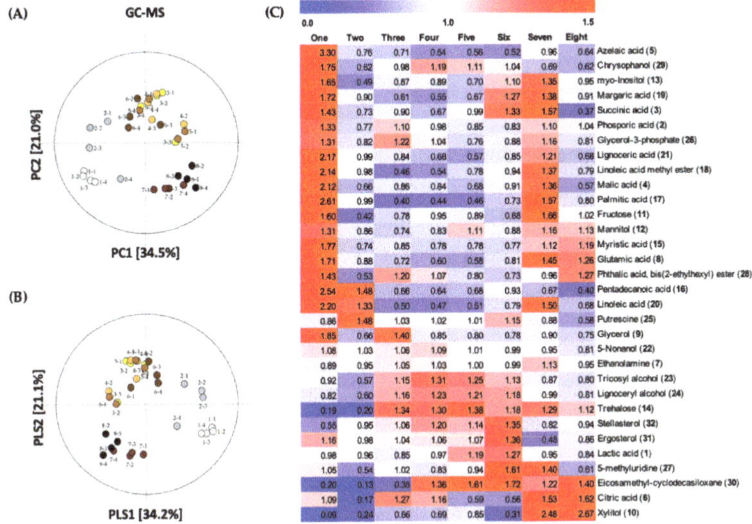

Figure 2. (**A**) PCA score plot and (**B**) PLS-DA score plot of each *G. lingzhi* samples at the eight developmental stages analyzed by GC/MS; (**C**) Heat maps of significantly different metabolites at the eight developmental stages of *G. lingzhi* from GC/MS analysis. Each column represents the developmental stage, and the fold change of average peak area denoted by the number and color of heat scale.

A total of 32 metabolites were selected and putatively identified as significant markers to discriminate between the eight developmental stages of *G. lingzhi*. A heat map analysis was applied to compare the relative levels of discriminant metabolites, on the basis of the corresponding peak area (Figure 2C). The discriminant markers were mostly primary metabolites including six organic acids, two amino acids, six sugars and polyols, seven fatty acids, three fatty alcohols, a phospholipid, and a nucleobase. In addition, six metabolites including two steroids were also selected (Table S1).

A vast majority of significantly different metabolites in the different developmental stages of *G. lingzhi* were observed in relatively greater quantities at stage one, including a sugar, fructose (**15**), polyols such as glycerol (**6**), mannitol (**16**), and myo-inositol (**19**); a phospholipid, glycerol 3-phosphate (**26**); organic acids such as phosphoric acid (**5**), succinic acid (**7**), malic acid (**8**), glutamic acid (**9**), and citric acid (**13**); fatty acids such as myristic acid (**14**), pentadecanoic acid (**17**), palmitic acid (**18**), linoleic acid methyl ester (**20**), margaric acid (**21**), linoleic acid (**22**), and lignoceric acid (**27**); a fatty alcohol, 5-nonanol (**2**); a nucleoside, 5-methyluridine (**23**); and other metabolites such as azelaic acid (**12**), chrysophanol (**26**), ergosterol (**31**), and phthalic acid ester (**24**) (Figure 2B). Pentadecanoic acid, linoleic acid, and 5-nonanol were also found in relatively higher level at stage two in addition to a polyamine, putrescine (**3**).

The following metabolites were detected in relatively higher-levels during stages three to six: a sugar, trehalose (**25**) which increased until stage eight; fatty alcohols such as tricosyl alcohol (**28**) and lignoceryl alcohol (**29**); steroids such as stellasterol (**32**) and ergosterol; and putrescine. Moreover, the levels of citric acid, 5-nonanol, glycerol 3-phosphate, and ethanolamine (**4**) were higher at stages three to four. Additionally, the quantities of glycerol and phosphoric acid peaked at stage three, while the levels of chrysophanol and eicosamethylcyclodecasiloxane (**30**) peaked at stage four.

The quantities of eicosamethylcyclodecasiloxane escalated at stages four to eight, while lactic acid (**1**) was found to be slightly higher at stages five to six. Xylitol (**10**), a sugar alcohol, and citric acid were found to be present at great quantities during stages seven to eight, compared to other developmental stages of *G. lingzhi*. Furthermore, the level of 5-methyluridine was found to be highest at stages six to seven. Interestingly, all of the metabolites which were found in greater quantities at stage one, except azelaic acid, chrysophanol, and phthalic acid ester, were also present in higher levels at stage seven.

2.2. Secondary Metabolite Profiling of *G. lingzhi* at the Eight Developmental Stages by LC/IT-TOF-MS

The changes of secondary metabolites at the eight developmental stages of *G. lingzhi* were profiled by means of LC/IT-TOF-MS data, in combination with multivariate analysis. As shown in Figure 3A, stage one was clearly separated from the other stages in PC1 by 58.8% of variance; while stage two, seven, and eight alienated from stages three to six in PC2 by 9.17% of variance ($R2X_{(cum)} = 0.680$, $Q2_{(cum)} = 0.615$). Moreover, PLS-DA analysis was performed to select secondary metabolites that gave significant differences on the basis of VIP value (>1.0) and p-value (<0.05) ($R2X_{(cum)} = 0.724$, $R2Y_{(cum)} = 0.412$, $Q2_{(cum)} = 0.325$; Figure 3B). The identification of secondary metabolites was elucidated by comparing the retention time, molecular weight, and MS/MS fragmentation pattern of samples, to that of standard compounds or through references [29–32].

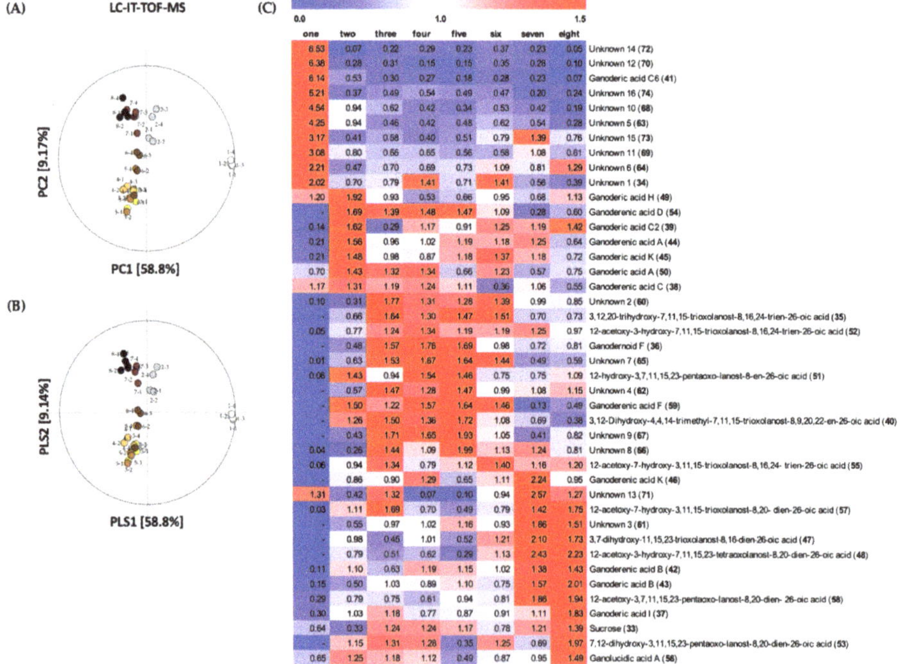

Figure 3. (**A**) PCA and (**B**) PLS-DA score plot of each *G. lingzhi* samples at the eight developmental stages analyzed by LC/IT-TOF-MS; (**C**) Heat maps of significantly different metabolites at the eight developmental stages of *G. lingzhi* from LC/IT-TOF-MS analysis. Each column represents the developmental stage, and the fold change of average peak area denoted by the number and color of heat scale.

A total of 42 metabolites, including 25 triterpenoids, sucrose, and 16 unidentified metabolites were selected for discrimination of *G. lingzhi* at eight developmental stage (Table S1). The variations of the quantity of secondary metabolites during the developmental stages of *G. lingzhi* has been illustrated in the heat map, as shown in Figure 3C. Ten metabolites were found to be present at their greatest quantities in stage one: unknown 1 (**34**), ganoderic acid C6 (**41**), unknown 5 (**63**), unknown 6 (**64**), unknown 10 (**68**), unknown 11 (**69**), unknown 12 (**70**), unknown 14 (**72**), unknown 15 (**73**), and unknown 16 (**74**). Furthermore, at stage two: ganoderenic acid C (**38**), ganoderic acid C2 (**39**), ganoderenic acid A (**44**), ganoderic acid K (**45**), ganoderic acid H (**49**), ganoderic acid A (**50**), and ganoderenic acid D (**54**) were found to be present in their highest levels. During stages three to six, the following metabolites were found in relatively higher levels: 3,12,20-trihydroxy-7,11,15-trioxolanost-8,16,24-trien-26-oic acid (**35**), 3,12-dihydroxy-4,4,14-trimethyl-7,11,15-trioxolanost-8,9,20,22-en-26-oic acid (**40**), 12-acetoxy-3-hydroxy-7,11,15-trioxolanost-8,16,24-trien-26-oic acid (**52**), ganoderenic acid F (**59**), unknown 2 (**60**), unknown 7 (**65**), unknown 8 (**66**), and unknown 9 (**67**). The levels of ganodernoid F (**36**), 12-hydroxy-3,7,11,15,23-pentaoxo-lanost-8-en-26-oic acid (**51**), unknown 4 (**62**) were present in relatively higher quantities during stages three to five, except **51** which was slightly lower at stage three. Furthermore, the quantities of sucrose (**33**), ganoderic acid I (**37**), ganoderenic acid B (**42**), ganoderic acid B (**43**), 3,7-dihydroxy-11,15,23-trioxolanost-8,16-dien-26-oic acid (**47**), 12-acetoxy-3-hydroxy-7,11,15,23-tetraoxolanost-8,20-dien-26-oic acid (**48**), 12-acetoxy-7-hydroxy-3,11, 15-trioxolanost-8,20-dien-26-oic acid (**57**), 12-acetoxy-3,7,11,15,23-pentaoxo-lanost-8,20 -dien-26-oic acid (**58**), unknown 3 (**61**), and unknown 13 (**71**) were found in higher levels at stage seven to eight. In addition, the levels of ganoderenic acid K (**46**) and 12-acetoxy-7-hydroxy-3,11, 15-trioxolanost-8,16,24-trien-26-oic acid (**55**) at stage six and at stage seven were found to be present in their highest levels, along with 7,12-dihydroxy-3,11,15,23-pentaoxo-lanost-8,20-dien-26-oic acid (**53**) and ganolucidic acid A (**56**) at stage eight. The structures of the discriminant metabolites can be seen in Figure 4.

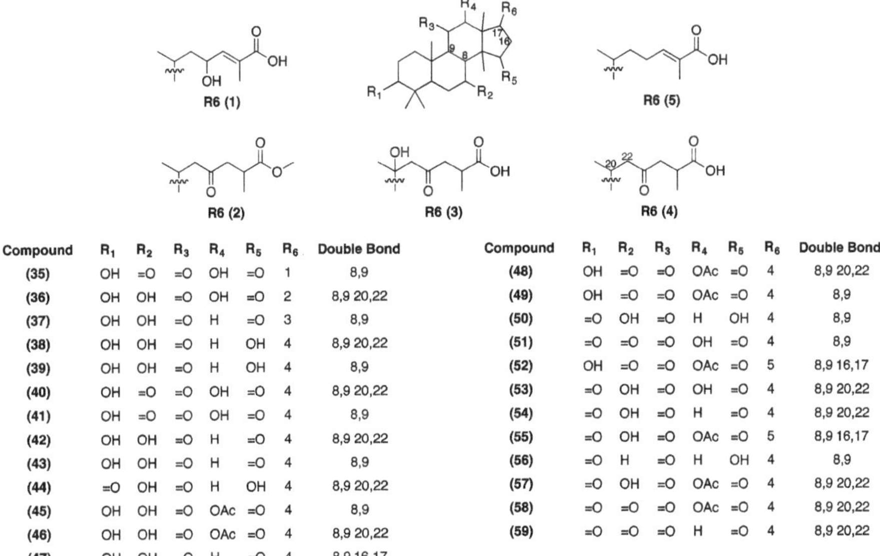

Compound	R1	R2	R3	R4	R5	R6	Double Bond
(35)	OH	=O	=O	OH	=O	1	8,9
(36)	OH	OH	=O	OH	=O	2	8,9 20,22
(37)	OH	OH	=O	H	=O	3	8,9
(38)	OH	OH	=O	H	OH	4	8,9 20,22
(39)	OH	OH	=O	H	OH	4	8,9
(40)	OH	=O	=O	OH	=O	4	8,9 20,22
(41)	OH	=O	=O	OH	=O	4	8,9
(42)	OH	OH	=O	H	=O	4	8,9 20,22
(43)	OH	OH	=O	H	=O	4	8,9
(44)	=O	OH	=O	H	OH	4	8,9 20,22
(45)	OH	OH	=O	OAc	=O	4	8,9
(46)	OH	OH	=O	OAc	=O	4	8,9 20,22
(47)	OH	OH	=O	H	=O	4	8,9 16,17
(48)	OH	=O	=O	OAc	=O	4	8,9 20,22
(49)	OH	=O	=O	OAc	=O	4	8,9
(50)	=O	OH	=O	H	OH	4	8,9
(51)	=O	=O	=O	OH	=O	4	8,9
(52)	OH	=O	=O	OAc	=O	5	8,9 16,17
(53)	=O	OH	=O	OH	=O	4	8,9 20,22
(54)	=O	OH	=O	H	=O	4	8,9 20,22
(55)	=O	OH	=O	OAc	=O	5	8,9 16,17
(56)	=O	H	=O	H	OH	4	8,9
(57)	=O	OH	=O	OAc	=O	4	8,9 20,22
(58)	=O	=O	=O	OAc	=O	4	8,9 20,22
(59)	=O	=O	=O	H	=O	4	8,9 20,22

Figure 4. Chemical structures of selected discriminant metabolites from LC/IT-TOF-MS analysis.

2.3. Comparison of α-Glucosidase Inhibitory Activity of *G. lingzhi* at the Eight Developmental Stages

The inhibitory activity of *G. lingzhi* at different developmental stages on α-glucosidase was investigated (Figure 5). Overall, ethanol extracts of *G. lingzhi* at every developmental stage showed stronger inhibitory effect on α-glucosidase enzymatic activity, compared to acarbose, a clinically approved α-glucosidase inhibitor which was used as a positive control in the present study, and which had an IC_{50} value of 347 ± 53.3 μg/mL (the data not shown in the figure). The average IC_{50} values of each stages in decreasing order were as follows: stage six (40 ± 0.9 μg/mL), stage three (47 ± 1.5 μg/mL), stage seven (50 ± 1.7 μg/mL), stage four (51 ± 0.8 μg/mL), stage five (52 ± 1.2 μg/mL), stage one (59 ± 0.9 μg/mL), stage two (71 ± 3.8 μg/mL), and stage eight (109 ± 4.6 μg/mL). Amongst all the developmental stages of *G. lingzhi*, the sample 6-3 which was harvested 31-34 weeks after the inoculation of the spawn was the strongest inhibitor, with an IC_{50} value of 27 ± 1.4 μg/mL.

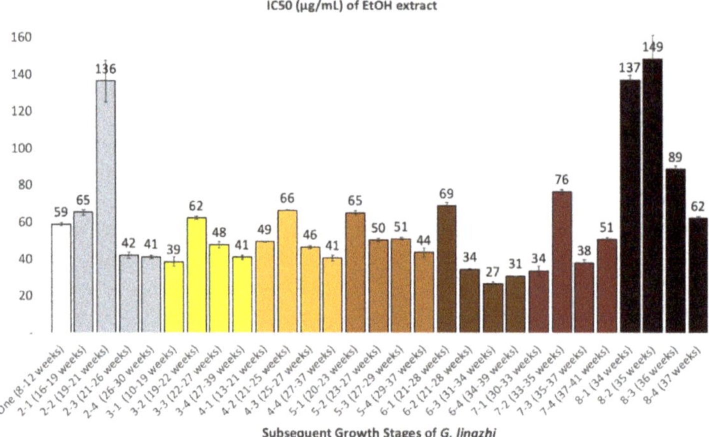

Figure 5. α-Glucosidase inhibitory activity (IC_{50} in μg/mL) of ethanol extract of *G. lingzhi* at different developmental stages.

3. Discussion

The untargeted metabolites profiling was performed for developmental stages of *G. lingzhi* using GC/MS and LC/IT-TOF-MS. From the PCA score plot (Figures 2A and 3A), the profile of metabolites has been clustered into four clusters as follows: stage one, stage two, stages three-six, and stages seven-eight which corresponded to mycelial stage, primordia stage, bud-breaking of primordia stage to developmental stage, and mature stage to post-mature stage respectively.

Mycelia stage is a vegetative state in the initial formation of fruiting bodies of Basidiomycetes [33]. In this stage, we found great levels of sugars and polyols, organic acids, fatty acids, and fatty alcohols that might be related to the carbon source and the production of energy during exponential rate of mycelium expansion, for the formation of fruiting bodies [34,35]. Another reason could be because the mycelium greedily absorbs nutrient from the substrate, stores the nutrients, and releases it only when they are invaded [36]. Fructose and glycerol were reported as the best carbon source for mycelial development [34]. In addition to these sources of carbon, mannitol and myo-inositol have also been reported to exist in *Ganoderma* species at the mycelial stage in relatively higher concentrations [37]. Glycerol-3-phosphate is an intermediate related to glycolysis pathway; while malic acid, succinic acid, citric acid, and glutamic acid are well-known organic acids that are involved in TCA cycle for energy production in the mushroom cells [34,38]. The fatty acids such as myristic acid, pentadecanoic acid, palmitic acid, margaric acid, linoleic acid, and lignoceric acid are essential components of cell membrane and cell wall of mushroom, which functions as food reserves and cell protectors [34,39].

The relatively high quantity of 5-methyluridine, a nucleobase, at mycelial stage might be associated with its reproductive function in preparation for primordia formation [40]. As for the secondary metabolites, azelaic acid was reported as a weak competitive tyrosinase inhibitor which might contribute to the tyrosinase inhibitory activity of *G. lingzhi* extract [5,41]. For this reason, we suggest that azelaic acid also contributed to the white color of mycelia. Chrysophanol has been reported as a highly active inhibitor of pathogenic fungi on plants, by suppression of mycelial growth [42]. Similarly, ergosterol, a principle sterol which can be found only in fungi, was reported to have the ability to activate defense enzyme in plants, against pathogenic fungi [43]. Therefore, we imply that the biosynthesis of these compounds might be related to the survival function of *G. lingzhi* mycelia, and to provide a competitive advantage against other fungi. As for triterpenes, in agreement with the previous reports about its biosynthesis in *G. lingzhi* [44,45], our metabolomic data confirmed that only a few triterpenoids are produced at mycelial stage. Several lanostane-type triterpenoids such as ganoderic acid C6 (**41**), ganoderic acid H (**49**), and other unidentified metabolites were produced at highest quantities in this stage. Ganoderic acid A (**50**), ganoderic acid B (**43**), **41**, and **49** isolated from fruiting bodies of *G. lingzhi* were reported as anti-nociceptive components [46]. Moreover, the strong effect of acetone extract of *G. lingzhi* mycelium on anti-nociceptive activity has also been reported [47]. These facts show that there is a strong correlation between secondary metabolites contained in crude extracts of natural products and their biological activity. In addition, it is worthy to note that phthalic acid ester, which is usually used as plasticizer, was detected at all stages; particularly at the mycelial stage with the highest intensities. This might be related to the degradation of cultivation bags during the expansive growth of mycelia.

At the primordia stage, we observed that the levels of almost all of the selected markers of primary metabolites were decreased, compared to the mycelial stage, apart from pentadecanoic acid and linoleic acid, which were still above average levels. These fatty acids might be essential to support the formation of primordia. On the other hand, the levels of ethanolamine and putrescine were increased. The latter, at primordia stage, was noticed at its highest levels in comparison to other developmental stages. Previous studies about the role of putrescine in *G. lingzhi* shows that it is responsible for mushroom-like smell [48], and it influences the biosynthesis of ganoderic acid [49]. Thus, at this stage we found that more triterpenoids were produced i.e., ganoderenic acid C (**38**), ganoderic acid C2 (**39**), ganoderenic acid A (**44**), ganoderic acid K (**45**), ganoderic acid H (**49**), ganoderic acid A (**50**), and ganoderenic acid D (**54**). All of them except **54** showed strong inhibitory activity on angiotensin-converting enzyme inhibition assay [50]. **39** and **50** were reported to have inhibitory effect on the induction of Epstein–Barr Virus Early Antigen (EBV-EA) [51], while **39** and **44** exhibited aldose reductase inhibitory activity [10]. As mentioned earlier, several triterpenoids such as **43**, **49**, and **50** have anti-nociceptive activity at the mycelial stage, but, their relative levels were not as high as their quantities at the primordia stage. Hence, we suggest that the anti-nociceptive effect of the primordia stage must be due to their triterpenoid contents. Since the triterpenoids at primordia stage are capable of exhibiting various biological activities, specific *G. lingzhi* products can be prepared simply from the primordia itself.

Primordia eventually developed into immature fruiting bodies, and this period of development was divided into four stages (stages three to six), as can be seen in Figure 1. As for the bud-breaking stages, several changes of metabolites were observed, such as increased levels of sucrose, trehalose, glycerol, glycerol-3-phosphate, citric acid, and phosphoric acid from primordia stage. On the other hand, the levels of pentadecanoic acid and linoleic acid were decreased intensely. Trehalose has been reported to be the best source of carbohydrate for the formation of fruiting bodies, which is converted from glucose inside hyphae, as a food reserve [34]. Regarding the secondary metabolites at bud-breaking stage, rapid increase in the quantities of some metabolites such as ganodernoid F, **35**, **52**, **55**, **60**, **62**, **65**, **66**, **67**, and **71** were observed, while ganoderic acid C2 was reduced. We examined that the level of eicosamethylcyclodecasiloxane increased rapidly when the bud-breaking stage developed to early cap formation, whereas, glycerol intensities contrarily decreased. However, the correlation between the changes of eicosamethylcyclodecasiloxane or glycerol and the early cap formation remains unclear. Concerning secondary metabolites produced at this stage, great changes were observed as the increasing quantities of ganoderic acid C2, ganoderenic acid B, **34**, and **51**, and decreasing of ganoderic acid H, ganoderic acid I, **47**, **55**, **57**, and **71**. Furthermore, we checked the metabolites alteration during cap formation which was assigned by the observation of a white edge on the cap. At this stage, the level of citric acid declined intensively. Also, ganoderic acid A, ganoderenic acid K, **34**, **47**, **48**, **53**, and **56** were reduced immensely, but, a sharp increase was observed for **66**. The immature fruiting bodies was assigned by the yellow color of the edge of the cap. We noticed the sharp inclination of margaric acid at this stage, as well as 5-methyluridine, myo-inositol, and succinic acid. The increasing inclination of margaric acid might be related to the yellow color of the edge of the cap. As for 5-methyluridine, we suggest that it might be associated with the elevation of reproductive function prior to maturity stage [40].

In general, the levels of metabolites were increased at the mature stage and some of them such as ganoderenic acid K, **47**, **48**, **61**, and **71** were at their highest levels. In contrast, pentadecanoic acid, putrescine, tricosyl alcohol, lignoceryl alcohol, chrysophanol, stellasterol, and ergosterol were decreased. Ergosterol in particular could be considered as a suitable discriminant marker for mature stage, since its quantities declined greatly in comparison to immature stage. As for putrescine, due to its role in influencing the biosynthesis of triterpenes, its declining levels at the mature stage indicates the maturity of fruiting bodies. In mature stage, the spores will accumulate until the dispersion. We observed that at this stage, when the spores have been dispersed, the levels of all metabolites decreased, except eicosamethylcyclodecasiloxane, myristic acid, citric acid, and xylitol. Especially, xylitol was observed at its highest levels in this stage, compared to other developmental stages.

In order to demonstrate the correlation between biological activity and variability of metabolites during developmental stages of *Ganoderma lingzhi*, we performed alpha-glucosidase inhibitory assay as shown in Figure 5. We observed that immature stage of *G. lingzhi* gave the highest activity (lowest IC50 value) especially for the sample which was harvested 31–34 weeks after the inoculation of spawn. This inhibitory activity might be associated with the accumulation of active compounds at this stage. To find the responsible metabolites that gave higher activity, we performed Pearson's correlation test between the $1/IC_{50}$ of each developmental stage and the peak area of the discriminant metabolites. As a result, several metabolites such as 34 and 35 gave medium positive correlation. Our current results show that immature stage correlates with the highest α-glucosidase inhibitory activity. This result also proves that the mature fruiting bodies are not always necessary for good quality *G. lingzhi* products; and the harvesting time of the fruiting bodies should depend on the specific purpose of *G. lingzhi* product.

4. Materials and Methods

4.1. Chemicals and Reagents

Deionized water was purified using an Ultrapure Water System RFU424CA (Advantec, Tokyo, Japan). LC-MS grade of methanol, acetonitrile, formic acid, dehydrated pyridine, sucrose, and acarbose were purchased from Fujifilm Wako Pure Chemical Corporation, Osaka, Japan. Methoxyamine hydrochloride was obtained from Aldrich (St Louis, Mo., USA). *N*-methyl-*N*-(trimethylsilyl)-trifluoroacetamide (MSTFA) + 1% trimethylchlorosilane (TMCS) was purchased from Thermo Scientific, Rockford, IL, USA. C_8–C_{20} alkane analytical standard solution containing ~40 mg/L of each C_8–C_{20} alkane in hexane was obtained from Fluka (Menlo Park, CA, USA). Triterpenoid standards of ganoderic acid A, ganoderic acid B, ganoderenic acid C, ganoderic acid C1, ganoderic acid C2, ganoderic acid C6, ganoderenic acid D, ganoderic acid I, and ganoderic acid K were purchased from Chem Faces Co., Ltd. (Wuhan, China). Ultrol grade HEPES was purchased from Calbiochem (San Diego, CA, USA). α-Glucosidase derived from yeast was obtained from Oriental Yeast Co. Ltd. (Tokyo, Japan).

4.2. Mushroom Materials and Sampling

G. lingzhi samples were received every week since February 14 to September 5, 2018 from Aso Biotech, a mushroom company located in Kumamoto, Japan. The spawn of *G. lingzhi* strain BMC9049 was inoculated into the substrate packed in heat-sealed cultivation bags, with microfilter windows. The substrate in each bag consisted of wood chips (hardwood) 1563 g, rice bran 93 g, wheat bran 69 g, and water 775 g, resulting in final moisture of 55% in the mushroom bed. The environmental condition of the cultivation site was maintained at the temperature 23–25 °C, humidity 80%, and air circulation for more than 12 h/day. The sample harvests were done randomly.

For the identification of different developmental stages, fruiting body samples were sent to Dr. Hiroto Suhara, a mushroom expert from Miyazaki Prefectural Wood Utilization Research Center, Japan. Eight developmental stages of *G. lingzhi* fruiting bodies were categorized on the basis of morphological changes; starting from mycelium stage to post-mature stage. *G. lingzhi* samples were freeze dried at approximately −50 °C temperature and 50 mmHg pressure. Sampling of each developmental stages were made by pooling the samples based on morphological observations into four levels, according to their age (in weeks), wherein each level consists of more than three fruiting bodies ($n > 3$). The detailed information regarding *G. lingzhi* samples used are listed in Table 1. The dried samples were pulverized and then keep in a dark and dry place at room temperature prior to extraction.

Table 1. List of harvested *G. lingzhi* (BMC9049) at different developmental stages after spawn inoculation.

Developmental Stages	Stage One [a]	Stage Two [b]	Stage Three [c]	Stage Four [d]	Stage Five [e]	Stage Six [f]	Stage Seven [g]	Stage Eight [h]
Sample label (age in weeks)	1-1 (8–12)	2-1 (16–19)	3-1 (10–19)	4-1 (13–21)	5-1 (20–23)	6-1 (21–28)	7-1 (30–33)	8-1 (34)
	1-2 (8–12)	2-2 (19–21)	3-2 (19–22)	4-2 (21–25)	5-2 (23–27)	6-2 (28–31)	7-2 (33–35)	8-2 (35)
	1-3 (8–12)	2-3 (21–26)	3-3 (22–27)	4-3 (25–27)	5-3 (27–29)	6-3 (31–34)	7-3 (35–37)	8-3 (36)
	1-4 (8–12)	2-4 (26–30)	3-4 (27–39)	4-4 (27–37)	5-4 (29–37)	6-4 (34–39)	7-4 (37–41)	8-4 (37)

[a] mycelia; [b] primordia; [c] bud-breaking, brown stipe; [d] early cap formation; [e] cap formation, white edge; [f] immature stage, yellow edge; [g] mature stage, spore not dispersed; [h] post-mature stage, after spore dispersed.

4.3. Metabolite Extraction and Sample Preparation

Pulverized fruiting bodies of *G. lingzhi* from each stage (500 mg) were transferred to a 15 mL High-Clarity Polypropylene Conical Tube (Falcon, Corning Science Mexico SA de CV, Tamaulipas, Mexico), and extracted by sonication for 30 min using 7 mL of ethanol. After centrifugation for 10 min at 2058× *g*, the supernatant was separated. The same extraction process was repeated by adding 3 mL of ethanol. The supernatants from both extractions were combined respectively, and then filtered, and evaporated in a rotary evaporator under vacuum at 45 °C. The dried extract was then reconstituted in methanol to make a 10 mg/mL solution. After being vortexed, the solution was filtered through 0.20 μm PTFE syringe driven filter unit (Millex-LG, Millipore, Tokyo, Japan), and then analyzed by LC/IT-TOF-MS and GC/MS. The sample oximation for GC/MS analysis was performed by incubating the dried sample extracts with 50 μL of methoxyamine hydrochloride in dry pyridine (20 mg/mL) at 30 °C for 90 min. Subsequently, the silylation step was carried out by adding 50 μL of MSTFA + 1%TMCS to the reaction mixture with 30 min incubation at 37 °C.

4.4. GC/MS Analysis

A GC/MS analysis was performed using a 7890A GC-5975C inert MSD (with Triple Axis Detector) system (Agilent Technologies, Palo Alto, CA). One microliter of sample was injected into the HP-5MS Agilent capillary column (30 m × 0.25 mm i.d.; 0.25 μm thickness) using an auto-injector with a split ratio of 1:10. Helium was used as a carrier gas at a flow rate of 1.5 mL/min. The inlets and MS source temperatures were maintained at 250 and 230 °C respectively. The oven temperature was maintained at 75 °C for 2 min and ramped to 300 °C at a rate of 10 °C/min, then held at 300 °C for 10 min. Data were acquired in full scan from 50 to 800 m/z. The metabolites were putatively identified based on library searches in the NIST database. The level of confidence for the identification of metabolites was based upon Metabolomics Standard Initiative (MSI) of the Metabolomics Society [52].

4.5. LC/IT-TOF-MS Analysis

Liquid chromatographic separation and mass spectrometric detection were achieved by employing UFLC coupled with IT-TOF-MS via electrospray ionization (ESI) interface (Shimadzu, Kyoto, Japan). The analytical column was ZORBAX Eclipse plus C18 RRHT (3.0 mm i.d. × 100 mm, 1.8 μm) equipped with Eclipse plus C18 guard column (3.0 mm i.d. × 5 mm, 1.8 μm). The automatic sampler was maintained at 4 °C and the injection volume was 2 μL. The gradient elution was performed using a 40 min method, at a flow rate of 0.43 mL/min, with the mobile phase containing water with 0.1% formic acid (mobile phase A) and acetonitrile with 0.1% formic acid (mobile phase B). In the first 26 min, mobile phase B was increased linearly from 5% to 100%. Then, mobile phase B was kept at 100% for 9 min. At 31–32 min, sodium trifluoroacetic (NaTFA) was injected into the mass system for mass calibration. At 35.1 min, mobile phase B was adjusted to 5% for equilibration for 5 min. Mass spectra in both positive and negative ionization modes were obtained simultaneously in a full-scan operation with a scan range of 100–1000, 50–650, and 50–300 *m/z* for MS^1, MS^2, and MS^3 respectively by switching the interface voltage between 4.5 kV and 3.5 kV in each 0.1 s. The CID collision energy of MS^2 and MS^3 were set to 50 and 75%, respectively. The flow rate of the nebulizing gas (N_2) was 1.5 L/min. The temperatures of the curved desorption line and the heat block were both 200 °C, and the microchannel plate detector voltage was set to 1.70 kV. The pressure of the drying gas (N_2) was 198 kPa, and the ion accumulation time was set to 10 ms. Mass spectra and chromatograms were acquired and processed with LC/MS solution version 3.0 (Shimadzu). The metabolites were positively identified using standard compounds. In the absence of the standard compounds, a tentative identification was made by comparing the MS and MS^n spectra of metabolites, reported in previously published research. The confidence level of metabolite annotations was also based upon MSI [52].

4.6. α-Glucosidase Inhibitory Activity

The α-glucosidase activity was carried out according to a previously described method with minor modifications [9]. A 100 μL portion of α-glucosidase (5 units/mL) in 0.15 M HEPES buffer was added to the mixture of 100 μL DMSO with or without (control) samples and 100 μl of sucrose in 0.15 M HEPES buffer, and then incubated at 37 °C for 30 min. After incubation, the reaction was ended by heating at 100 °C for 10 min. The formation of glucose was determined by means of the glucose oxidase method using a BF-5S Biosensor (Oji Scientific Instrument, Hyogo, Japan).

4.7. Data Processing and Multivariate Analysis

GC/MS raw data files were converted to computable document form (CDF) format (*.cdf) using the Enhanced Data analysis (Agilent). The converted GC/MS data files were pre-processed for peak alignment, retention time correction, and peak intensity calculation using a web-based software, XCMS Online (xcmsonline.scripps.edu). A default Centwave method for single quadrupole, "GC/Single Quad (centWave)", was selected as the parameter set [53]. The parameters were as follows: maximal tolerated m/z deviation, 100 ppm; signal/noise threshold, 6; mzdiff, 0.01; integration methods, 1; prefilter peaks, 3; prefilter intensity, 100; mzwid, 0.25; minfrac, 0.5; and bandwidth, 10. After processing, a data matrix containing retention times, masses, and normalized peak intensities were obtained in an Excel (Microsoft, Redmond, WA, USA) format and was analyzed further with multivariate software.

The obtained data from LC/IT-TOF-MS were processed by Profiling Solution version 1.1 (Shimadzu) for peak deconvolution and alignment. The method parameters were as follows: width (5 s), slope (2000/min), retention time range (0.5–30 min), ion *m/z* tolerance (20 mDa), ion retention time tolerance (0.5 min), and ion intensity threshold (10,000 counts). After completing the integration parameters, a report of peaks based on areas, retention time and *m/z* was generated for each sample. Signals of different samples were considered to be similar when they simultaneously fulfilled both retention time (0.5 min tolerance) and *m/z* value (30 mDa tolerance) criteria. The resulting data were exported to Excel for multivariate analysis.

Multivariate statistical analysis was performed using SIMCA (version 15.0.2, Umetrics, Umea, Sweden). PCA and PLS-DA models were used to compare the changes of metabolites during developmental stages of *G. lingzhi*. The data sets were pareto scaled without transformation prior to PCA and PLS-DA modelling. The PCA model was evaluated with statistical parameters, including the R2X which explained variance of X-data, and the Q2 which illustrated the predicted variance. In PLS-DA model, the R2X and R2Y represent fraction of the variance of the X and Y matrix variables explained by the model. Q2 represented the predictive capacity of the model. R2X(cum) and R2Y(cum) is cumulative fraction of sum of squares of X and Y explained by the current component, respectively. Metabolites with variable importance as determined by a projection (VIP) value of >1.0 and a *p*-value of <0.05 were selected. Significant differences were tested by one-way analysis of variance (ANOVA) using SPSS Statistics 24 (SPSS Inc., Chicago, IL, USA). Differences in alpha-glucosidase inhibitory activity were analyzed by one-way ANOVA and Duncan's multiple-range test using SPSS Statistics 24 (SPSS Inc.). Pairwise correlations between secondary metabolites and alpha-glucosidase inhibitory activity were calculated by Pearson's correlation coefficient test using Excel.

5. Conclusions

MS-based untargeted metabolomics was carried out on the developmental stages of *G. lingzhi*. We confirmed that there are significant changes in the primary and secondary metabolites during the developmental stages of *G. lingzhi*. Moreover, we found that the state of the metabolites during various developmental stages of *G. lingzhi* was associated with the morphological changes. Two metabolites, **34** and **35**, were found to have potential contribution to the α-glucosidase inhibitory activity. The list of selected marker metabolites might be useful for discrimination of developmental stages. It might not only be helpful in explaining the variability in the quality of *G. lingzhi* products, but can also be applicable for increasing cultivation efficiency through selection of optimum harvesting time of *G. lingzhi*. Our current findings may provide practical information for more comprehensive targeted analysis of selected markers in *G. lingzhi*, for prospective studies in the future.

Supplementary Materials: The following are available online. Figure S1: Principle Component Analysis (PCA) loading plot of metabolites derived from GC/MS data sets of *Ganoderma lingzhi* at eight developmental stages, Figure S2: Principle Component Analysis (PCA) loading plot of metabolites derived from LC/IT-TOF-MS data sets of *Ganoderma lingzhi* at eight developmental stages, Table S1: Tentatively identified chemical markers at eight developmental stages of *G. lingzhi* analyzed by GC/MS and LC/IT-TOF-MS.

Author Contributions: Conceptualization, D.S. and K.S.; Data curation, D.S.; Formal analysis, D.S. and S.T.; Funding acquisition, K.S.; Investigation, D.S., H.S. and S.K.; Methodology, D.S.; Resources, K.S.; Software, D.S.; Supervision, K.S.; Writing – original draft, D.S.; Writing – review & editing, S.T.

Funding: This work was supported by JSPS KAKENHI Grant-in-Aid for Scientific Research [17H03845].

Acknowledgments: We would like to thank Bisoken and Co. Ltd., Oita, Japan for providing the spawn of *G. lingzhi* BMC9049. We are greatly indebted to Aso Biotech, Kumamoto, Japan for cultivating and providing of *G. lingzhi* samples. We appreciate the technical assistance from the Center for Advanced Instrumental and Educational Supports, Faculty of Agriculture, Kyushu University. D.S. gratefully acknowledges an Indonesia Endowment Fund for Education, the Ministry of Finance of Indonesia, for scholarship support.

Conflicts of Interest: The authors declare no conflict of interest. The funding sponsors had no role in the design of the study; in the collection, analyses, interpretation of data; in the writing of the manuscript, and in the decision to publish the results.

References

1. Cheung, P.C.K. Mini-review on edible mushrooms as source of dietary fiber: Preparation and health benefits. *Food Sci. Hum. Wellness* **2013**, *2*, 162–166. [CrossRef]

2. Wachtel-Galor, S.; Tomlinson, B.; Benzie, I.F.F. *Ganoderma lucidum* ('Lingzhi'), a Chinese medicinal mushroom: Biomarker responses in a controlled human suppl. study. *Br. J. Nutr.* **2004**, *91*, 263–269. [CrossRef]
3. Wachtel-galor, S.; Szeto, Y.; Tomlinson, B.; Benzie, I.F. *Ganoderma lucidum* ('Lingzhi'); acute and short-term biomarker response to supplementation. *Int. J. Food Sci. Nutr.* **2004**, *55*, 75–83. [CrossRef]
4. Satria, D.; Amen, Y.; Niwa, Y.; Ashour, A.; Allam, A.E.; Shimizu, K. Lucidumol D, a new lanostane-type triterpene from fruiting bodies of Reishi (*Ganoderma lingzhi*). *Nat. Prod. Res.* **2019**, *33*, 189–195. [CrossRef] [PubMed]
5. Chien, C.-C.; Tsai, M.-L.; Chen, C.-C.; Chang, S.-J.; Tseng, C.-H. Effects on tyrosinase activity by the extracts of *Ganoderma lucidum* and related mushrooms. *Mycopathologia* **2008**, *166*, 117. [CrossRef] [PubMed]
6. Kim, J.-W.; Kim, H.-I.; Kim, J.-H.; Kwon, O.-C.; Son, E.-S.; Lee, C.-S.; Park, Y.-J. Effects of ganodermanondiol, a new melanogenesis inhibitor from the medicinal mushroom *Ganoderma lucidum*. *Int. J. Mol. Sci.* **2016**, *17*, 1798. [CrossRef] [PubMed]
7. Fatmawati, S.; Shimizu, K.; Kondo, R. Ganoderic acid Df, a new triterpenoid with aldose reductase inhibitory activity from the fruiting body of *Ganoderma lucidum*. *Fitoterapia* **2010**, *81*, 1033–1036. [CrossRef]
8. Fatmawati, S.; Shimizu, K.; Kondo, R. Inhibition of aldose reductase in vitro by constituents of *Ganoderma lucidum*. *Plant. Med.* **2010**, *76*, 1691–1693. [CrossRef]
9. Fatmawati, S.; Shimizu, K.; Kondo, R. Ganoderol B: A potent α-glucosidase inhibitor isolated from the fruiting body of *Ganoderma lucidum*. *Phytomedicine* **2011**, *18*, 1053–1055. [CrossRef] [PubMed]
10. Fatmawati, S.; Shimizu, K.; Kondo, R. Structure–activity relationships of ganoderma acids from *Ganoderma lucidum* as aldose reductase inhibitors. *Bioorg. Med. Chem. Lett.* **2011**, *21*, 7295–7297. [CrossRef]
11. Wachtel-Galor, S.; Yuen, J.; Buswell, J.A.; Benzie, I.F.F. Ganoderma lucidum (Lingzhi or Reishi): A Medicinal Mushroom. In *Herbal Medicine: Biomolecular and Clinical Aspects*; Benzie, I.F.F., Wachtel-Galor, S., Eds.; CRC Press/Taylor & Francis: Boca Raton, FL, USA, 2011; ISBN 978-1-4398-0713-2.
12. Hapuarachchi KK, K. Current status of global Ganoderma cultivation, products, industry and market. *Mycosphere* **2018**, *9*, 1025–1052. [CrossRef]
13. Liu, J.; Kurashiki, K.; Fukuta, A.; Kaneko, S.; Suimi, Y.; Shimizu, K.; Kondo, R. Quantitative determination of the representative triterpenoids in the extracts of *Ganoderma lucidum* with different growth stages using high-performance liquid chromatography for evaluation of their 5α-reductase inhibitory properties. *Food Chem.* **2012**, *133*, 1034–1038. [CrossRef]
14. O'Gorman, A.; Barry-Ryan, C.; Frias, J.M. Evaluation and identification of markers of damage in mushrooms (*Agaricus bisporus*) postharvest using a GC/MS metabolic profiling approach. *Metabolomics* **2012**, *8*, 120–132. [CrossRef]
15. Woldegiorgis, A.Z.; Abate, D.; Haki, G.D.; Ziegler, G.R. LC-MS/MS based metabolomics to identify biomarkers unique to *Laetiporus sulphureus*. *Int. J. Nutr. Food Sci.* **2015**, *4*, 141. [CrossRef]
16. Putri, S.P.; Fukusaki, E. *Mass Spectrometry-Based Metabolomics: A Practical Guide*; CRC Press: Boca Raton, FL, USA, 2014; ISBN 978-1-4822-2376-7.
17. Lei, Z.; Huhman, D.V.; Sumner, L.W. Mass spectrometry strategies in metabolomics. *J. Biol. Chem.* **2011**, *286*, 25435–25442. [CrossRef]
18. t'Kindt, R.; Morreel, K.; Deforce, D.; Boerjan, W.; Van Bocxlaer, J. Joint GC–MS and LC–MS platforms for comprehensive plant metabolomics: Repeatability and sample pre-treatment. *J. Chromatogr. B* **2009**, *877*, 3572–3580. [CrossRef]
19. Lee, S.; Do, S.-G.; Kim, S.Y.; Kim, J.; Jin, Y.; Lee, C.H. Mass spectrometry-based metabolite profiling and antioxidant activity of aloe vera (*Aloe barbadensis Miller*) in different growth stages. *J. Agric. Food Chem.* **2012**, *60*, 11222–11228. [CrossRef]
20. Lombardo, V.A.; Osorio, S.; Borsani, J.; Lauxmann, M.A.; Bustamante, C.A.; Budde, C.O.; Andreo, C.S.; Lara, M.V.; Fernie, A.R.; Drincovich, M.F. Metabolic profiling during peach fruit development and ripening reveals the metabolic networks that underpin each developmental stage. *Plant Physiol.* **2011**, *157*, 1696–1710. [CrossRef]
21. Nakagawa, T.; Zhu, Q.; Tamrakar, S.; Amen, Y.; Mori, Y.; Suhara, H.; Kaneko, S.; Kawashima, H.; Okuzono, K.; Inoue, Y.; et al. Changes in content of triterpenoids and polysaccharides in *Ganoderma lingzhi* at different growth stages. *J. Nat. Med.* **2018**, *72*, 734–744. [CrossRef]

22. Isidorov, V.A.; Lech, P.; Żółciak, A.; Rusak, M.; Szczepaniak, L. Gas chromatographic–mass spectrometric investigation of metabolites from the needles and roots of pine seedlings at early stages of pathogenic fungi. *Armillaria ostoyae attack. Trees* **2008**, *22*, 531. [CrossRef]
23. Birkemeyer, C.; Kopka, J. Design of Metabolite Recovery by Variations of the Metabolite Profiling Protocol. In Proceedings of the Concepts in Plant Metabolomics; Nikolau, B.J., Wurtele, E.S., Eds.; Springer: Dortrecht, The Netherlands, 2007; pp. 45–69.
24. Erxleben, A.; Gessler, A.; Vervliet-Scheebaum, M.; Reski, R. Metabolite profiling of the moss *Physcomitrella patens* reveals evolutionary conservation of osmoprotective substances. *Plant Cell Rep.* **2012**, *31*, 427–436. [CrossRef] [PubMed]
25. Isidorov, V.A.; Kotowska, U.; Vinogorova, V.T. GC identification of organic compounds based on partition coefficients of their TMS derivatives in a hexane-acetonitrile system and retention indices. *Anal. Sci.* **2005**, *21*, 1483–1489. [CrossRef]
26. Kim, J.-S.; Chung, H.Y. GC-MS analysis of the volatile components in dried boxthorn (*Lycium chinensis*) fruit. *J. Korean Soc. Appl. Biol. Chem.* **2009**, *52*, 516–524. [CrossRef]
27. Yaşar, A.; Üçüncü, O.; Güleç, C.; İnceer, H.; Ayaz, S.; Yayl, N. GC-MS analysis of chloroform extracts in flowers, stems, and roots of *Tripleurospermum callosum*. *Pharm. Biol.* **2005**, *43*, 108–112. [CrossRef]
28. Radulović, N.S.; Đorđević, N.D. Steroids from poison hemlock (*Conium maculatum* L.): A GC-MS analysis. *J. Serbian Chem. Soc.* **2011**, *76*, 1471–1483. [CrossRef]
29. Qian, Z.; Zhao, J.; Li, D.; Hu, D.; Li, S. Analysis of global components in *Ganoderma* using liquid chromatography system with multiple columns and detectors. *J. Sep. Sci.* **2012**, *35*, 2725–2734. [CrossRef] [PubMed]
30. Cheng, C.-R.; Yang, M.; Wu, Z.-Y.; Wang, Y.; Zeng, F.; Wu, W.-Y.; Guan, S.-H.; Guo, D.-A. Fragmentation pathways of oxygenated tetracyclic triterpenoids and their application in the qualitative analysis of *Ganoderma lucidum* by multistage tandem mass spectrometry. *Rapid Commun. Mass Spectrom.* **2011**, *25*, 1323–1335. [CrossRef] [PubMed]
31. Wu, L.; Liang, W.; Chen, W.; Li, S.; Cui, Y.; Qi, Q.; Zhang, L. Screening and analysis of the marker components in *Ganoderma lucidum* by HPLC and HPLC-MSn with the aid of chemometrics. *Molecules* **2017**, *22*, 584. [CrossRef]
32. Yang, M.; Wang, X.; Guan, S.; Xia, J.; Sun, J.; Guo, H.; Guo, D. Analysis of triterpenoids in *Ganoderma lucidum* using liquid chromatography coupled with electrospray ionization mass spectrometry. *J. Am. Soc. Mass Spectrom.* **2007**, *18*, 927–939. [CrossRef] [PubMed]
33. Kües, U.; Liu, Y. Fruiting body production in basidiomycetes. *Appl. Microbiol. Biotechnol.* **2000**, *54*, 141–152. [CrossRef]
34. Miles, P.G.; Chang, S.-T.; Chang, S.-T. *Mushrooms: Cultivation, Nutritional Value, Medicinal Effect, and Environmental Impact*; CRC Press: Boca Raton, FL, USA, 2004; ISBN 978-0-203-49208-6.
35. Trinci, A.P.J. A study of the kinetics of hyphal extension and branch initiation of fungal mycelia. *Microbiology* **1974**, *81*, 225–236. [CrossRef]
36. Boddy, L.; Watkinson, S.C. Wood decomposition, higher fungi, and their role in nutrient redistribution. *Can. J. Bot.* **1995**, *73*, 1377–1383. [CrossRef]
37. Tseng, Y.-H.; Lee, Y.-L.; Li, R.-C.; Mau, J.-L. Non-volatile flavour components of *Ganoderma tsugae*. *Food Chem.* **2005**, *90*, 409–415. [CrossRef]
38. Moore, D. *Fungal Morphogenesis*; Cambridge University Press: Cambridge, UK, 2002; ISBN 978-0-521-52857-3.
39. Etten, J.L.V.; Gottlieb, D. Biochemical changes during the growth of fungi. *J. Bacteriol.* **1965**, *89*, 6.
40. Park, Y.J.; Jung, E.S.; Singh, D.; Lee, D.E.; Kim, S.; Lee, Y.W.; Kim, J.-G.; Lee, C.H. Spatial (cap & stipe) metabolomic variations affect functional components between brown and white beech mushrooms. *Food Res. Int.* **2017**, *102*, 544–552.
41. Kim, Y.-J.; Uyama, H. Tyrosinase inhibitors from natural and synthetic sources: Structure, inhibition mechanism and perspective for the future. *CmlsCell. Mol. Life Sci.* **2005**, *62*, 1707–1723. [CrossRef]
42. Choi, G.J.; Lee, S.-W.; Jang, K.S.; Kim, J.-S.; Cho, K.Y.; Kim, J.-C. Effects of chrysophanol, parietin, and nepodin of *Rumex crispus* on barley and cucumber powdery mildews. *Crop Prot.* **2004**, *23*, 1215–1221. [CrossRef]
43. Shao, S.; Hernandez, M.; Kramer, J.K.G.; Rinker, D.L.; Tsao, R. Ergosterol profiles, fatty acid composition, and antioxidant activities of button mushrooms as affected by tissue part and developmental stage. *J. Agric. Food Chem.* **2010**, *58*, 11616–11625. [CrossRef]

44. Liang, W.-Q.; Zhang, D.-B.; Wang, N.; Wang, C.-G.; Pan, Y.-J. Cloning and characterization of squalene synthase (SQS) gene from *Ganoderma lucidum*. *J. Microbiol. Biotechnol.* **2007**, *17*, 1106–1112.
45. Zhao, M.W.; Zhong, J.Y.; Liang, W.Q.; Wang, N.; Chen, M.J.; Zhang, D.B.; Pan, Y.J.; Jong, S.C. Analysis of squalene synthase expression during the development of *Ganoderma lucidum*. *J. Microbiol. Biotechnol.* **2004**, *14*, 116–120.
46. Koyama, K.; Imaizumi, T.; Akiba, M.; Kinoshita, K.; Takahashi, K.; Suzuki, A.; Yano, S.; Horie, S.; Watanabe, K.; Naoi, Y. Antinociceptive components of *Ganoderma lucidum*. *Plant. Med.* **1997**, *63*, 224–227. [CrossRef]
47. Han, C. A Comparison of antinociceptive activity of mycelial extract from three species of fungi of basidiomycetes. *Open Complement. Med. J.* **2009**, *1*, 73–77. [CrossRef]
48. Costa, R.; De Grazia, S.; Grasso, E.; Trozzi, A. Headspace-Solid-Phase Microextraction-Gas Chromatography as Analytical Methodology for the Determination of Volatiles in Wild Mushrooms and Evaluation of Modifications Occurring during Storage. Available online: https://www.hindawi.com/journals/jamc/2015/951748/abs/ (accessed on 21 April 2019).
49. Wu, C.-G.; Tian, J.-L.; Liu, R.; Cao, P.-F.; Zhang, T.-J.; Ren, A.; Shi, L.; Zhao, M.-W. Ornithine decarboxylase-mediated production of putrescine influences ganoderic acid biosynthesis by regulating reactive oxygen species in *Ganoderma lucidum*. *Appl. Env. Microbiol.* **2017**, *83*, e01289-17. [CrossRef]
50. Hai-Bang, T.; Shimizu, K. Structure–activity relationship and inhibition pattern of reishi-derived (*Ganoderma lingzhi*) triterpenoids against angiotensin-converting enzyme. *Phytochem. Lett.* **2015**, *12*, 243–247. [CrossRef]
51. Akihisa, T.; Nakamura, Y.; Tagata, M.; Tokuda, H.; Yasukawa, K.; Uchiyama, E.; Suzuki, T.; Kimura, Y. anti-inflammatory and anti-tumor-promoting effects of triterpene acids and sterols from the fungus *Ganoderma lucidum*. *Chem. Biodivers.* **2007**, *4*, 224–231. [CrossRef] [PubMed]
52. Blaženović, I.; Kind, T.; Ji, J.; Fiehn, O. Software tools and approaches for compound identification of LC-MS/MS data in metabolomics. *Metabolites* **2018**, *8*, 31. [CrossRef] [PubMed]
53. Chi, L.; Bian, X.; Gao, B.; Tu, P.; Lai, Y.; Ru, H.; Lu, K. Effects of the artificial sweetener neotame on the gut microbiome and fecal metabolites in mice. *Molecules* **2018**, *23*, 367. [CrossRef] [PubMed]

Sample Availability: Samples of *G. lingzhi* at different developmental stages are available from the authors.

Article

Identification of Azoxystrobin Glutathione Conjugate Metabolites in Maize Roots by LC-MS

Giuseppe Dionisio [1,*,†], Maheswor Gautam [2,†,‡] and Inge Sindbjerg Fomsgaard [2]

1 Department of Molecular Biology and Genetics, Research Center Flakkebjerg, Aarhus University, Forsøgsvej 1, 4200 Slagelse, Denmark

2 Department of Agroecology, Research Center Flakkebjerg, Aarhus University, Forsøgsvej 1, 4200 Slagelse, Denmark; maheswor.gautam@gmail.com (M.G.); inge.fomsgaard@agro.au.dk (I.S.F.)

* Correspondence: giuseppe.dionisio@mbg.au.dk

† These authors have contributed equally to the development of this manuscript.

‡ Current address: Department of Biological Sciences, University of Alberta, Canada.

Academic Editor: Thomas Letzel
Received: 9 June 2019; Accepted: 3 July 2019; Published: 5 July 2019

Abstract: Xenobiotic detoxification in plant as well as in animals has mostly been studied in relationship to the deactivation of the toxic residues of the compound that, surely for azoxystrobin, is represented by its β-methoxyacrylate portion. In maize roots treated for 96 h with azoxystrobin, the fungicide accumulated over time and detoxification compounds or conjugates appeared timewise. The main detoxified compound was the methyl ester hydrolysis product (azoxystrobin free acid, 390.14 *m*/*z*) thought to be inactive followed by the glutathione conjugated compounds identified as glutathione conjugate (711.21 *m*/*z*) and its derivative lacking the glycine residue from the GSH (654.19 *m*/*z*). The glycosylated form of azoxystrobin was also found (552.19 *m*/*z*) in a minor amount. The identification of these analytes was done by differential untargeted metabolomics analysis using Progenesis QI for label free spectral counting quantification and MS/MS confirmation of the compounds was carried out by either Data Independent Acquisition (DIA) and Data Dependent Acquisition (DDA) using high resolution LC-MS methods. Neutral loss scanning and comparison with MS/MS spectra of azoxystrobin by DDA and MSe confirmed the structures of these new azoxystrobin GSH conjugates.

Keywords: azoxystrobin; glutathione; glutathione conjugate; untargeted metabolomics

---

1. Introduction

Azoxystrobin is one of the widely used fungicides in plants. The metabolism of azoxystrobin in plants is well described [1] and recently our laboratory has reported mass spectrometric methods to identify and quantify azoxystrobin metabolites in plants [2–4]. However, glutathione conjugation of azoxystrobin in plants has not been reported yet [1,5]. The only known report on GSH conjugation of azoxystrobin comes from an animal metabolism study where azoxystrobin was reported to conjugate with glutathione in the cyanophenyl aromatic ring with the electrophilic carbon in the –ortho position to the cyano group [6]. However, a later in vitro experiment undertaken to conjugate azoxystrobin with glutathione by plant-derived glutathione transferase was unsuccessful [5]. Glutathione (GSH) conjugation is a crucial step in the detoxification process of xenobiotics in living organisms [5–7]. GSH conjugation occurs in the cytoplasm of plant cells when electrophilic centers in xenobiotics are attacked by the nucleophilic thiol (–SH) group of glutathione [5]. This process increases the solubility of xenobiotics, which can thus be transported from cytoplasm to vacuoles. As increasing concentration of GSH-xenobiotic could be potentially harmful for plant cells, the conjugates are sequentially degraded in vacuoles by enzymatic actions thereby enabling plants to tackle harmful effects of xenobiotics [7]. During the degradation of GSH-xenobiotic conjugate in vacuoles, glycine residue from GSH is cleaved off enzymatically, which results in glutamic acid and cysteine residue of GSH attached to the xenobiotic [5,7].

GSH detoxification has been shown to occur in xenobiotics compound by conjugation, at a structural level, to aliphatic, aromatic, benzylic, disulfide, and thioester groups [8]. However, no other conjugation than aromatic conjugation of azoxystrobin with GSH has been reported or studied so far. We hypothesized that the carbon in the –CN group of azoxystrobin is a favorable candidate for the nucleophilic attack by the –SH group in GSH. The synthesis of analytical standards of such conjugates of GSH with azoxystrobin was out of the scope of the present study. Thus, we aimed to verify the presence of GSH-azoxystrobin conjugates at the –CN group by suspect screening of such metabolites and others possible ones by using high-resolution mass spectrometry. Our initial candidate structure for GSH conjugates were aromatic and cyano conjugation with azoxystrobin. However, the algorithm used here to detect GSH conjugates of azoxystrobin, a priori of its structure determination, is to detect at MS/MS level the simultaneous presence of GSH fragments neutral loss and the main azoxystrobin MS/MS fragments. Untargeted LC-MS/MS analysis was automated by using Progenesis QI for metabolomics where the candidate MS scans were compared also over time for increased appearance due to the plant detoxification of azoxystrobin. Once we had identified by MS/MS spectral matching we did confirm its reproducibility in detection by setting up a method to use in future for targeted detection by quadrupole-linear ion trap mass spectrometry to develop sensitive triple-quadrupole mass spectrometry-based methods for quantification of GSH-azoxystrobin conjugates.

The conjugation of GSH to xenobiotics is operated enzymatically by plant Glutathione Transferases (GSTs), which is a multi-functional superfamily of detoxification enzymes [9]. They possess a hydrophobic binding domain (H-site) and the ability to bind glutathione (G-site). The greatest similarity between members of the GST family is observed in the GSH binding domain containing a motif of four highly conserved amino acids [10]. Here, a catalytically essential tyrosine or serine activates glutathione by lowering the pKa of the thiol group from around pH 9 to approximately pH 6, thus enhancing the rate of nucleophilic attack by the resulting thiolate anion toward electrophilic substrates at physiological pH [11–13]. The H site is more variable, accounting for the broad range of endogenous and xenobiotic substrates utilized by this family of GSH dependent enzymes [13].

Despite the fact that reduced glutathione (GSH) is capable of undergoing spontaneous conjugation with compounds containing electrophilic centers for the detoxification of several xenobiotics compounds, different classes of GSTs are, in fact, involved because they can lower by several fold the Km related to the nucleophilic thiolate anion attack of GSH to the electrophilic substrates, with concomitant displacement of a leaving group (i.e., a proton) which more rarely takes part in addition reactions [14].

In our former study, we established a maize root culture model system to study the biotransformation of azoxystrobin in plants [3]. In the current study, we report on more results from the use of maize root culture as a model system to study not only the demethylated metabolite, azoxystrobin free acid (AzFA), which represents the main plant metabolite but also, in lower extent of abundancy, the GSH-azoxystrobin conjugates.

2. Results

The first observation to note in treated maize roots samples as compared to control is the accumulation of azoxystrobin (monoisotopic mass 403.1168), as expected, judged by looking at the appearance of the peak at 404.1197 *m/z* representing MH+ of azoxystrobin in the MS survey profile (Figure 1), that is accumulating after 96h as compared to the control. Further evidence that the peak of 404.1197 *m/z* eluting at about 28.5 min represents for real the MH+ of azoxystrobin is given by its fragmentation pattern when observing its MS/MS profile (Figure 2). Furthermore, the major methyl ester hydrolysis product of azoxystrobin, azoxystrobin free acid, was differentially present in the treated samples as compared to the controls (Supplementary Figure S1).

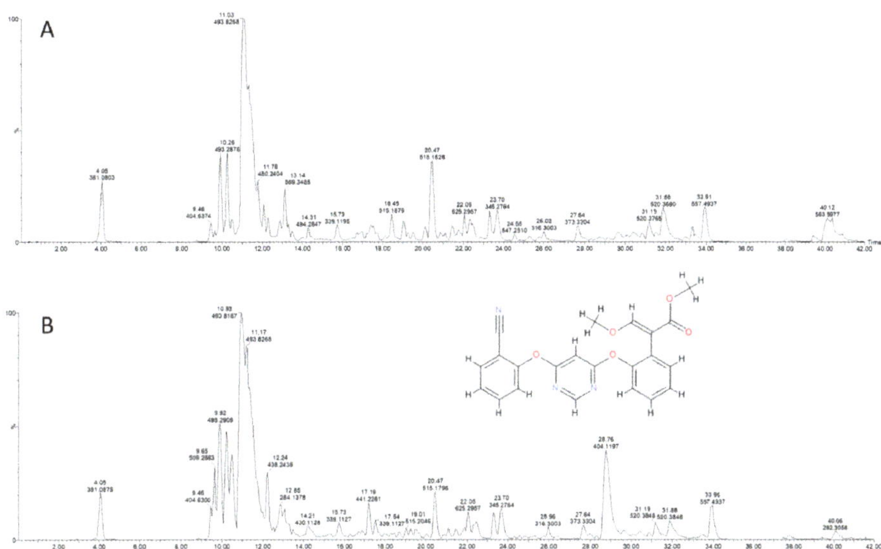

Figure 1. Q-TOF survey profile of the total metabolites extracted from maize root in control (panel **A**) and azoxystrobin treated ones (Panel **B**). Untargeted MS of roots methanol extracts at 96h has revealed the presence of a differential peak occurring at about 28.5 min with 404.1197 as *m*/*z* which is correspond to MH+ of azoxystrobin (exact mass, 403.1168; IUPAC Name, methyl (*E*)-2-[2-[6-(2-cyanophenoxy)pyrimidin-4-yl]oxyphenyl]-3-methoxy-prop-2-enoate). In the ordinate is shown the signal intensity in percentage; in the abscissa the retention time (RT) in minutes.

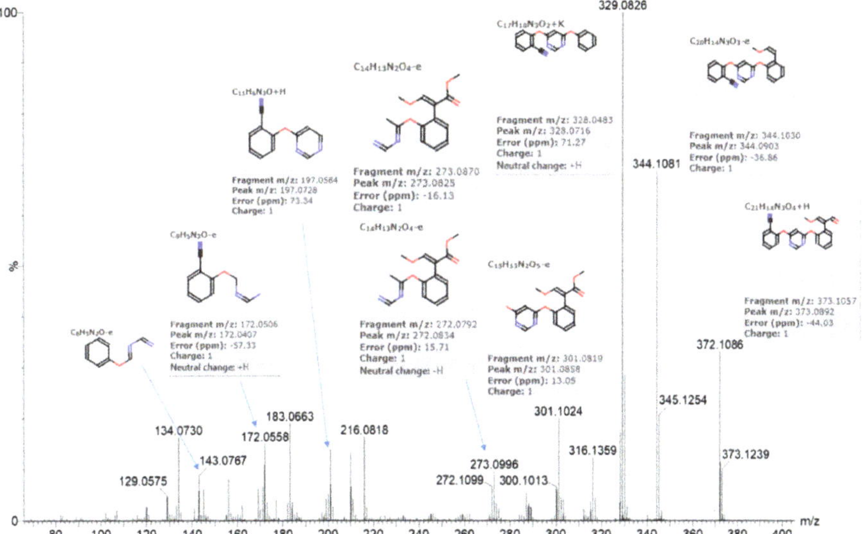

Figure 2. Q-TOF MS/MS DDA profile of the root metabolite extracted from maize root at 28.667 min as retention time. The precursor specie isolated by the quadrupole at 404.09 *m*/*z* was subjected to high energy ramping from 10 to 40 eV and its MS/MS profile shown here clearly indicates the presence of fragments of azoxystrobin as assessed by comparing the peak with similar fragmentation pattern from massbank (http://www.massbank.jp/RecordDisplay.jsp?id=AU324504&dsn=Athens_Univ). The peak annotation has extracted by Progenesis QI Metascope built-in fragmentation and annotation system.

The conjugated azoxystrobin metabolite analysis relied on the fact that the MS/MS spectra of the main peaks of azoxystrobin-conjugates should contain, at least, one of the most abundant MS/MS peak as shown in Figure 2. Using Masslynx function for extracting selected peaks from a chromatogram, it is possible to create one extracted chromatogram profile at survey as well as at MS/MS level. By scanning the MS/MS fragment in order to filter out peaks with 372 *m*/*z* from the MS/MS spectra of the treated versus untreated controls it was possible to visualize the presence of four main peaks at MS/MS level that contains the species 372 *m*/*z* (Figure 3). Four potential azoxystrobin metabolites containing at MS/MS level the 372 *m*/*z* species were found at retention time (RT) respectively: (a) 18.36 min; (b) 18.68 min; (c) 19.72 min; and (d) 30.5 min, while RT 28.79 min was related to the azoxystrobin itself.

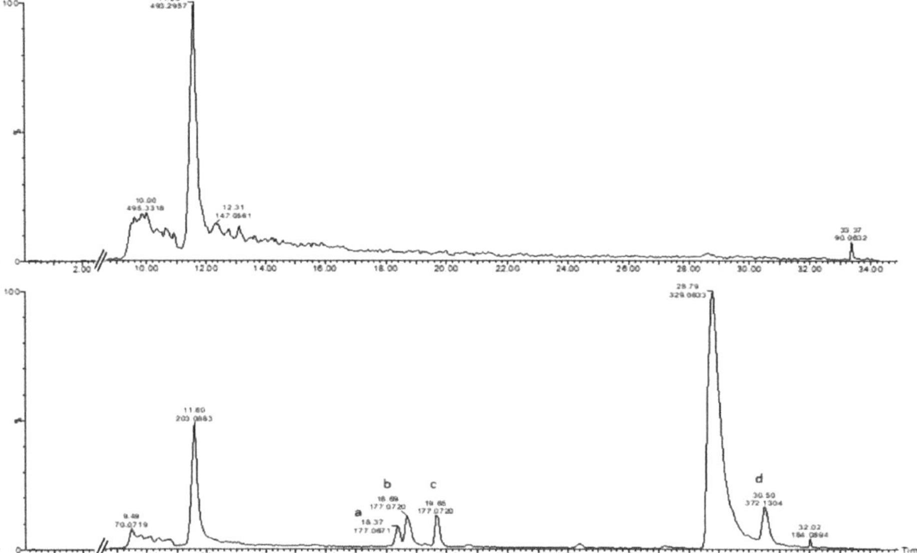

Figure 3. Extracted MS chromatograms at MS/MS (MSe) level of the precursor specie 372 *m*/*z*. A comparative extracted chromatogram for control untreated (**top** MS/MS trace) and treated with azoxystrobin (**lower** MS/MS trace) extracted and analyzed after 96 h. Four potential azoxystrobin metabolites are differentially displayed containing the 372 *m*/*z* species as visualized by the related extracted chromatogram at *m*/*z* 372 with different retention times: (a) RT 18.36 min; (b) RT 18.68 min; (c) RT 19.72 min; and (d) RT 30.5 min, being RT 28.79 min related to the azoxystrobin itself.

Matches were found in relation to the metabolites proposed to be azoxystrobin-glutathione conjugates at the cyano group due to the bonding of sulphur with the electrophilic carbon with monoisotopic mass 710.200 (MH+ 711.30 *m*/*z*) and after glycil neutral loss 653.1792 (MH+ 654.19 *m*/*z*) (Figure 4). The peak eluting at retention time (RT) 18.36 min is therefore compatible with precursor ion 711.30 *m*/*z* (Figure 5) and the peak eluting at RT 18.67 min could be represented by the same azoxystrobin-glutathione conjugate with glycil neutral loss, represented by the precursor ion 654.19 *m*/*z* (Supplementary Figure S2). The correct identification of potential azoxystrobin conjugate at RT 19.72 min was indicated by the neutral loss of 162 Da (glucosyl) and corresponds to azoxystrobin free acid glucoside with a monoisotopic mass of 551.1540 Da (MH+ 552.19 *m*/*z*) (Supplementary Figure S3). An analysis with QTRAP IDA method revealed similarity in MS/MS spectra for 711.3 *m*/*z* including the evidence for a neutral loss of 129 corresponding to the loss of glutamic acid residue as shown in Supplementary Figure S4. Finally, the correct identification of potential azoxystrobin conjugate at RT 30.55 min was revealed by the neutral loss discovery of demethylated azoxystrobin-GSH conjugate of the precursor specie 697.48 *m*/*z* (Supplementary Figure S5).

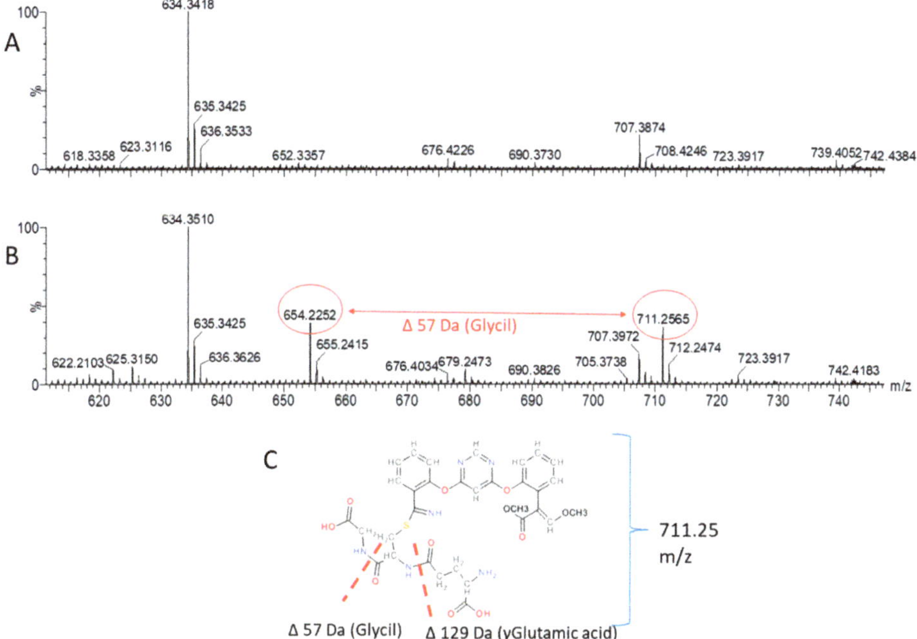

Figure 4. Neutral loss discovery of azoxystrobin-GSH conjugate at MS survey conditions of the precursor specie 711 *m/z*. Possible identification of potential azoxystrobin conjugate at RT 18.10–18.36 min. The analysis of MS survey profile for untreated (panel **A**) and treated sample (panel **B**) reveals that both 711.25 and 654.22 *m/z* species are present only in the treated sample. The glycil neutral loss (654.22 *m/z*) is already occurring at +5eV low energy level (MS1 or MS survey). The azoxystrobin-glutathione conjugate (panel **C**) at the cyano group has the MS/MS signature of azoxystrobin alongside the evidence of neutral loss of the glycil residue (neutral loss 57).

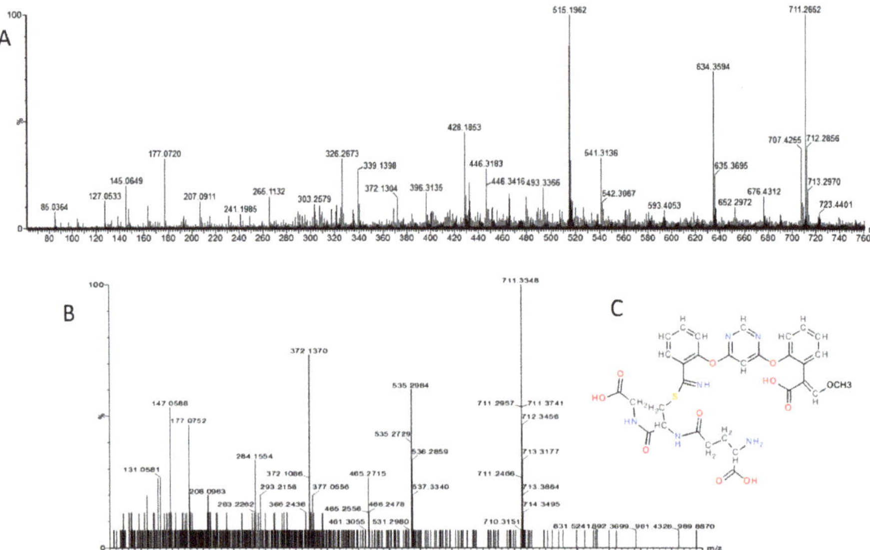

Figure 5. Analysis of the precursor specie 711 *m*/*z*. The identification of potential azoxystrobin conjugate at RT 18.36 min was possible by the analysis of MS survey profile (panel **A**) of azoxystrobin treated roots and its DDA MS/MS profile (panel **B**). The isolated MS/MS fragmentation profile reveals that the precursor of this azoxystrobin-conjugate/metabolite fits well with the azoxystrobin glutathione conjugate (panel **C**) at the cyano group compatible with the compound 711.30 *m*/*z* as confirmed by the presence of 372 *m*/*z* in MS/MS of the isolated peak at 711.30 *m*/*z* (panel B). Further confirmation of this compound has been done by studying the neutral losses.

The presence of these metabolites of azoxystrobin in maize roots was also investigated by differential untargeted metabolomics using Progenesis QI ver. 2.0. A database of SDF structures to search such metabolites was custom made from known and possible metabolites and validation of such metabolites was performed by the internal Metascope theoretical fragmentation engine using DDA MS/MS fragmentation profile both at MS1 and the most abundant MS/MS fragments.

The top relative maximum abundance is shown in Table 1 and it was recorded respectively for compounds at *m*/*z* 404.1068, 372.0798, 549.1369, 549.1715, 654.1935, 711.2189, 557.1727, and 552.1740. The validation of each peak was performed by looking at the control versus the treated root MS/MS profile for each *m*/*z* species above which a signature of azoxystrobin MS/MS spectra could be found.

Table 1. Possible azoxystrobin containing peaks differentially present in the treated maize root as compared untreated roots over 96 h. The searched azoxystrobin SDF structures and related metabolites present here were not all validated since most of them were also present in the control roots. Their relative max abundance as compared to the control and spectral counting is shown in the last column, where clearly the relative max abundance was recorded respectively for compounds at *m/z* 404.1068, 372.0798, 549.1369, 549.1715, 654.1935, 711.2189, 557.1727, and 552.1740. In bold are reported the compounds which have been confirmed being azoxystrobin metabolites.

Compound (RT_*m/z*)	*m/z*	z	RT	PW	Anova (p)	q Value	FC	HM	LM	Max Abundance
18.18_711.2189 *m/z*	**711.2189**	**1**	**18.18**	**0.61**	**$<1.1 \times 10^{-16}$**	**$<1.1 \times 10^{-16}$**	**18.7**	**AZO_96h**	CTRL_24h	139.9
19.33_552.1740 *m/z*	**552.1740**	**1**	**19.33**	**0.69**	**1.11×10^{-16}**	**4.58×10^{-15}**	**27.8**	**AZO_96h**	CTRL_48h	50.7
18.47_654.1935 *m/z*	**654.1935**	**1**	**18.47**	**0.61**	**2.22×10^{-16}**	**8.10×10^{-15}**	**110**	**AZO_96h**	CTRL_3h	162.0
28.70_403.0995 *m/z*	**404.1068**	**1**	**28.70**	**3.12**	**3.33×10^{-16}**	**1.13×10^{-14}**	**2.98×10^{-3}**	**AZO_3h**	CTRL_48h	8260.1
28.70_372.0798 *m/z*	**372.0798**	**1**	**28.70**	**2.94**	**5.33×10^{-16}**	**1.28×10^{-13}**	**8.05×10^{3}**	**AZO_3h**	CTRL_48h	6041.6
16.10_534.1777 *m/z*	534.1777	1	16.10	0.64	2.12×10^{-10}	1.60×10^{-9}	3.91	AZO_96h	CTRL_3h	57.2
23.22_557.1621 *m/z*	557.1621	1	23.22	1.25	4.47×10^{-9}	2.41×10^{-8}	6.76	CTRL_3h	AZO_96h	3340.4
13.32_743.2417 *m/z*	743.2417	1	13.32	0.59	5.68×10^{-9}	3.01×10^{-8}	3.46	AZO_96h	CTRL_3h	63.9
11.19_549.1369 *m/z*	549.1369	1	11.19	0.96	4.20×10^{-6}	1.27×10^{-5}	4.68	AZO_96h	CTRl_3h	229.8
21.12_557.1727 *m/z*	557.1727	1	21.12	1.17	7.97×10^{-6}	2.30×10^{-5}	2.39	AZO_3h	CTRl_72h	104.5
34.25_404.1264 *m/z*	404.1264	1	34.25	0.80	1.02×10^{-5}	2.89×10^{-5}	Infinity	AZO_3h	CTRL_3h	12.9
3.90_1103.2994 *m/z*	1103.2994	1	3.90	0.45	2.50×10^{-5}	6.57×10^{-5}	4.87	CTRL_3h	CTRL_48h	20.4
21.25_663.2184 *m/z*	663.2184	1	21.25	0.59	9.41×10^{-5}	0.00022	2.78	AZO_3h	AZO_72h	26.5
13.21_426.1160 *m/z*	426.1160	1	13.21	0.85	0.000114	0.000263	1.74	CTRL_72h	CTRL_3h	94.7
13.37_549.1715 *m/z*	549.1715	1	13.37	2.59	0.0258	4	1.42	AZO_3h	CTRL_24h	330.3
9.32_367.0946 *m/z*	367.0946	2	9.32	0.51	0.0296	456	2.84	AZO_3h	CTRL_3h	57.5
1.98_404.1227 *m/z*	404.1227	1	1.98	0.24	0.143	0.197	Infinity	AZO_24h	CTRL_24h	14.2
3.82_370.0885 *m/z*	370.0885	2	3.82	0.51	0.208	0.278	2.36	CTRL_96h	AZO_96h	5.3
1.52_404.1226 *m/z*	404.1226	1	1.52	0.35	0.31	0.4	Infinity	CTRL_3h	CTRL_24h	3.6
9.32_355.0933 *m/z*	355.0933	2	9.32	0.51	0.336	0.432	4.08	AZO_3h	CTRL_3h	7.5
38.31_404.1245 *m/z*	404.1245	1	38.31	0.45	0.508	0.629	Infinity	AZO_48h	CTRL_3h	13.4
37.53_404.1205 *m/z*	404.1205	1	37.53	0.56	0.607	0.74	Infinity	AZO_72h	CTRL_3h	16.6
24.29_743.2418 *m/z*	743.2418	1	24.29	0.61	0.704	0.838	1.38	AZO_3h	AZO_96h	11.4
3.71_380.1017 *m/z*	380.1017	2	3.71	0.32	0.978	>0.9995	4.22	CTRL_96h	CTRL_24h	18.3

Z = charge; RT = retention time; PW = peak width; FC = Fold change; HM = highest mean; LM = lowest mean.

Not all the metabolites of azoxystrobin were found by the Progenesis Metascope engine but at least the main azoxystrobin conjugates were detected in order of significance (ANOVA) and abundance. The peaks were presented by precursor ions 654.1935 *m*/*z*, 711.2189 *m*/*z*, and 552.1740 *m*/*z*, corresponding respectively to azoxystrobin-GSH conjugate with glycil loss, azoxystrobin-GSH conjugate, and azoxystrobin free acid glucoside conjugate (Figure 6). Time course accumulation of selected metabolites have been chosen for visualization in Progenesis QI where the untreated control (CTRL) and the azoxystrobin treated samples (AZO) have been analyzed at 3 h, 24 h, 48 h, and 96 h (Figure 6). The typical representation of the trend of *m*/*z* over time and its isotope distribution can be visualized in 2D or 3D and the related area quantified by relative units which are normalized by abundance related to a standard (i.e., CTRL) in Figure 6.

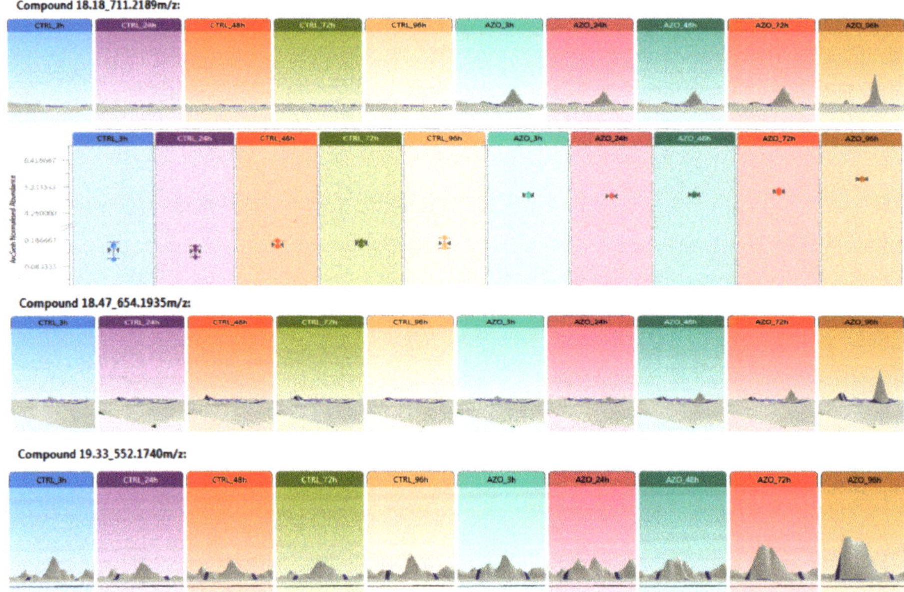

Figure 6. Relative Ion abundance distribution of azoxystrobin-GSH (glutathione) conjugate at MS survey conditions as quantified by the Progenesis QI software for metabolomics. The label free relative spectral counting quantification analysis (here depicted as 3D isotope abundance) has revealed the differential presence of the *m*/*z* 711.2189, accumulating over time (max at 96 h) in azoxystrobin treated roots versus the untreated control. The same trend has been detected for the peaks 654.1935 *m*/*z* and 552.1740 *m*/*z*.

All the rest of Progenesis identified peaks at MS1 containing at MS/MS level the 372 *m*/*z* species, shown in Table 1, were not validated because they were also present in the control roots extract and lacked full similarity to all MS/MS spectra of azoxystrobin.

The QTRAP MS was hence used to setup a quantification method to be further implemented in future screening of the 711.3 *m*/*z* metabolite in plant once it would be commercially available to be used as standards. The MS conditions used for the future QQQ detection of 711.3 *m*/*z* metabolite are summarized in Table 2.

Table 2. Mass spectrometric parameters used for the Multiple Reaction Monitoring-Information Dependent Acquisition (MRM-IDA) spectra acquisition of glutathione conjugate of azoxystrobin.

Scan type	Q1	Q3	DP (V)	EP (V)	CE (V)	CXP (V)	CUR (psi)	CAD	IS (V)	TEM (°C)	GS 1 (psi)	GS 2 (psi)
	711	404	60	10	30	13	35	High	4300	450	90	50
MRM	711	372	60	10	30	13	35	High	4300	450	90	50
	711	582	60	10	30	13	35	High	4300	450	90	50

3. Discussion

Azoxystrobin is a strobilurin fungicide that was developed by Zeneca UK Ltd., in 1992. Its mechanism of action in fungi and bacteria rely on inhibiting the electron transport system by binding the Qo site of mitochondrial cytochrome bcl complex to inhibit microbial respiration [15]. The toxophore (chemical group that produces the toxic effect) portion of the molecule is actually not the cyano group, but the β-methoxyacrylate portion that is double methylated and that, hence, should represent the immediate sensitive region of the molecules for enzymatic deactivation by conjugation or demethylation. One of the azoxystrobin environmental degradation products, azoxystrobin free acid, was present in the methanol extracts of maize roots for the treated samples as compared to the controls (Supplementary Figure S1) indicating an active demethylation mechanism of detoxification occurring *in vivo*. This initial demethylation opens the molecule to a definitive inactivation of its toxophore by glycosylation as was shown by the presence of the peak with 552.1740 *m/z* (Supplementary Figure S3).

The conjugation of azoxystrobin with glutathione by enzymatic action of glutathione S-transferase enzyme has not been described yet in plants, although a glutathione (GSH) conjugate of azoxystrobin has been identified in rats where sulphur is conjugated to the phenolic ring in the ortho position relative to the cyano group where the conjugate species have shown *m/z* values of 708.18 and 694.16 respectively for the fully methylated and demethylated GSH azoxystrobin conjugate [6] not found here in our experiments. Here, we present mass spectrometric evidences that GSH was conjugated with azoxystrobin through the –CN group thereby potentially allowing detoxification of azoxystrobin. We have also proposed chemical structures of the corresponding precursors of *m/z* 711.2189 (Figures 4 and 5) and 654.1935 (Supplementary Figure S2) as being full azoxystrobin-GSH and azoxystrobin-cysteine-γglutamic acid conjugates. During the MS survey at low energy, the glycine neutral loss is already present in the MS/MS spectra of *m/z* 711.2189 (Figure 4), but the elution of the peak with *m/z* 654.1935 is at different time than that with *m/z* 711.2189 meaning that the loss could be happening *in vivo*. In barley, Glutathione S-conjugates accumulates in the plant vacuole and that the first step of its degradation, the formation of the respective T-glutamylcysteinyl-S-conjugate, is catalyzed by a vacuolar carboxypeptidase [16]. Following the action of a vacuolar carboxypeptidase that degrades GS-conjugates by cleaving the glycine, it is also reported a removal of the amino terminal glutamic acid residue, cleaved also successively enzymatically [17]. Besides, a membrane associated γ-glutamyl transpeptidase (γ-GTase) has been identified [18], as part of the transport from the vacuole to the cytosol, that then removes the terminal glutamic acid leaving the S-cysteinyl derivative [7]. The loss of the glycine residue from glutathione conjugates of a xenobiotic is well documented [5,19] and is known to occur during transport of xenobiotic glutathione conjugates from cytosol to vacuoles (for review see [5]). Alongside the glycine neutral loss (Δ57) (Figure 4), γ-Glutamic acid neutral loss (Δ129) has been detected (Supplementary Figure S4) confirming the nature of the analyte with precursor ion *m/z* 711.2189. The neutral loss of 129 Da corresponding to the loss of glutamic acid from GSH conjugates is previously described [20] and validated by the method for simultaneous screening of GSH and CN adducts using precursor ion (PI) and neutral loss (NL) scans dependent product ion spectral acquisition and datamining tools on QTRAP mass spectrometry [21]. Here, hence, we have demonstrated that the 711.3 *m/z* metabolite possesses a 129 Da neutral loss (Supplementary Figure S4) and that can be quantified by QQQ MS methods (Table 2) once the standard is available as pure compound.

Naturally occurring isothiocyanate compounds present in cruciferous vegetables, e.g., sulforaphane, are powerful inducers of glutathione S-transferase activities in laboratory animals, meaning that the GSH conjugation at the cyano group is operated enzymatically and the reactive group is represented by the carbonyl portion of the isothiocyanate [22]. Similar results have been shown for phenyl isocyanate detoxification by glutathione conjugation [23]. Here we have proposed and have shown evidence that glutathione is conjugated to the electrophilic carbon of the cyano group (–CN) of azoxystrobin. Further experiments are currently underway in order to identify possible enzymes, i.e., glutathione S-transferases, that might mediate in vivo GSH conjugation and detoxification of azoxystrobin in maize roots.

4. Materials and Methods

4.1. Chemicals

Analytical grade azoxystrobin (purity 99.4%) and sucrose (≥99.5%) were purchased from Sigma-Aldrich (Copenhagen, Denmark). Azoxystrobin free acid ((*E*)-2-(2-((6-(2-cyanophenoxy) pyrimidin-4-yl)oxy)phenyl)-3-methoxyacrylic acid) (10 ng/μL in acetonitrile) was procured from Dr. Ehrenstorfer GmbH (Augsburg, Germany). Acetonitrile and methanol (TOF/MS grade) were obtained from Fisher Chemical (Roskilde, Denmark). Milli-Q water was generated from a Milli-Q system (Millipore, MA, USA). Murashige and Skoog (M&S) media was purchased from Duchefa Biochemie (Harlem, The Netherlands).

4.2. Maize Root Culture Growth

Maize seeds (*cv.* Kaspian) were purchased from KWS Scandinavia (Vejle, Denmark) and maize roots were cultured as explained in our earlier study [3]. Briefly, maize seeds were surface sterilized with ethanol and sodium hypochlorite and rinsed with sterilized Milli-Q water. The seeds were transferred into petri plates containing sucrose and full-strength M&S media that had been solidified with phytagel. The pH of the media was maintained at 5.8. The plates were incubated under light/dark condition for 16/8 h at 23 °C for seven days. The roots from germinated seeds were detached and transferred into Erlenmeyer flasks containing half-strength M&S and sucrose where the pH was maintained at 5.8. The flasks were covered with aluminum foil to maintain a dark condition within the flasks. The flasks were shaken at 100 rpm at 25 °C for seven days and the maize root culture was ready for azoxystrobin uptake and metabolism study.

4.3. Azoxystrobin Uptake

Azoxystrobin was dissolved in methanol and added to sterile half-strength liquid M&S growth medium (pH 5.8) contained in a beaker to yield a final azoxystrobin concentration 100 μM in 500 mL medium as explained in Gautam et al. (2018). Maize roots were transferred into the beaker and swirled gently to mix azoxystrobin thoroughly. A beaker with similar content except azoxystrobin was used as control. Azoxystrobin treatment continued for three hours after which a portion of roots (5–10 g) was cut-off with sterile scissors, washed thoroughly with sterile water, wiped with tissue paper, wrapped in aluminum foil, and flash frozen in liquid nitrogen. Control maize roots was sampled similarly after three hours. The samples were labelled as 3 h sample.

The treatment solution was decanted off, maize roots were first washed with sterilized Milli-Q water and then with M&S medium, and finally transferred to sterile liquid M&S medium devoid of azoxystrobin. The control was also washed and transferred similarly. The flasks were wrapped with aluminum foil to prevent exposure to light and allowed to stand in a sterile bench throughout the uptake experiment period. Samples were collected after 3, 24, 48, 72, and 96 h of initial exposure with azoxystrobin.

4.4. Sample Preparation and Extraction

Frozen maize roots were freeze dried on a DryWinner 6-85 freeze drier (Holm & Halby, Brøndby, Denmark) while still wrapped in aluminum foil. The freeze-dried roots were ground in GenoGrinder 100 (Spex, Metuchen, NJ, USA). 20.0 mg of homogenized root tissue was weighed in 2-mL Eppendorf tube and suspended in 1 mL of 50% methanol (*v/v*) containing 0.1% formic acid. The tube was shaken vigorously, ultrasonicated for 15 min, and centrifuged at 16,000× *g* at 20 °C for 15 min in a Sigma 1-14K microcentrifuge (Buch and Holm, Herlev, Denmark). The supernatant was filtered with 0.22 μm KX Syringe PTFE filter (Mikrolab, Aarhus, Denmark) and analyzed immediately with or stored at −20 °C before analysis.

4.5. LC-MS/MS Analysis

4.5.1. Q-TOF Analysis

Methanol content in the extracts were adjusted to 25% before injecting. The liquid chromatography system was composed of a nano Acquity UPLC (Waters, Milford, CT, USA) equipped with a nano Trap column (Waters XSelect™ HHS T3, 5 μm beads diameter, 180 μm × 20 mm) for accumulating the injected samples and an analytical capillary column with the same chemistry (Waters nano RP-C18 HHS T3 type, 1.8μm, 150μm × 100 mm) for sample separation. The nano capillary column was attached with a nano ESI source to a Waters Q-TOF premiere (Waters, Milford, USA). The Q-TOF operational conditions were the following: 3.2 kV for capillary cone voltage and extraction, 30 kV as sampling cone voltage, and 3.4 kV as ion guide voltage, and the source temperature was set to 120 °C. The instrument was calibrated in V positive mode using the MS/MS profile of M2H + 785.8426 *m/z* GluFib-B ([Glu1]-Fibrinopeptide B human, F3261, Sigma Aldrich, Merck, Germany) as Z-lock mass (reference calibration) at 23 eV as collision energy obtaining a calibration range from 72 to 1286 Da, and with a ppm error of +/− 2.

The entire length of the LC run was 42 min. Mobile phase A was 0.1% formic acid (FA, Sigma Aldrich, Germany) and phase B was acetonitrile (ACN, Thermo-Fischer Scientific, Hvidovre, Denmark) with 0.1% FA. The gradient conditions were from 15% phase B to 70% phase B in 25 min and from 25 min to 27 min from 70% phase B to 85% phase B; from 27 to 30 min, the percentage of mobile phase B raised from 85% to 95%. After 30 min the percentage of phase B was constant set to 95%. Data acquisition was performed in Masslynx version 4.1 (Waters, Milford, USA) either as data-independent acquisition (DIA) mode by all ion fragmentation (MSe) with MS1 and MS/MS (MS2) range from 50 to 2000 *m/z*, or data dependent (DDA). For the DDA mode the MS1 survey was acquired from 50 to 1000 *m/z* and MS/MS was acquired from 500 Da above after peak deisotoping: scan time 1.5 s with variable collision energy from 5 to 40 V.

Differential untargeted azoxystrobin metabolite, by the means of spectral counting comparison, was performed by Progenesis QI, version 2.0 (Nonlinear Dynamics, a Waters company, Newcastle upon Tyne, UK). The import of raw MSe data acquired by Masslynx was transformed into centroid data by M2H + 785.8426 *m/z* GluFib-B lock mass data calibration and refinement including dead time correction and charge deconvolution. Data runs, in triplicate, were hence layered into 2D graphs obtained by the retention time versus the *m/z* indexing of the MS1 and MS/MS peaks. After the multirun 2D alignment the peptide3D Apex3D algorithm was used to peak picking and quantification [24]. Label-free ion/adducts quantitation was performed by general all ion compound normalization using built in LOWESS (locally weighted scatterplot smoothing) algorithm and weighted spectral counting alongside the ion/adduct isotopic peaks [25]. The search algorithm was the built-in Progenesis Metascope using a custom Structure Data File (SDF) databases created by joining together many single chemical 2D structures of azoxystrobin precursors as SDF files by Progenesis SDF studio. Chemical 2D structures of azoxystrobin precursors were downloaded from ChemSpider website or manually drawn either by ChemDraw Pro (CambridgeSoft, PerkinElmer, Waltham, MA, USA) or BIOVIA Draw 2018 (BIOVIA,

Accelrys, San Diego, CA, USA). The Progenesis Metascope search of the custom SDF database was carried out assuming 50 ppm error for precursor ions and 30 ppm for theoretical fragments.

4.5.2. QTRAP Analysis

The LC-MS/MS analysis was done with a QTRAP 4500 mass spectrometer (AB Sciex, Framingham, MA, USA) coupled to an Agilent 1260 Infinity HPLC system (Santa Clara, CA, USA). Reversed phase HPLC was done with a BDS Hypersil C18 column (250 × 2.1; 5 μm) (Thermo Scientific, Waltham, MA, USA) using a gradient with eluent A 0.1% formic acid in Milli-Q water and B 0.1% formic acid in acetonitrile. Column compartment was maintained at 30 °C. The column was equilibrated with 20% B for 10 min in the beginning of each run. The linear gradient was increased from 20% to 95% B (0–15 min) and maintained at 95% B (15–25 min). The mobile phase flow rate was maintained at 300 μL/min with 5 μL injection volume. Samples were filtered through a 0.22 μm KX Syringe PTFE filters (Mikrolab, Aarhus, Denmark) before LC-MS/MS analysis.

The QTRAP MS was used in Information Dependent Acquisition (IDA) mode. After the suspect screening of glutathione conjugates with Q-TOF analysis, MS/MS spectra of the selected hits were acquired using IDA.

The MS/MS spectra from one to two of the most intense peaks matching the survey MRM transitions outlined in Table 1 were acquired if the peak intensity was above IDA threshold of 3000 counts per second (cps). Dynamic exclusion was applied to avoid acquiring MS/MS spectra from the same transition for the next 15 s after 3 occurrences of a transition, thus allowing for the detection of co-eluting peaks. Two EPI experiments were done for the two most intense peaks of each transition meeting the IDA criteria. EPI spectra were acquired between m/z 80 to 720 in positive ionization mode with a scan rate at 20,000 amu per second. Two EPI spectra from each transition was summed to give a final spectrum which resulted in a dwell time of 1.6 s for a complete MRM-EPI scan. During the acquisition of EPI spectra, DP was ramped between 50 to 70 V. Collision energy of 35 V and a spread of 15 V was used thus acquiring EPI at CE 20, 35, and 50 V.

Supplementary Materials: The supplementary materials are available online.

Author Contributions: Conceptualization, G.D., I.S.F., M.G.; methodology, G.D. and M.G.; software, (Q-TOF) G.D., (QTRAP) M.G.; data analysis, G.D. I.S.F., M.G.; data curation, G.D. I.S.F., M.G.; writing—original draft preparation, G.D. and M.G.; writing—review and editing, I.S.F., G.D. and M.G.; supervision, I.S.F.

Funding: The study was funded by the PhD project (Project no. 17638) of Maheswor Gautam at the Graduate School of Science and Technology, Aarhus University (GSST, AU).

Acknowledgments: The GSST, AU is thanked for the funding for the PhD project. Bente Laursen, Kirsten Heinrichson and Elena Claudia Jensen from the Team Natural Products Chemistry and Environmental Chemistry at AU are thanked for their assistance in the laboratory. Agnes Corbin and Juliet Evans from Nonlinear Dynamics, a Waters company (Newcastle upon Tyng, United Kingdom), are thanked a lot for their assistance during the data processing in Progenesis QI for our experiments.

Conflicts of Interest: The authors declare no conflict of interest.

Abbreviations

MSe	Waters® acquisition method synonymous of "all ions fragmentation" and data independent acquisition (DIA)
GSH	reduced glutathione
LC-MS	Liquid Chromatography followed by Mass Spectroscopy
Q-TOF	Quadrupole-Time of Flight
Q-TRAP	Triple Quadrupole (QQQ) Linear Ion Traps
MRM-IDA	Multiple Reaction Monitoring using Information Dependent Acquisition (IDA)
MRM-EPI	Multiple Reaction Monitoring (MRM) with Enhanced Product Ion (EPI)

References

1. Joseph, R.S. Metabolism of Azoxystrobin in Plants and Animals. *Special Publication-Royal Soc. Chem.* **1999**, *233*, 265–278. [CrossRef]
2. Gautam, M.; Fomsgaard, I.S. Liquid chromatography-tandem mass spectrometry method for simultaneous quantification of azoxystrobin and its metabolites, azoxystrobin free acid and 2-hydroxybenzonitrile, in greenhouse-grown lettuce. *Food Addit. Contam. A* **2017**, *34*, 2173–2180. [CrossRef] [PubMed]
3. Gautam, M.; Elhiti, M.; Fomsgaard, I.S. Maize root culture as a model system for studying azoxystrobin biotransformation in plants. *Chemosphere* **2018**, *195*, 624–631. [CrossRef] [PubMed]
4. Gautam, M.; Etzerodt, T.; Fomsgaard, I.S. Quantification of azoxystrobin and identification of two novel metabolites in lettuce via liquid chromatography–quadrupole-linear ion trap (QTRAP) mass spectrometry. *Int. J. Environm. Anal. Chem.* **2017**, *97*, 419–430. [CrossRef]
5. Bryant, D. Glutathione conjugation of herbicides and fungicides in plants and fungi: Functional characterization of glutathione transferases from phytopathogens. Ph.D. Thesis, Durham University, Durham, UK, 2004.
6. Laird, W.J.D.; Gledhill, A.J.; Lappin, G.J. Metabolism of methyl-(*E*)-2-{2-[6-(2-cyanophenoxy)pyrimidin-4-yloxy]phenyl}-3-methoxyacrylate (azoxystrobin) in rat. *Xenobiotica* **2003**, *33*, 677–690. [CrossRef] [PubMed]
7. Ohkama-Ohtsu, N.; Zhao, P.; Xiang, C.; Oliver, D.J. Glutathione conjugates in the vacuole are degraded by gamma-glutamyl transpeptidase GGT3 in Arabidopsis. *Plant. J.* **2007**, *49*, 878–888. [CrossRef] [PubMed]
8. Xie, C.; Zhong, D.; Chen, X. A fragmentation-based method for the differentiation of glutathione conjugates by high-resolution mass spectrometry with electrospray ionization. *Anal. Chim. Acta* **2013**, *788*, 89–98. [CrossRef] [PubMed]
9. Marrs, K.A. THE FUNCTIONS AND REGULATION OF GLUTATHIONE S-TRANSFERASES IN PLANTS. *Annual Rev. Plant Biol.* **1996**, *47*, 127–158. [CrossRef]
10. Edwards, R.; Dixon, D.P.; Walbot, V. Plant glutathione S-transferases: Enzymes with multiple functions in sickness and in health. *Trends Plant. Sci.* **2000**, *5*, 193–198. [CrossRef]
11. Dixon, D.P.; Lapthorn, A.; Edwards, R. Plant glutathione transferases. *Genome Biol.* **2002**, *3*, reviews3004. [CrossRef]
12. Caccuri, A.M.; Lo Bello, M.; Nuccetelli, M.; Nicotra, M.; Rossi, P.; Antonini, G.; Federici, G.; Ricci, G. Proton release upon glutathione binding to glutathione transferase P1-1: Kinetic analysis of a multistep glutathione binding process. *Biochemistry* **1998**, *37*, 3028–3034. [CrossRef] [PubMed]
13. Sheehan, D.; Meade, G.; Foley, V.M.; Dowd, C.A. Structure, function and evolution of glutathione transferases: Implications for classification of non-mammalian members of an ancient enzyme superfamily. *Biochem. J.* **2001**, *360*, 1–16. [CrossRef] [PubMed]
14. Caccuri, A.M.; Antonini, G.; Board, P.G.; Parker, M.W.; Nicotra, M.; Lo Bello, M.; Federici, G.; Ricci, G. Proton release on binding of glutathione to alpha, Mu and Delta class glutathione transferases. *Biochem. J.* **1999**, *344*, 419–425. [CrossRef] [PubMed]
15. Bartlett, D.W.; Clough, J.M.; Godwin, J.R.; Hall, A.A.; Hamer, M.; Parr-Dobrzanski, B. The strobilurin fungicides. *Pest. Manage. Sci.* **2002**, *58*, 649–662. [CrossRef] [PubMed]
16. Wolf, A.E.; Dietz, K.J.; Schroder, P. Degradation of glutathione S-conjugates by a carboxypeptidase in the plant vacuole. *FEBS Lett.* **1996**, *384*, 31–34. [CrossRef]
17. Beck, A.; Lendzian, K.; Oven, M.; Christmann, A.; Grill, E. Phytochelatin synthase catalyzes key step in turnover of glutathione conjugates. *Phytochemistry* **2003**, *62*, 423–431. [CrossRef]
18. Martin, M.N.; Slovin, J.P. Purified gamma-glutamyl transpeptidases from tomato exhibit high affinity for glutathione and glutathione S-conjugates. *Plant. Physiol.* **2000**, *122*, 1417–1426. [CrossRef]
19. Huber, C.; Bartha, B.; Harpaintner, R.; Schroder, P. Metabolism of acetaminophen (paracetamol) in plants–two independent pathways result in the formation of a glutathione and a glucose conjugate. *Environ. Sci. Pollut. Res. Int.* **2009**, *16*, 206–213. [CrossRef]
20. Murphy, C.M.; Fenselau, C.; Gutierrez, P.L. Fragmentation characteristic of glutathione conjugates activated by high-energy collisions. *J. Am. Soc. Mass Spectr.* **1992**, *3*, 815–822. [CrossRef]
21. Jian, W.; Liu, H.-F.; Zhao, W.; Jones, E.; Zhu, M. Simultaneous Screening of Glutathione and Cyanide Adducts Using Precursor Ion and Neutral Loss Scans-Dependent Product Ion Spectral Acquisition and Data Mining Tools. *J. Am. Soc. Mass Spectr.* **2012**, *23*, 964–976. [CrossRef]

22. Kassahun, K.; Davis, M.; Hu, P.; Martin, B.; Baillie, T. Biotransformation of the Naturally Occurring Isothiocyanate Sulforaphane in the Rat: Identification of Phase I Metabolites and Glutathione Conjugates. *Chem. Res. Toxicol.* **1997**, *10*, 1228–1233. [CrossRef] [PubMed]
23. Mali'n, T.J.; Lindberg, S.; Åstot, C. Novel glutathione conjugates of phenyl isocyanate identified by ultra-performance liquid chromatography/electrospray ionization mass spectrometry and nuclear magnetic resonance. *J. Mass Spectr.* **2014**, *49*, 68–79. [CrossRef]
24. Mueller, L.N.; Brusniak, M.-Y.; Mani, D.R.; Aebersold, R. An Assessment of Software Solutions for the Analysis of Mass Spectrometry Based Quantitative Proteomics Data. *J. Proteome Res.* **2008**, *7*, 51–61. [CrossRef] [PubMed]
25. Vogel, C.; Marcotte, E.M. Label-Free Protein Quantitation Using Weighted Spectral Counting. In *Quantitative Methods in Proteomics*; Marcus, K., Ed.; Humana Press: Totowa, NJ, USA, 2012; pp. 321–341.

Sample Availability: Samples of the compounds are not available from the authors.

MDPI

Article

Metabolomics Approach on Non-Targeted Screening of 50 PPCPs in Lettuce and Maize

Weifeng Xue *, Chunguang Yang, Mengyao Liu, Xiaomei Lin, Mei Wang and Xiaowen Wang

Technical Center of Dalian Customs, Dalian 116000, China; 2004ycg51@163.com (C.Y.); laoyao1024@163.com (M.L.); lynn9857@163.com (X.L.); monkeywangcn@sina.com (M.W.); wxw0652@sina.com (X.W.)
* Correspondence: xwf526@163.com

Abstract: The metabolomics approach has proved to be promising in achieving non-targeted screening for those unknown and unexpected (U&U) contaminants in foods, but data analysis is often the bottleneck of the approach. In this study, a novel metabolomics analytical method via seeking marker compounds in 50 pharmaceutical and personal care products (PPCPs) as U&U contaminants spiked into lettuce and maize matrices was developed, based on ultrahigh-performance liquid chromatography-tandem mass spectrometer (UHPLC-MS/MS) output results. Three concentration groups (20, 50 and 100 ng mL^{-1}) to simulate the control and experimental groups applied in the traditional metabolomics analysis were designed to discover marker compounds, for which multivariate and univariate analysis were adopted. In multivariate analysis, each concentration group showed obvious separation from other two groups in principal component analysis (PCA) and orthogonal partial least squares discriminant analysis (OPLS-DA) plots, providing the possibility to discern marker compounds among groups. Parameters including S-plot, permutation test and variable importance in projection (VIP) in OPLS-DA were used for screening and identification of marker compounds, which further underwent pairwise *t*-test and fold change judgement for univariate analysis. The results indicate that marker compounds on behalf of 50 PPCPs were all discovered in two plant matrices, proving the excellent practicability of the metabolomics approach on non-targeted screening of various U&U PPCPs in plant-derived foods. The limits of detection (LODs) for 50 PPCPs were calculated to be 0.4~2.0 μg kg^{-1} and 0.3~2.1 μg kg^{-1} in lettuce and maize matrices, respectively.

Keywords: metabolomics; marker compounds; non-targeted screening; pharmaceutical and personal care products; plant-derived food

Citation: Xue, W.; Yang, C.; Liu, M.; Lin, X.; Wang, M.; Wang, X. Metabolomics Approach on Non-Targeted Screening of 50 PPCPs in Lettuce and Maize. *Molecules* **2022**, *27*, 4711. https://doi.org/10.3390/molecules27154711

Academic Editor: Thomas Letzel

Received: 23 June 2022
Accepted: 18 July 2022
Published: 23 July 2022

Publisher's Note: MDPI stays neutral with regard to jurisdictional claims in published maps and institutional affiliations.

1. Introduction

Pharmaceutical and personal care product (PPCP) contamination in animal-derived foods has attracted worldwide attention, and a series of formal regulatory documents on the maximum residue limits (MRLs) of PPCPs from different countries and organizations has been issued [1–4]. However, PPCPs-induced contamination in plant-derived foods has not been fully addressed [5]. Previous studies [6–14] indicate that some plant-derived foods (e.g., corn, barley, pea, wheat, carrot, potato, cucumber and lettuce) can easily absorb PPCPs from soil with animal manure used as a fertilizer, which contains several kinds of commonly used antibiotics, e.g., tetracyclines, quinolones, sulfonamides and β-lactam, with their total concentration from the μg kg^{-1} to the mg kg^{-1} level in the plants [9,15–18]. Due to the lack of evaluation standards of PPCPs in plant-derived foods, it is hard to directly judge whether the residue concentrations of PPCPs can induce adverse effects on human health. Referring to the regulatory files on MRLs of PPCPs in animal-derived foods [2,4], which proposed a concentration of 10 μg kg^{-1} as the threshold of safety for most PPCPs, it can be inferred that if the concentrations of PPCPs in plant-derived foods exceed 10 μg kg^{-1}, it triggers a food safety risk. Therefore, the top priority is to develop reliable analytical methods for the investigation of PPCP residues in plant-derived foods.

Previous studies [6,9,10,19,20] have proposed some analytical methods based on high-performance liquid chromatography-tandem mass spectrometer (HPLC-MS/MS) for PPCPs detection in plant matrices. These studies mainly focused on the contamination of antibiotics, especially for tetracyclines, quinolones and sulfonamides. Most methods are customized and showcase the excellent detection performance for specific PPCPs, but are invalid for detecting other PPCPs not in the customized database. With the rapid development of the modern chemical industry's ability to synthesize new compounds, there is reason to believe that more and more PPCPs will be produced and applied in animal husbandry; as a result, continuous uptake of PPCPs by plant-derived foods will probably lead to more complicated, serious and underlying food safety risk. The United States, China and Japan, as the world's top three economies, plus the European Union, have issued regulatory documents on MRLs of only 95, 128, 180 and 139 PPCPs in animal-derived foods, and some listed PPCPs are of repeated emergence [1–4]. The sticking point is that the existing technologies cannot meet the detection requirements for increasing unknown and unexpected (U&U) PPCPs, for which the most effective method is to develop non-targeted screening methods, as proposed by the NORMAN network (www.norman-network.net, accessed on 20 November 2021) founded in 2005 by the European Commission [21,22].

Non-targeted screening can be defined from the narrow and broad senses. The former is reliant on the established screening database to discern contaminants [23]. The contaminants in the database are known, but those existing in the matrix are obscure, thus the screening practicability depends on the database size. The latter sense is to employ omics-related approaches to complete U&U contaminant screening [24,25], which can be realized by high resolution mass spectrometry (HRMS) technology [26,27]. To date, LC-MS/MS-based metabolomics analytical methods have showed good practicability on non-targeted screening of some pesticides in food matrices, e.g., orange juice [28], milk [24] and tea [29], with the screening ratio of pesticides depending on their contents. These studies have obtained desirable outcomes, but the methods they proposed are so sophisticated that they are not favorable for wide application. Nowadays, the development of non-targeted metabolomics analysis still encounters many great obstacles, especially for data analysis, which is the bottleneck to be urgently solved through the advancement of data processing tools and improvement of HRMS data quality.

In view of this, we developed a novel metabolomics-based analytical method via seeking marker compounds on behalf of 50 PPCPs as U&U contaminants spiked in lettuce and maize matrices to achieve non-targeted screening. Ultrahigh-performance liquid chromatography-tandem mass spectrometer (UHPLC-MS/MS) was used to obtain output results for metabolomics analysis. Herein, 14 sulfonamides, 12 quinolones, 10 nitroimidazoles, 7 agonists, 4 steroids and 3 tetracyclines were selected as target PPCPs, in which quinolones, sulfonamides and tetracyclines are of relatively high detection frequency in plant-derived foods [9,15–18]. Lettuce and maize are consumed in high quantities worldwide, and have proved to easily absorb PPCPs from the soil [6,19]. Lack of formal documents to regulate the MRLs of PPCPs in plant-derived foods makes it difficult to directly evaluate whether the contents of PPCPs in the foods are in the safety range. According to the guidelines of GB 31650-2019 [4] and Commission Regulation (EU) No 37/2010 [2], the MRLs of most PPCPs in animal-derived foods are no lower than 10 $\mu g\ kg^{-1}$, which was used as the test concentration of 50 PPCPs in our study to perform screening analysis. The goal of this study is to develop an applicable analytical method on the basis of metabolomics, which can accurately, rapidly and comprehensively achieve the screening and identification of potential non-targeted contaminants in plant-derived foods.

2. Materials and Methods

2.1. Chemicals and Materials

The lettuce was bought from a local market in Dalian City. Ethylenediamine tetraacetic acid disodium salt (Na_2EDTA), citric acid, sodium hydrogen phosphate (Na_2HPO_4), anhydrous sodium sulfate (Na_2SO_4), sodium chloride (NaCl), sodium hydroxide (NaOH), hydrochloric acid (HCl) and C18 powder (Sinopharm Chemical Reagent Co., Ltd., Shanghai, China); methanol and acetonitrile (HPLC grade, Merck, Darmstadt, Germany); formic acid (HPLC grade, Shanghai ANPEL Laboratory Technologies Inc., Shanghai, China); filter membrane (0.22 μm, Agilent Technologies, Singapore, MI, USA); ultrapure water (Milli-Q ultrapure water system, Merck, Darmstadt, Germany); ciprofloxacin-d8 hydrochloride solution (100 μg mL^{-1} in methanol, First Standard, Ridgewood, NY, USA). Analytical standard compounds for 50 PPCPs (purity > 98.3%) were obtained from First Standard (Ridgewood, NY, USA), Sigma (Alexandria, VA, USA), TRC (Toronto, ON, Canada) and Dr. Ehrenstorfer (Augsburg, Germany). More details on the 50 PPCPs are shown in Table 1.

Table 1. Basic information on the 50 PPCPs.

No.	Compound	CAS No.	Category	Adduct	Parent Ion (*m*/*z*)	Retention Time (min)
1	Clorprenaline	3811-25-4	Agonist	M + H	214.09932	7.03
2	Terbutaline	23031-25-6			226.14377	2.65
3	Tolobuterol	41570-61-0			228.11496	8.23
4	Cimbuterol	54239-39-3			234.16009	4.47
5	Propranolol	5051-22-9			260.16451	9.53
6	Sotalol	959-24-0			273.12674	2.47
7	Nadolol	42200-33-9			310.20128	6.47
8	5-Chloro-1-methyl-4-nitroimidazole	4897-25-0	Nitroimidazoles	M + H	162.00649	4.43
9	Ipronidazole	14885-29-1			170.09241	7.79
10	Metronidazole	443-48-1			172.07167	2.58
11	Metronidazole-hydroxy	4812-40-2			188.06658	1.69
12	Ronidazole	7681-76-7			201.06183	2.96
13	Thiabendazole	148-79-8			202.04334	5.83
14	Ornidazole	16773-42-5			220.04835	6.41
15	Tinidazole	19387-91-8			248.06996	4.96
16	Albendazole	54965-21-8			266.09577	10.48
17	Fenbendazole	43210-67-9			300.08012	11.06
18	Oxolinic acid	14698-29-4	Quinolones	M + H	262.07100	9.16
19	Cinoxacin	28657-80-9			263.06625	8.69
20	Norfloxacin	70458-96-7			320.14051	6.50
21	Enoxacin	74011-58-8			321.13575	6.35
22	Ciprofloxacin	85721-33-1			332.14052	6.73
23	Lomefloxacin	98079-51-7			352.14672	7.12
24	Danofloxacin	112398-08-0			358.15615	7.00
25	Enrofloxacin	93106-60-6			360.17183	7.00
26	Ofloxacin	82419-36-1			362.15106	6.39
27	Marbofloxacin	11550-35-1			363.14631	5.93
28	Sparfloxacin	110871-86-8			393.17327	8.34
29	Difloxacin	98106-17-3			400.14672	7.42
30	Trenbolone	10161-33-8	Steroid	M + H	271.16926	10.98
31	Boldenone	846-48-0			287.20056	11.02
32	Testosterone propionate	57-85-2			345.24242	12.30
33	Deflazacort	14484-47-0			442.22241	11.06
34	Tetracycline	60-54-8	Tetracyclines	M + H	445.13444	6.56
35	Oxytetracycline	79-57-2			461.13391	6.41
36	Chlorotetracycline	57-62-5			479.38028	7.88
37	Sulphacetamide	144-80-9	Sulfonamides	M + H	215.04849	1.99
38	Sulfapyridine	144-83-2			250.06447	4.68
39	Sulfadiazine	68-35-9			251.05972	3.37
40	Sulfathiazole	72-14-0			256.02089	4.32
41	Sulfamerazine	127-79-7			265.07537	5.11
42	Sulfamoxole	729-99-7			268.07504	6.10
43	Sulfamethizole	144-82-1			271.03179	6.11
44	Sulfabenzamide	127-71-9			277.06414	8.05
45	Sulfmethazine	57-68-1			279.09102	3.76
46	Sulfisomidine	515-64-0			279.09102	6.24
47	Sulfachloropyridazine	80-32-0			285.02075	6.80
48	Trimethoprim	738-70-5			291.14517	6.01
49	Sulfaquinoxaline	59-40-5			301.07537	9.26
50	Sulfanitran	122-16-7			336.06486	10.08

2.2. Solution Preparation

A total of 50 PPCPs were separately prepared with methanol at 100 μg mL^{-1}, 1 mL of which was withdrawn, mixed together and further diluted with methanol to obtain a 1 μg mL^{-1} solution. Then, 100 ng mL^{-1} ciprofloxacin-d8 methanol solution was prepared by diluting its 100 μg mL^{-1} solution. A 0.1 mol L^{-1} Na_2EDTA-Mcllvaine buffer solution was prepared with Na_2HPO_4 (5.5 g), citric acid (12.9 g) and Na_2EDTA (37.2 g) dissolved in 1 L pure water, which was further adjusted to pH 4.0 with 0.1 mol L^{-1} HCl or NaOH solution.

2.3. Sample Preparation and Pretreatment Process

(a) Lettuce sample was cut into small pieces, then ground into batter by tissue homogenizer; (b) 2.0, 5.0 and 10.0 g lettuce batters, together with one-to-one corresponding 20, 50 and 100 μL of 50 PPCPs mixed solutions (1 μg mL^{-1}) were poured into 50 mL polypropylene centrifuge tubes. To calibrate the recovery during the sample pretreatment process, ciprofloxacin-d8 methanol solution (0.5 mL, 100 ng mL^{-1}) as recovery internal standard was further added, as adopted in previous studies [30–32]; (c) 5 mL Na_2EDTA-Mcllvaine buffer solution (0.1 mol L^{-1}) was dumped into the tube, vortexed for 1 min, then 20 mL 1% (*V*/*V*) formic acid/acetonitrile solution was added further, stirring for 1 min. An extraction salt package (10.0 g Na_2SO_4 + 2.0 g NaCl) was added for stratification under salting out after the solution standing for 10 min, centrifuging at 4500 r min^{-1} for 5 min; (d) then, after transferring all the supernatant into new 50 mL polypropylene centrifuge tubes, adding 100 mg C18 powder, vortexing for 1 min, centrifuging at 4500 r min^{-1} for 3 min, the solution was extracted to another 50 mL centrifuge tube, dried with N_2 blowing by nitrogen blowing apparatus (N-EVAP-112, Organomation, Berlin, MA, USA), and redissolved in 1 mL 40% (*V*/*V*) methanol 0.1% formic acid/water solution, vortexed for 1 min; (e) then, filtered with a 0.22 μm filter membrane, the sample solutions of 50 PPCPs at the theoretical concentrations of 20, 50 and 100 ng mL^{-1} were prepared. Each concentration experiment was repeated nine times.

2.4. Sample Grouping and Naming

Samples of 20 ng mL^{-1}–1~20 ng mL^{-1}–9, 50 ng mL^{-1}–1~50 ng mL^{-1}–9 and 100 ng mL^{-1}–1~100 ng mL^{-1}–9 were employed to label samples from three concentration groups. Each sample provided a 30 μL solution as a quality control (QC) sample [29,33,34], which experienced 3 injections before and after each concentration group. As a result, 12 samples marked as QC-1, QC-2, and QC-12 were obtained to evaluate the stability of LC-MS/MS.

2.5. Analytical Method

The 50 PPCPs and ciprofloxacin-d8 were analyzed on a quadrupole/electrostatic field orbitrap LC-MS/MS system (Q Exactive Plus, Thermo Fisher Scientific Inc., Waltham, MA, USA) under the positive mode of electrospray ion (ESI) source. Components in the sample solution underwent separation within an Accucore RP-MS column (100 × 2.1 mm, 2.6 μm particle diameter, Thermo Fisher Scientific Inc., Waltham, MA, USA), with injection volume of 10 μL. Next, 0.1% (*V*/*V*) formic acid/water and 0.1% (*V*/*V*) formic acid/methanol solutions were prepared as the mobile phase A and B, respectively, with flow rate of 0.3 mL min^{-1}. In consideration of the matrix complexity of lettuce and maize, there may be some impurities not eluted from the LC-MS/MS system in a relatively short time (738 s for the last eluted target PPCP in this study) designed only for 50 PPCPs, leading to the potential disruption for the elution and analysis of the next sample. Therefore, a longer elution program was designed as follows: gradient started from 5% B, kept for 2 min, then increased to 30% B in 1 min, at a duration of 7 min, further increased to 90% B in 1 min, holding on 25 min, finally decreased to 5% B in 1 min, equilibrating for 16 min. The oven temperature was set at 40 °C. Other parameter settings were as follows: heating and capillary temperature 320 °C; lens and spray voltage 50 and 3200 V, respectively; auxiliary and sheath gas N_2, with flow rate at 10 and 40 arb, respectively; scan mode: full-scan/data-dependent two-stage scanning; MS parameters: full-scan resolution 70,000, maximum dwell time 100 ms, AGC target 1×10^6, m/z scan range 100~1000; MS/MS parameters: resolution 17,500, maximum dwell time 50 ms, AGC target 2×10^5.

LC-MS/MS output results of 50 PPCPs and ciprofloxacin-d8 were analyzed by Trace Finder 3.3 software, with screening conditions as follows: (a) for primary parent ion, signal to noise ratio 5.0, response intensity threshold 10,000, and mass error 5 ppm; (b) for secondary fragment ions, minimum matching number of ion 1, response intensity threshold 10,000, and mass error 5 ppm. On the basis of the peak area of the primary parent ion, ciprofloxacin-d8 was quantified with standard curve for recovery calculation.

2.6. Metabolomics Data Processing

LC-MS/MS was operated in full scan mode with RAW-formatted files as the direct output, which underwent conversion to corresponding mzXML-formatted files via the ProteoWizard software [35]. These new files are adaptable to the upload to the Workflow4Metabolomics (W4M) platform (https://workflow4metabolomics.usegalaxy.fr/, accessed on 20 November 2021) for metabolomics analysis [36]. After peak detection, alignment and retention time calibration, plus data normalization, centralization, scaling and transformation performed on the W4M platform, the data matrix was obtained in the format of variable and sample named as abscissa and ordinate, respectively [36,37]. Variable contains a series of information, e.g., molecular weight and retention time, with every marker compound corresponding to its unique variable, that is to say, the process to pursue marker compounds is actually a process to pursue eligible variables. Multivariate statistical analysis including principal component analysis (PCA) [38–40] and orthogonal partial least squares discriminant analysis (OPLS-DA) [41,42] was performed in SIMCA 14.1 software [43] after importing the data matrix. A permutation test with 200 iterations was employed for over-fitting judgement of the OPLS-DA model [43,44]. Other parameters to screen marker compound candidates include the absolute value of variable confidence in the S-plot plot [45] and variable importance in projection (VIP) [43,44,46], with the threshold above 0.9 and 1, respectively. After this, eligible marker compound candidates from 20 and 100 ng mL^{-1} groups can both be obtained, and only overlapped candidates in two groups, representing their significantly low and high concentration in the corresponding 20 and 100 ng mL^{-1} groups, were further investigated by pairwise t-test [47–49] in SPSS Statistics V17.0 software and fold change judgement for the univariate analysis. Univariate analysis is simple, intuitive and easy to be understood. It was used to quickly investigate the differences of marker compound candidates in different groups. To more rapidly verify the identity of marker compounds on behalf of 50 PPCPs, we directly compared the precise molecular weight (<5 ppm in absolute value of error), retention time and the adduct structure of marker compounds with that of the authentic 50 PPCPs (Table 1).

3. Results and Discussion

3.1. Data Preprocessing

As indicated in Figure 1, although only part of the total ion chromatograms at the retention time of 0~900 s is shown, during which all 50 PPCPs were eluted, obvious differences in peak intensity have already been observed in three concentration groups, implying the possibility to seek marker compounds among groups. The principle for relative standard deviation of peak intensity above 30% was employed to filter out invalid variables in QC and three concentration groups [50], with a final 6512×39 data matrix obtained for further analysis.

3.2. Multivariate Analysis

3.2.1. PCA Analysis

As Taguchi [51] pointed out, PCA can make a natural classification for sample groups and eliminate the extreme data without knowing their categories, thus PCA can be used in metabolomics to assess the data quality and to identify outliers [38–40]. As indicated in

Figure 2, no extreme data and outliers were observed. Samples at the same concentration gathered together, indicating the good classification of groups. Obvious separation among three concentration groups indicates the existence of major discrepancies, further paving the way to seek marker compounds from different groups.

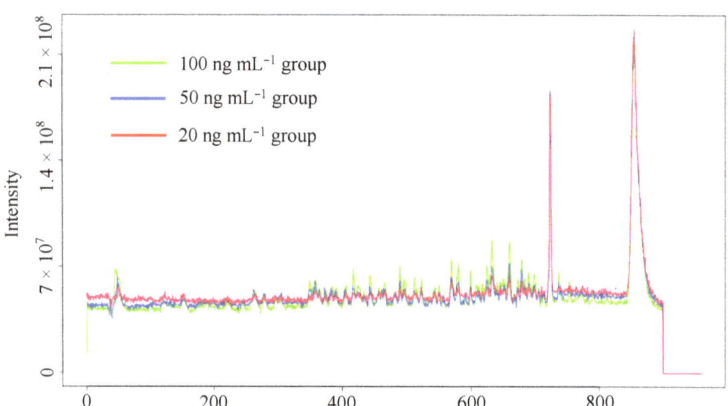

Figure 1. Total ion chromatograms (0~900 s) of spiked lettuce sample groups on the W4M platform.

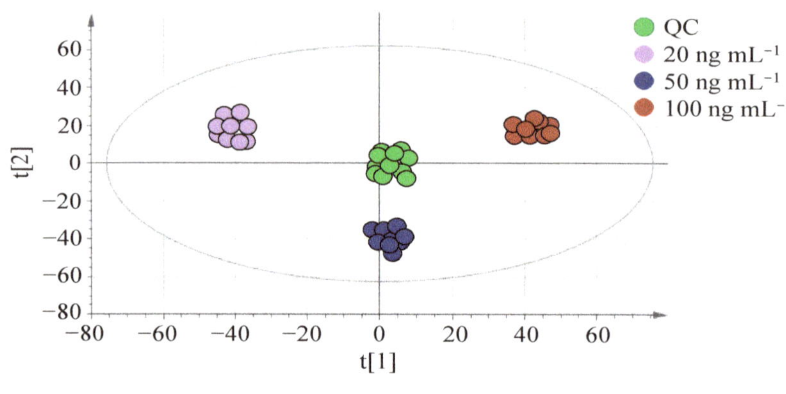

Figure 2. PCA score plot of spiked lettuce sample groups.

3.2.2. OPLS-DA Analysis

Theoretically speaking, the peak intensities of variables ought to increase with their rising concentrations, i.e., 20 and 100 ng mL^{-1} groups should present the minimum and maximum peak intensities, respectively. However, the reality may be different, due to the discrepancies in sample recoveries. Previous studies [30–32] proposed deuterated antibiotics as recovery internal standards to correct losses of PPCPs during sample preparation. In consideration of this, ciprofloxacin-d8 (parent ion m/z 340.19132; fragment ions m/z 296.20156, 253.15933 and 239.14367; retention time 6.73 min) was employed here to eliminate the peak intensity errors of variables induced by disparate recoveries of PPCPs during the pretreatment process. As shown in Table S1 (Supplementary Materials), the recoveries of ciprofloxacin-d8 were calculated to be 80.1~85.9%, 80.3~86.2% and 81.6~87.7% in the 20, 50 and 100 ng mL^{-1} groups, respectively, based on the ciprofloxacin-d8 standard curve solutions (100, 50, 25, 10 and 5 ng mL^{-1}) prepared in blank lettuce extract solution. After this, the recoveries of ciprofloxacin-d8 were all calibrated to 100% by multiplying a corresponding calibration coefficient, with which the peak intensities of ciprofloxacin-d8 were also calibrated, together with peak intensities for all the variables.

As shown in Figure 3, we can observe the separation of two camps on the first principal component axis. One camp represents the specific concentration group (green part), and the other camp is on behalf of the remaining two groups (blue part), indicating the existence of variables with significant differences between the two camps. Each point in the S-plot plots (Figure 4) represents a variable, which keeps away from the origin along *X*- and *Y*-axis, implying more contribution and higher confidence level of the variable to the difference. Therefore, the points at the two ends of 'S' can be deemed the most differentiating components. In the S-plot analysis, absolute value of confidence > 0.9 has been proposed to screen variables as marker compound candidates [45], which at the significantly low and high concentration should be searched at the right and left ends of S-plot plots in Figure 4a,b, respectively.

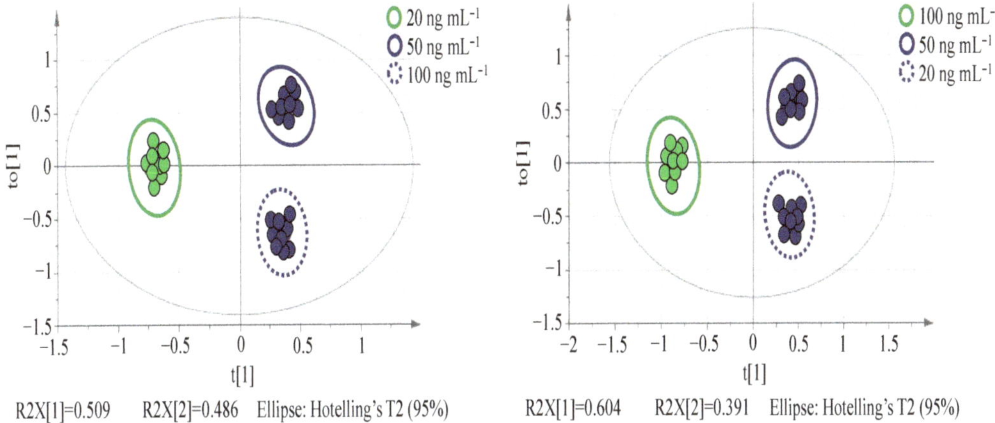

Figure 3. OPLS-DA score plots of spiked lettuce sample groups.

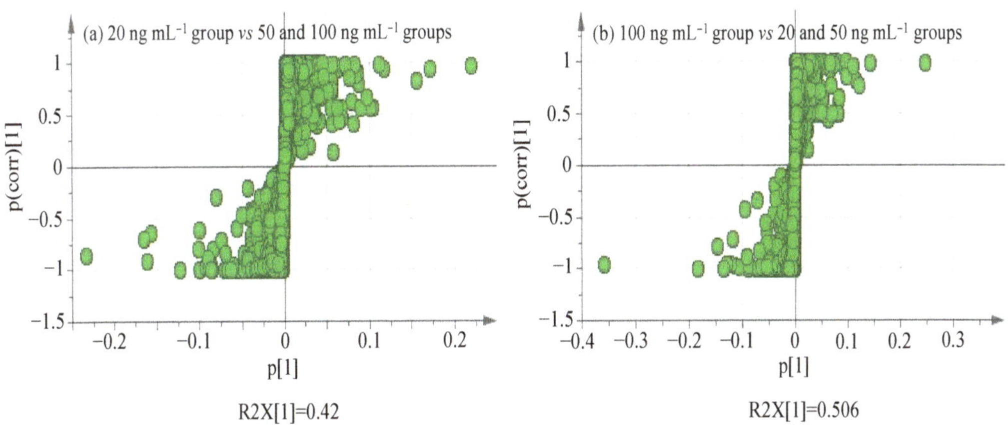

Figure 4. S-plot plots of spiked lettuce sample groups.

SIMCA 14.1 software performed permutation tests with 200 iterations to investigate whether the OPLS-DA models underwent data over-fitting, for which R^2Y and Q^2 are two common parameters to describe the interpretation level of the model in the Y-axis direction and the prediction level of the model [52,53], respectively. If R^2Y and Q^2 are both close (or equal) to 1, the OPLS-DA models are not susceptible to over-fitting. As can be seen from Figure 5, R^2Y and Q^2 values were no less than 0.991, indicating the good reliability, predictability and no over-fitting for all OPLS-DA models. VIP > 1 principle continues to screen marker compounds. Eventually, marker compounds on behalf of 50 PPCPs were all screened out as shown in Table 2. Negligible concentrations (<0.1 ng mL^{-1}) of 50 PPCPs in the blank lettuce extract solution were obtained by the metabolomics analysis, which eliminates the interference of inherent (rather than spiked) 50 PPCPs residues in lettuce matrix to seek marker compounds.

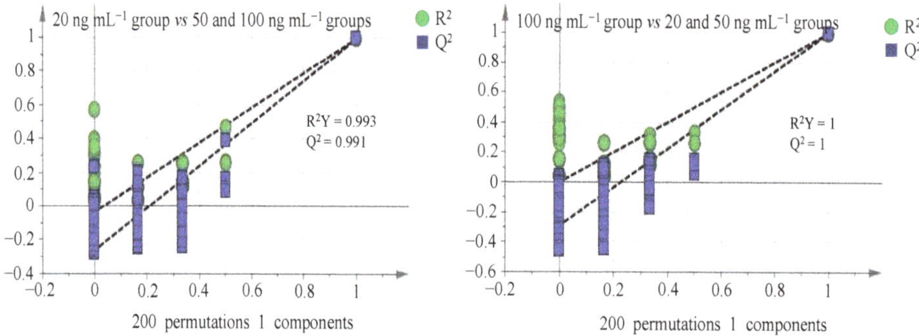

Figure 5. Permutation test plots of spiked lettuce sample groups.

Table 2. Marker compounds screened in lettuce sample groups.

Var ID (Primary)	Marker Compounds	VIP Pred [a]	Coordinate in S-Plot [b]	Mass Error (ppm) [c]	LOD (μg kg^{-1})
M214T418	Clorprenaline	3.300/3.057	(−0.077, −0.911)/(0.057, 0.934)	−0.225	1.3
M226T148	Terbutaline	2.297/2.467	(−0.049, −0.969)/(0.049, 0.945)	−1.751	2.0
M228T491	Tolobuterol	4.045/3.311	(−0.098, −0.922)/(0.063, 0.929)	−0.294	0.6
M234T261	Cimbuterol	3.226/4.071	(−0.075, −0.934)/(0.082, 0.965)	−2.061	1.1
M260T571	Propranolol	3.710/4.505	(−0.089, −0.927)/(0.099, 0.945)	0.617	0.5
M273T138	Sotalol	1.513/1.453	(−0.025, −0.923)/(0.025, 0.916)	0.563	0.7
M310T388	Nadolol	4.165/2.141	(−0.101, −0.981)/(0.038, 0.933)	−2.351	0.7
M162T266	5-Chloro-1-methyl-4-nitroimidazole	1.854/1.675	(−0.038, −0.971)/(0.032, 0.948)	−0.394	0.6
M170T467	Ipronidazole	4.537/4.102	(−0.111, −0.921)/(0.087, 0.927)	2.920	0.4
M172T152	Metronidazole	2.838/2.675	(−0.064, −0.928)/(0.052, 0.967)	2.916	1.7
M188T101	Metronidazole-hydroxy	2.178/2.756	(−0.046, −0.964)/(0.053, 0.956)	−0.353	1.7
M201T179	Ronidazole	1.957/1.673	(−0.040, −0.961)/(0.032, 0.907)	−1.134	2.0
M202T351	Thiabendazole	3.879/4.056	(−0.093, −0.946)/(0.078, 0.986)	−1.654	0.9
M220T383	Ornidazole	3.627/1.928	(−0.087, −0.932)/(0.035, 0.914)	−1.476	1.4
M248T298	Tinidazole	2.977/3.588	(−0.069, −0.938)/(0.070, 0.952)	−2.146	0.4
M266T625	Albendazole	4.193/3.237	(−0.102, −0.918)/(0.060, 0.911)	0.143	1.7
M300T661	Fenbendazole	3.050/2.891	(−0.071, −0.946)/(0.054, 0.961)	−0.272	1.1
M262T553	Oxolinic acid	2.091/2.258	(−0.044, −0.932)/(0.040, 0.955)	4.435	0.5
M263T524	Cinoxacin	2.481/2.115	(−0.055, −0.986)/(0.037, 0.973)	2.106	0.7
M320T375	Norfloxacin	2.684/2.113	(−0.059, −0.966)/(0.037, 0.949)	0.600	0.6
M321T381	Enoxacin	2.479/3.480	(−0.055, −0.937)/(0.067, 0.924)	1.102	0.7
M332T406	Ciprofloxacin	3.110/3.589	(−0.072, −0.934)/(0.070, 0.948)	4.829	1.0
M352T427	Lomefloxacin	3.052/2.676	(−0.071, −0.928)/(0.052, 0.933)	1.175	1.6
M358T420	Danofloxacin	2.842/2.469	(−0.065, −0.929)/(0.049, 0.937)	2.019	1.8
M360T422	Enrofloxacin	2.836/2.397	(−0.064, −0.957)/(0.045, 0.959)	0.770	1.9
M362T386	Ofloxacin	1.553/1.672	(−0.028, −0.925)/(0.031, 0.966)	0.908	0.8
M363T386	Marbofloxacin	1.617/1.678	(−0.030, −0.923)/(0.032, 0.929)	−4.510	0.6
M393T500	Sparfloxacin	1.827/2.946	(−0.036, −0.911)/(0.056, 0.945)	−3.455	0.8
M400T446	Difloxacin	1.535/2.317	(−0.028, −0.937)/(0.042, 0.949)	−2.715	0.8
M271T659	Trenbolone	3.077/1.586	(−0.071, −0.947)/(0.029, 0.945)	1.869	0.7
M287T661	Boldenone	2.071/2.398	(−0.043, −0.944)/(0.045, 0.936)	−0.921	0.8
M345T738	Testosterone propionate	2.698/1.676	(−0.061, −0.928)/(0.032, 0.927)	0.122	0.8
M442T663	Deflazacort	1.535/1.545	(−0.026, −0.977)/(0.027, 0.919)	−4.677	1.0
M445T393	Tetracycline	4.047/4.059	(−0.098, −0.911)/(0.079, 0.941)	−0.539	1.0
M461T385	Oxytetracycline	2.708/2.677	(−0.061, −0.928)/(0.052, 0.934)	1.062	0.5
M479T473	Chlorotetracycline	2.180/2.392	(−0.046, −0.945)/(0.044, 0.928)	1.502	0.5

Table 2. *Cont.*

Var ID (Primary)	Marker Compounds	VIP Pred [a]	Coordinate in S-Plot [b]	Mass Error (ppm) [c]	LOD (μg kg^{-1})
M215T120	Sulphacetamide	1.615/1.468	(−0.030, −0.915)/(0.026, 0.933)	4.277	0.7
M250T278	Sulfapyridine	3.108/4.087	(−0.072, −0.956)/(0.085, 0.927)	−0.686	0.7
M251T200	Sulfadiazine	2.075/1.893	(−0.043, −0.912)/(0.034, 0.921)	0.422	0.7
M256T260	Sulfathiazole	3.880/2.893	(−0.093, −0.945)/(0.054, 0.948)	0.280	0.9
M265T304	Sulfamerazine	2.415/3.483	(−0.053, −0.987)/(0.067, 0.987)	−2.799	0.4
M268T345	Sulfamoxole	3.301/3.312	(−0.077, −0.922)/(0.063, 0.909)	0.002	1.1
M271T366	Sulfamethizole	2.704/1.412	(−0.061, −0.956)/(0.023, 0.919)	−2.372	1.8
M277T482	Sulfabenzamide	1.883/2.417	(−0.038, −0.944)/(0.046, 0.951)	2.561	0.6
M279T222	Sulfmethazine	4.119/3.117	(−0.100, −0.927)/(0.058, 0.978)	−4.681	1.3
M279T374	Sulfisomidine	2.711/3.661	(−0.061, −0.946)/(0.072, 0.923)	−3.077	0.5
M285T407	Sulfachloropyridazine	1.639/1.569	(−0.031, −0.982)/(0.028, 0.958)	−2.350	0.5
M291T360	Trimethoprim	1.952/1.585	(−0.040, −0.945)/(0.029, 0.928)	0.465	0.6
M301T557	Sulfaquinoxaline	1.829/2.115	(−0.036, −0.934)/(0.037, 0.922)	−0.561	0.4
M336T672	Sulfanitran	1.535/1.568	(−0.027, −0.928)/(0.028, 0.927)	2.996	1.6

Note: [a] two VIP values from 100 and 20 ng mL^{-1} groups, respectively; [b] two-group coordinate values from 100 and 20 ng mL^{-1} groups, respectively; [c] Mass error (ppm) = (extracted molecular weight from W4M platform—extracted molecular weight from LC-MS/MS) × 10^6/extracted molecular weight from LC-MS/MS.

3.3. *Univariate Analysis*

After multivariate analysis, a pairwise *t*-test [47–49] was firstly employed to examine whether marker compounds from a specific concentration group presented significant differences in peak intensity with those from other two groups. Pairwise *t*-test, as a reliable statistical test method, was performed to calculate *p* values between the two concentration groups and the $p < 0.05$ observed in this study indeed showed the existence of significant differences among groups. Previous studies [29,54] also adopted fold change of concentration > 2 to discern variables with high contrast among groups as marker compounds. Herein, marker compounds on behalf of 50 PPCPs all presented fold change values above 2, supporting the validity of marker compounds obtained with our analytical strategy.

The limits of detection (LODs) for 50 PPCPs were also considered here. Firstly, a 2.0 g blank lettuce sample was used to prepare an extract solution (1 mL) after the same pretreatment mentioned above. Then, a 20 ng mL^{-1} PPCPs solution was obtained by diluting their mixed methanol solution (20 μL, 1 μg mL^{-1}) with 1 mL blank lettuce extract solution. The experiments were repeated in septuplicate to obtain seven samples, which underwent the same metabolomics analysis to obtain the peak intensities of 50 PPCPs. For each PPCP, a 20 ng mL^{-1} concentration level was deemed to correspond to average values of seven samples in peak intensity; therefore, the concentration (unit: ng mL^{-1}) of each PPCP in a sample was calculated by its own peak intensity × 20/average peak intensity for the standard deviation measurement of the seven samples. According to the method proposed by US Environmental Protection Agency [55], the LOD values for 50 PPCPs were calculated to be 0.4~2.0 μg kg^{-1}, as shown in Table 2.

3.4. *Method Applicability in Maize Matrix*

Maize as the primary food crop in China has proved to easily absorb PPCPs from the soil [19]; therefore, it was selected as another plant matrix different from vegetables to investigate the applicability of the developed metabolomics-based screening method. Maize sample was purchased from the local market and turned into a powder by a grinder. Then, it underwent the same above-mentioned pretreatment process after 50 PPCPs spiked at 10 μg kg^{-1} as well. Ciprofloxacin-d8 methanol solution (0.5 mL, 100 ng mL^{-1}) was added for recovery calibration, with the results shown in Table S2. The same metabolomics analysis was performed as indicated in Figures S1–S5 (Supplementary Materials). Marker compounds to represent 50 PPCPs were also discovered (Table S3), proving the good applicability of the metabolomics analytical method to non-targeted screening of various PPCPs residues in different plant matrices. As can be seen from Table S3, the LOD values for 50 PPCPs in maize matrix were calculated to be 0.3~2.1 μg kg^{-1}.

3.5. Real Sample Test

We collected lettuce and maize samples from six administrative districts including Zhongshan, Xigang, Shahekou, Gaoxin, Ganjingzi and Jinpu affiliated to Dalian City, each district with two sampling points. A total of 12 fresh lettuce samples were purchased from the local farmer's market and immediately delivered to the laboratory for testing. The above process was also applied to the maize samples. After pretreatment experiments and metabolomics analysis, only one lettuce sample from Jinpu District was found to contain enrofloxacin and its content was 17.4 $\mu g \ kg^{-1}$. Other samples had no detection of PPCPs. Although the detection rate of PPCPs in all the samples is only 1/24, and seemingly only one district is vulnerable to PPCPs contamination, the results are enough to show that our proposed method is competent for the screening of PPCPs in plant-derived foods. These spot check results alert us to the fact that PPCP-induced safety risk of plant-derived foods is on the horizon.

Previous studies have successfully applied non-targeted screening methods on the basis of metabolomics to pesticide residues in plant matrices, e.g., orange juice [28] and tea [29], providing the feasibility to screen PPCPs residues in plant-derived foods. In light of the otherness of analytes, the reported methods may not be completely applied to our study. Herein, we firstly considered spiked contaminants to be marker compounds and then implemented a marker compound-seeking analytical strategy of metabolomics to finish the non-targeted screening of contaminants in plant-derived foods, which is the biggest difference from previous studies [24,28,29]. Despite only 50 PPCPs and two plant matrices considered here, the developed method still has wide applicability due to the representation of these PPCPs and universal consumption of lettuce and maize.

Extensive use of PPCPs in livestock farming raises the risk that these compounds end up in soil where animal waste is used as fertilizer [9,56], which leads to the uptake of PPCPs by plant-derived foods from the soil [57–64]. Compared with other plants, leafy vegetables generally show higher detection ratio and concentrations of PPCPs [60,64] and therefore deserve more attention in their food safety risk. Although there are no official documents to explicitly clarify the MRLs of PPCPs in plant-derived foods, we can still deduce their safety thresholds from their corresponding MRLs in animal-derived foods [1–4]. Relative to the colossal number of analytical methods for PPCPs in animal-derived foods [65–69], the methods for PPCPs detection in plant matrices are in short supply. To better cope with the complicated PPCPs contamination in plants, the top priority is to develop a high-throughput screening method that can accurately, rapidly and comprehensively determine which PPCPs exist in the foods. With this consideration, we developed this novel metabolomics-based analytical method to achieve non-targeted screening of PPCPs in plant-derived foods.

4. Conclusions

The newly developed metabolomics analytical method was successfully applicable to non-targeted screening of 50 PPCPs residues in lettuce and maize matrices. We intentionally designed three concentration groups of PPCPs (20, 50 and 100 ng mL^{-1}) to simulate the experimental and control groups adopted in the traditional metabolomics analytical procedures to search for marker compounds on behalf of 50 PPCPs. The process to perform metabolomics analysis has less artificial interference, a more concise workflow and higher screening efficiency. It is worth mentioning that this is the first implemented analytical strategy of metabolomics for non-targeted screening of PPCPs in plant-derived foods through seeking marker compounds. Due to the lack of binding legal documents on MRLs of PPCPs in plant matrices, together with constant development and application of new PPCPs in animal husbandry, it is urgent to compile legal rules to control MRLs of PPCPs in plant-derived foods, otherwise it may evolve as a serious food safety issue. To date, plant uptake from PPCP-contaminated soil is a known source of PPCP residues in plant-derived foods. It is not yet clear whether other ways can also induce the accumulation of PPCPs in the foods, potentially increasing the complexity of PPCPs contamination. Even worse, this increases the exposure risk of PPCPs to human health via the food chain. Therefore, we advocate that early attention to this issue would help defuse the potential crisis.

Supplementary Materials: The following supporting information can be downloaded at: https://www.mdpi.com/article/10.3390/molecules27154711/s1, Figure S1: Total ion chromatograms (0~900 s) of spiked maize sample groups on the W4M platform; Figure S2: PCA score plot of spiked maize sample groups; Figure S3: OPLS-DA score plots of spiked maize sample groups; Figure S4: S-plot plots of spiked maize sample groups; Figure S5: Permutation test plots of spiked maize sample groups; Table S1: Recovery (%) of spiked ciprofloxacin-d8 in lettuce sample groups (n = 9); Table S2: Recovery (%) of spiked ciprofloxacin-d8 in maize sample groups (n = 9); Table S3: Marker compounds screened in maize sample groups.

Author Contributions: Conceptualization, W.X.; Data curation, W.X.; Investigation, C.Y., M.L., X.L., M.W. and X.W.; Methodology, W.X.; Writing—original draft, W.X.; Writing—review and editing, W.X. All authors have read and agreed to the published version of the manuscript.

Funding: This work was funded by the National Natural Science Foundation of China (No. 21777014), Natural Science Foundation of Liaoning Province of China (No. 2019-BS-008), Special Foundation for Basic Research Program of Dalian Customs (No. 2021DK11).

Institutional Review Board Statement: Not applicable.

Informed Consent Statement: Not applicable.

Data Availability Statement: The data presented in this study are available in this article and Supplementary Materials.

Conflicts of Interest: The authors declare no conflict of interest.

Sample Availability: Samples of the compounds are available from the authors.

References

1. Code of Federal Regulations (CFR). Tolerances for Residues of New Animal Drugs in Food. 2022. Available online: https://www.ecfr.gov/current/title-21/chapter-I/subchapter-E/part-556 (accessed on 15 November 2021).
2. European Union (EU). Commission Regulation (EU) No 37/2010 of 22 December 2009. 2010. Available online: https://www.legislation.gov.uk/eur/2010/37 (accessed on 15 November 2021).
3. The Japan Food Chemical Research Foundation (JFCRF). Maximum Residue Limits (MRLs) List of Agricultural Chemicals in Foods. 2021. Available online: https://www.ffcr.or.jp/en/zanryu/the-japanese-positive/the-japanese-positive-list-system-for-agricultural-chemical-residues-in-foods-enforcement-on-may-29-.html (accessed on 15 November 2021).
4. *GB 31650-2019*; National Food Safety Standard-Maximum Residue Limits for Veterinary Drugs in Foods. Ministry of Agriculture and Rural Affairs of the People's Republic of China (MARAPRC): Beijing, China, 2019.
5. Qi, H.; Yan, H.; Zhang, L.; Zhang, Z.; Cui, F. Research progress in determination methods of antibiotics residues in plants. *J. Food Saf. Qual.* **2019**, *10*, 5098–5103. (In Chinese)
6. Ahmed, M.B.M.; Rajapaksha, A.U.; Lim, J.E.; Vu, N.T.; Kim, S.; Kang, H.M.; Lee, S.S.; Ok, Y.S. Distribution and accumulative pattern of tetracyclines and sulfonamides in edible vegetables of cucumber, tomato and lettuce. *J. Agric. Food Chem.* **2015**, *63*, 398–405. [CrossRef] [PubMed]
7. Boxall, A.B.A.; Johnson, P.; Smith, E.J.; Sinclair, C.J.; Stutt, E.; Levy, L.S. Uptake of veterinary medicines from soils into plants. *J. Agric. Food Chem.* **2006**, *54*, 2288–2297. [CrossRef] [PubMed]
8. Hawker, D.W.; Cropp, R.; Boonsaner, M. Uptake of zwitterionic antibiotics by rice (*Oryza sativa* L.) in contaminated soil. *J. Hazard. Mater.* **2013**, *263*, 458–466. [CrossRef]
9. Hu, X.; Zhou, Q.; Luo, Y. Occurrence and source analysis of typical veterinary antibiotics in manure, soil, vegetables and groundwater from organic vegetable bases, northern China. *Environ. Pollut.* **2010**, *158*, 2992–2998. [CrossRef]
10. Jones-Lepp, T.L.; Sanchez, C.A.; Moy, T.; Kazemi, R. Method development and application to determine potential plant uptake of antibiotics and other drugs in irrigated crop production systems. *J. Agric. Food Chem.* **2010**, *58*, 11568–11573. [CrossRef]
11. Tanoue, R.; Sato, Y.; Motoyama, M.; Nakagawa, S.; Shinohara, R.; Nomiyama, K. Plant uptake of pharmaceutical chemicals detected in recycled organic manure and reclaimed wastewater. *J. Agric. Food Chem.* **2012**, *60*, 10203–10211. [CrossRef]
12. Tasho, R.P.; Cho, J.Y. Veterinary antibiotics in animal waste, its distribution in soil and uptake by plants: A review. *Sci. Total Environ.* **2016**, *563–564*, 366–376. [CrossRef]
13. Wang, C.; Luo, Y.; Mao, D. Sources, fate, ecological risks and mitigation strategies of antibiotics in the soil environment. *Environ. Chem.* **2014**, *33*, 19–29.
14. Zhang, M.; Xu, Q. Review on pollution and behavior of veterinary antibiotics in agricultural system. *Acta Agric. Zhejiang* **2013**, *25*, 416–424. (In Chinese)

15. Bao, Y.; Li, Y.; Mo, C.; Yao, Y.; Tai, Y.; Wu, X.; Zhang, Y. Determination of six sulfonamide antibiotics in vegetables by solid phase extraction and high performance liquid chromatography. *Environ. Chem.* **2010**, *29*, 513–518. (In Chinese)
16. Wu, X.; Xiang, L.; Mo, C.; Li, Y.; Jiang, Y.; Yan, Q.; Lv, X.; Huang, X. Determination of quinolones in vegetables using ultra performance liquid chromatography-electrospray ionization tandem mass spectrometry. *Chin. J. Anal. Chem.* **2013**, *41*, 876–881. (In Chinese) [CrossRef]
17. Wu, X.; Bao, Y.; Xiang, L.; Yan, Q.; Jiang, Y.; Li, Y.; Huang, X.; Mo, C. Simultaneous sulfonamide antibiotics in vegetables using solid phase extraction and high performance liquid chromatography coupled with mass spectrometry. *Environ. Chem.* **2013**, *32*, 1038–1044. (In Chinese)
18. Yao, Y.; Mo, C.; Li, Y.; Bao, Y.; Tai, Y.; Wu, X. Determination of tetracyclines in vegetables using solid phase extraction and HPLC with fluorescence detection. *Environ. Chem.* **2010**, *29*, 536–541. (In Chinese)
19. Pan, M.; Wong, C.K.C.; Chu, L.M. Distribution of antibiotics in wastewater-irrigated soils and their accumulation in vegetable crops in the Pearl River Delta, Southern China. *J. Agric. Food Chem.* **2014**, *62*, 11062–11069. [CrossRef] [PubMed]
20. Sallach, J.B.; Snow, D.; Hodges, L.; Li, X.; Bartelt-Hunt, S. Development and comparison of four methods for the extraction of antibiotics from a vegetative matrix. *Environ. Toxicol. Chem.* **2015**, *35*, 889–897. [CrossRef]
21. Brack, W.; Dulio, V.; Slobodnik, J. The NORMAN network and its activities on emerging environmental substances with a focus on effect-directed analysis of complex environmental contamination. *Environ. Sci. Eur.* **2012**, *24*, 29–33. [CrossRef]
22. Fu, Y.; Zhou, Z.; Kong, H.; Lu, X.; Zhao, X.; Chen, Y.; Chen, J.; Wu, Z.; Xu, Z.; Zhao, C.; et al. Nontargeted screening method for illegal additives based on ultrahigh-performance liquid chromatography-high-resolution mass spectrometry. *Anal. Chem.* **2016**, *88*, 8870–8877. [CrossRef]
23. Hakme, E.; Lozano, A.; Gómez-Ramos, M.M.; Hernando, M.D.; Fernández-Alba, A.R. Non-target evaluation of contaminants in honey bees and pollen samples by gas chromatography time-of-flight mass spectrometry. *Chemosphere* **2017**, *184*, 1310–1319. [CrossRef]
24. Kunzelmann, M.; Winter, M.; Aberg, M.; Hellenäs, K.E.; Rosén, J. Non-targeted analysis of unexpected food contaminants using LC-HRMS. *Anal. Bioanal. Chem.* **2018**, *410*, 5593–5602. [CrossRef]
25. Shepherd, L.V.; Fraser, P.; Stewart, D. Metabolomics: A second-generation platform for crop and food analysis. *Bioanalysis* **2011**, *3*, 1143–1159. [CrossRef] [PubMed]
26. Bueno, M.J.M.; Díaz-Galiano, F.J.; Rajski, L.; Cutillas, V.; Fernández-Alba, A.R. A non-targeted metabolomic approach to identify food markers to support discrimination between organic and conventional tomato crops. *J. Chromatogr. A* **2018**, *1546*, 66–76. [CrossRef] [PubMed]
27. Johanningsmeier, S.D.; Harris, G.K.; Klevorn, C.M. Metabolomic technologies for improving the quality of food: Practice and promise. *Annu. Rev. Food Sci. Technol.* **2016**, *7*, 413–438. [CrossRef] [PubMed]
28. Tengstrand, E.; Rosén, J.; Hellenäs, K.E.; Aberg, K.M. A concept study on non-targeted screening for chemical contaminants in food using liquid chromatography-mass spectrometry in combination with a metabolomics approach. *Anal. Bioanal. Chem.* **2013**, *405*, 1237–1243. [CrossRef]
29. Delaporte, G.; Cladière, M.; Jouan-Rimbaud Bouveresse, D.; Camel, V. Untargeted food contaminant detection using UHPLC-HRMS combined with multivariate analysis: Feasibility study on tea. *Food Chem.* **2019**, *277*, 54–62. [CrossRef] [PubMed]
30. Azanu, D.; Mortey, C.; Darko, G.; Weisser, J.J.; Styrishave, B.; Abaidoo, R.C. Uptake of antibiotics from irrigation water by plants. *Chemosphere* **2016**, *157*, 107–114. [CrossRef]
31. Du, J.; Zhao, H.; Chen, J. Simultaneous determination of 23 antibiotics in mariculture water using solid-phase extraction and high performance liquid chromatography-tandem mass spectrometry. *Chin. J. Chromatogr.* **2015**, *33*, 348–353. (In Chinese) [CrossRef]
32. Liu, S.; Du, J.; Chen, J.; Zhao, H. Determination of 19 antibiotic and 2 sulfonamide metabolite residues in wild fish muscle in mariculture areas of Laizhou Bay using accelerated solvent extraction and high performance liquid chromatography-tandem mass spectrometry. *Chin. J. Chromatogr.* **2014**, *32*, 1320–1325. (In Chinese) [CrossRef]
33. Dunn, W.B.; Broadhurst, D.; Begley, P.; Zelena, E.; Francis-McIntyre, S.; Anderson, N.; Brown, M.; Knowles, J.D.; Halsall, A.; Haselden, J.N.; et al. Procedures for large-scale metabolic profiling of serum and plasma using gas chromatography and liquid chromatography coupled to mass spectrometry. *Nat. Protoc.* **2011**, *6*, 1060–1083. [CrossRef]
34. Want, E.J.; Wilson, I.D.; Gika, H.; Theodoridis, G.; Plumb, R.S.; Shockcor, J.; Holmes, E.; Nicholson, J.K. Global metabolic profiling procedures for urine using UPLC-MS. *Nat. Protoc.* **2010**, *5*, 1005–1018. [CrossRef]
35. Chambers, M.C.; MacLean, B.; Burke, R.; Amodei, D.; Ruderman, D.L.; Neumann, S.; Gatto, L.; Fischer, B.; Pratt, B.; Egertson, J.; et al. A cross-platform toolkit for mass spectrometry and proteomics. *Nat. Biotechnol.* **2012**, *30*, 918–920. [CrossRef] [PubMed]
36. Giacomoni, F.; Le Corguille, G.; Monsoor, M.; Landi, M.; Pericard, P.; Petera, M.; Duperier, C.; Tremblay-Franco, M.; Martin, J.F.; Jacob, D.; et al. Workflow4Metabolomics: A collaborative research infrastructure for computational metabolomics. *Bioinformatics* **2015**, *31*, 1493–1495. [CrossRef] [PubMed]
37. Smith, C.A.; Want, E.J.; O'Maille, G.; Abagyan, R.; Siuzdak, G. XCMS: Processing mass spectrometry data for metabolite profiling using nonlinear peak alignment, matching, and identification. *Anal. Chem.* **2006**, *78*, 779–787. [CrossRef] [PubMed]
38. Gika, H.G.; Theodoridis, G.A.; Wilson, I.D. Liquid chromatography and ultra-performance liquid chromatography-mass spectrometry fingerprinting of human urine: Sample stability under different handling and storage conditions for metabonomics studies. *J. Chromatogr. A* **2008**, *1189*, 314–322. [CrossRef]

39. Winnike, J.H.; Busby, M.G.; Watkins, P.B.; O'Connell, T.M. Effects of a prolonged standardized diet on normalizing the human metabolome. *Am. J. Clin. Nutr.* **2009**, *90*, 1496–1501. [CrossRef]
40. Yin, P.; Peter, A.; Franken, H.; Zhao, X.; Neukamm, S.S.; Rosenbaum, L.; Lucio, M.; Zell, A.; Haring, H.U.; Xu, G.; et al. Preanalytical aspects and sample quality assessment in metabolomics studies of human blood. *Clin. Chem.* **2013**, *59*, 833–845. [CrossRef]
41. Pontes, T.A.; Barbosa, A.D.; Silva, R.D.; Melo-Junior, M.R.; Silva, R.O. Osteopenia-osteoporosis discrimination in postmenopausal women by 1H NMR-based metabonomics. *PLoS ONE* **2019**, *14*, e0217348. [CrossRef]
42. Schievano, E.; Morelato, E.; Facchin, C.; Mammi, S. Characterization of markers of botanical origin and other compounds extracted from unifloral honeys. *J. Agric. Food Chem.* **2013**, *61*, 1747–1755. [CrossRef]
43. Zhao, W.; Wang, G.; Xun, W.; Yu, Y.; Ge, C.; Liao, G. Selection of water-soluble compounds by characteristic flavor in Chahua Chicken muscles based on metabolomics. *Sci. Agric. Sin.* **2020**, *53*, 1627–1642. (In Chinese)
44. Dai, Y.; Lv, C.; He, L.; Yi, C.; Liu, X.; Huang, W.; Chen, J. Metabolic changes in the processing of Yunkang 10 sun-dried green tea based on metabolomics. *Sci. Agric. Sin.* **2020**, *53*, 357–370. (In Chinese)
45. Inoue, K.; Tanada, C.; Sakamoto, T.; Tsutsui, H.; Akiba, T.; Min, J.Z.; Todoroki, K.; Yamano, Y.; Toyo'oka, T. Metabolomics approach of infant formula for the evaluation of contamination and degradation using hydrophilic interaction liquid chromatography coupled with mass spectrometry. *Food Chem.* **2015**, *181*, 318–324. [CrossRef] [PubMed]
46. Yang, J.; Chen, T.; Sun, L.; Zhao, Z.; Wan, C. Potential metabolite markers of schizophrenia. *Mol. Psychiatry* **2013**, *18*, 67–78. [CrossRef] [PubMed]
47. Gika, H.G.; Theodoridis, G.A.; Plumb, R.S.; Wilson, D. Current practice of liquid chromatography-mass spectrometry in metabolomics and metabonomics. *J. Pharm. Biomed. Anal.* **2014**, *87*, 12–25. [CrossRef]
48. Rubert, J.; Righetti, L.; Stranska-Zachariasova, M.; Dzuman, Z.; Chrpova, J.; Dall'Asta, C.; Hajslova, J. Untargeted metabolomics based on ultra-high-performance liquid chromatography-high-resolution mass spectrometry merged with chemometrics: A new predictable tool for an early detection of mycotoxins. *Food Chem.* **2017**, *224*, 423–431. [CrossRef] [PubMed]
49. Thévenot, E.A.; Roux, A.; Xu, Y.; Ezan, E.; Junot, C. Analysis of the human adult urinary metabolome variations with age, body mass index, and gender by implementing a comprehensive workflow for univariate and OPLS statistical analyses. *J. Proteome Res.* **2015**, *14*, 3322–3335. [CrossRef]
50. Chen, Q.; Dai, W.; Lin, Z.; Xie, D.; Lv, M.; Lin, Z. Effects of shading on main quality components in tea (*Camellia Sinensis (L) O. Kuntze*) leaves based on metabolomics analysis. *Scientia Agricultura Sinica* **2019**, *52*, 1066–1077. (In Chinese)
51. Taguchi, Y.H. Principal component analysis based unsupervised feature extraction applied to budding yeast temporally periodic gene expression. *BioData Min.* **2016**, *9*, 22. [CrossRef]
52. Broadhurst, D.I.; Kella, D.B. Statistical strategies for avoiding false discoveries in metabolomics and related experiments. *Metabolomics* **2006**, *2*, 171–196. [CrossRef]
53. Slupsky, C.M.; Steed, H.; Wells, T.H.; Dabbs, K.; Schepansky, A.; Capstick, V.; Faught, W.; Sawyer, M.B. Urine metabolite analysis offers potential early diagnosis of ovarian and breast cancers. *Clin. Cancer Res.* **2010**, *16*, 5835–5841. [CrossRef]
54. Ortmayr, K.; Charwat, V.; Kasper, C.; Hann, S.; Koellensperger, G. Uncertainty budgeting in fold change determination and implications for non-targeted metabolomics studies in model systems. *Analyst* **2017**, *142*, 80–90. [CrossRef]
55. Environmental Protection Agency (EPA). Method 8061A Phthalate Esters by Gas Chromatography with Electron Capture detection (GC/ECD). 1996. Available online: https://www.epa.gov/hw-sw846/sw-846-test-method-8061a-phthalate-esters-gas-chromatography-electron-capture-detection (accessed on 15 November 2021).
56. Li, Y.; Wu, X.; Mo, C.; Tai, Y.; Huang, X.; Xiang, L. Investigation of sulfonamide, tetracycline, and quinolone antibiotics in vegetable farmland soil in the Pearl River delta area, Southern China. *J. Agric. Food Chem.* **2011**, *59*, 7268–7276. [CrossRef] [PubMed]
57. Goldstein, M.; Shenker, M.; Chefetz, B. Insights into the uptake processes of wastewater-borne pharmaceuticals by vegetables. *Environ. Sci. Technol.* **2014**, *48*, 5593–5600. [CrossRef]
58. Kang, D.H.; Gupta, S.; Rosen, C.; Fritz, V.; Singh, A.; Chander, Y.; Murray, H.; Rohwer, C. Antibiotic uptake by vegetable crops from manure-applied soils. *J. Agric. Food Chem.* **2013**, *61*, 9992–10001. [CrossRef] [PubMed]
59. Malchi, T.; Maor, Y.; Tadmor, G.; Shenker, M.; Chefetz, B. Irrigation of root vegetables with treated wastewater: Evaluating uptake of pharmaceuticals and the associated human health risks. *Environ. Sci. Technol.* **2014**, *48*, 9325–9333. [CrossRef] [PubMed]
60. Miller, E.L.; Nason, S.L.; Karthikeyan, K.G.; Pedersen, J.A. Root uptake of pharmaceuticals and personal care product ingredients. *Environ. Sci. Technol.* **2016**, *50*, 525–541. [CrossRef] [PubMed]
61. Sabourin, L.; Duenk, P.; Bonte-Gelok, S.; Payne, M.; Lapen, D.R.; Topp, E. Uptake of pharmaceuticals, hormones and parabens into vegetables grown in soil fertilized with municipal biosolids. *Sci. Total Environ.* **2012**, *431*, 233–236. [CrossRef] [PubMed]
62. Wu, X.; Ernst, F.; Conkle, J.L.; Gan, J. Comparative uptake and translocation of pharmaceutical and personal care products (PPCPs) by common vegetables. *Environ. Int.* **2013**, *60*, 15–22. [CrossRef] [PubMed]
63. Wu, X.; Conkle, J.L.; Ernst, F.; Gan, J. Treated wastewater irrigation: Uptake of pharmaceutical and personal care products by common vegetables under field conditions. *Environ. Sci. Technol.* **2014**, *48*, 11286–11293. [CrossRef]
64. Wu, X.; Dodgen, L.K.; Conkle, J.L.; Gan, J. Plant uptake of pharmaceutical and personal care products from recycled water and biosolids: A review. *Sci. Total Environ.* **2015**, *536*, 655–666. [CrossRef]
65. Berrada, H.; Molto, J.C.; Manes, J.; Font, G. Determination of aminoglycoside and macrolide antibiotics in meat by pressurized liquid extraction and LC-ESI-MS. *J. Sep. Sci.* **2015**, *33*, 522–529. [CrossRef]

66. Chung, S.W.C.; Lam, C. Development of a 15-class multiresidue method for analyzing 78 hydrophilic and hydrophobic veterinary drugs in milk, egg and meat by liquid chromatography-tandem mass spectrometry. *Anal. Methods* **2015**, *7*, 6764–6776. [CrossRef]
67. Desmarchelier, A.; Fan, K.; Minh Tien, M.; Savoy, M.C.; Tarres, A.; Fuger, D.; Goyon, A.; Bessaire, T.; Mottier, P. Determination of 105 antibiotic, anti-inflammatory, antiparasitic agents and tranquilizers by LC-MS/MS based on an acidic QuEChERS-like extraction. *Food Add. Contam. A* **2018**, *35*, 646–660. [CrossRef] [PubMed]
68. Schneider, M.J.; Lehotay, S.J.; Lightfield, A.R. Evaluation of a multi-class, multi-residue liquid chromatography-tandem mass spectrometry method for analysis of 120 veterinary drugs in bovine kidney. *Drug Test. Anal.* **2012**, *4*, 91–102. [CrossRef] [PubMed]
69. Zhang, N.; Fan, S.; Xue, Y.; Liu, P.; Liu, W.; Wu, G.; Zhao, R. Rapid detection of 8 veterinary drug residues in chicken by QuEChERS-ultra performance liquid chromatography tandem mass spectrometry. *J. Hyg. Res.* **2017**, *46*, 89–93. (In Chinese)

MDPI

Article

An Integrated Mass Spectrometry-Based Glycomics-Driven Glycoproteomics Analytical Platform to Functionally Characterize Glycosylation Inhibitors

Michael Russelle S. Alvarez [1,2], Qingwen Zhou [2], Sheryl Joyce B. Grijaldo [1], Carlito B. Lebrilla [2], Ruel C. Nacario [1], Francisco M. Heralde III [3], Jomar F. Rabajante [4] and Gladys C. Completo [1,*]

1 Institute of Chemistry, College of Arts and Sciences, University of the Philippines Los Baños, Los Baños 4031, Philippines; mralvarez@ucdavis.edu (M.R.S.A.); sbgrijaldo@up.edu.ph (S.J.B.G.); rcnacario@up.edu.ph (R.C.N.)
2 Department of Chemistry, University of California Davis, Davis, CA 95616, USA; qwzzhou@ucdavis.edu (Q.Z.); cblebrilla@ucdavis.edu (C.B.L.)
3 Lung Center of the Philippines, Quezon City 1100, Philippines; fmheralde1@up.edu.ph
4 Institute of Mathematical Sciences and Physics, College of Arts and Sciences, University of the Philippines Los Baños, Los Baños 4031, Philippines; jfrabajante@up.edu.ph
* Correspondence: gjcompleto@up.edu.ph

Abstract: Cancer progression is linked to aberrant protein glycosylation due to the overexpression of several glycosylation enzymes. These enzymes are underexploited as potential anticancer drug targets and the development of rapid-screening methods and identification of glycosylation inhibitors are highly sought. An integrated bioinformatics and mass spectrometry-based glycomics-driven glycoproteomics analysis pipeline was performed to identify an N-glycan inhibitor against lung cancer cells. Combined network pharmacology and in silico screening approaches were used to identify a potential inhibitor, pictilisib, against several glycosylation-related proteins, such as Alpha1-6FucT, GlcNAcT-V, and Alpha2,6-ST-I. A glycomics assay of lung cancer cells treated with pictilisib showed a significant reduction in the fucosylation and sialylation of N-glycans, with an increase in high mannose-type glycans. Proteomics analysis and in vitro assays also showed significant upregulation of the proteins involved in apoptosis and cell adhesion, and the downregulation of proteins involved in cell cycle regulation, mRNA processing, and protein translation. Site-specific glycoproteomics analysis further showed that glycoproteins with reduced fucosylation and sialylation were involved in apoptosis, cell adhesion, DNA damage repair, and chemical response processes. To determine how the alterations in N-glycosylation impact glycoprotein dynamics, modeling of changes in glycan interactions of the ITGA5–ITGB1 (Integrin alpha 5-Integrin beta-1) complex revealed specific glycosites at the interface of these proteins that, when highly fucosylated and sialylated, such as in untreated A549 cells, form greater hydrogen bonding interactions compared to the high mannose-types in pictilisib-treated A549 cells. This study highlights the use of mass spectrometry to identify a potential glycosylation inhibitor and assessment of its impact on cell surface glycoprotein abundance and protein–protein interaction.

Keywords: glycomics; glycoproteomics; glycosylation; proteomics; in silico docking; network pharmacology; non-small cell lung cancer

Citation: Alvarez, M.R.S.; Zhou, Q.; Grijaldo, S.J.B.; Lebrilla, C.B.; Nacario, R.C.; Heralde, F.M., III; Rabajante, J.F.; Completo, G.C. An Integrated Mass Spectrometry-Based Glycomics-Driven Glycoproteomics Analytical Platform to Functionally Characterize Glycosylation Inhibitors. *Molecules* **2022**, 27, 3834. https://doi.org/10.3390/molecules27123834

Academic Editor: Thomas Letzel

Received: 8 April 2022
Accepted: 11 June 2022
Published: 14 June 2022

1. Introduction

Lung cancer is the leading cause of cancer-related mortalities worldwide [1,2]. Cancer incidence and mortality are increasing worldwide, reflecting several factors: aging; population growth; cancer risk factors; and socioeconomic development. According to the GLOBOCAN 2020 database, there will be 19.3 million new cases and 10 million cancer deaths worldwide [3]. Out of these, 2,206,771 cases (11.4%) and 1,796,144 (18.0%) deaths for both sexes will be due to lung cancer.

Protein glycosylation is one of the most complex and most frequent post-translational modifications and is involved in many of the cellular interactions, such as host–pathogen interactions, cell differentiation and trafficking, and intra- and intercellular signaling [4–6]. Protein glycosylation is a complex process that starts at the endoplasmic reticulum and is continued in the Golgi apparatus, where the glycans are further processed to achieve the diversity and complexity of the final glycan structures, through a series of steps involving glycosyltransferases and glycosidases [7,8]. The overexpression of these glycan-processing enzymes is usually observed in cancer cells, resulting in an enhanced expression of the related glycan structures. For example, the enzymes Alpha1-6FucT, B4GALT2, MAN1A2, and MAN2A1 are overexpressed in lung cancer tissue samples [9]. Likewise, glycans synthesized by these enzymes are also overexpressed in lung cancer tissues [10,11]. Additionally, aberrant glycosylation leads to increased biosynthesis of various tumor antigens, such as Sialyl Lewis X (sLe^{X}), which serves as a ligand for the cell adhesion molecule, selectin. This antigen is also involved in the adhesion of cancer cells to the vascular endothelium and hematogenous metastasis. Cancer progression is also associated with changes in the glycosylation of cell-surface proteins involved in the loss of cell-to-cell adhesion and increased metastatic potential. Furthermore, altered glycosylation is also correlated with the other hallmarks of cancer, such as enhanced proliferation, angiogenesis potential, replicative immortality, metastatic potential, apoptosis, and tumor suppression [12,13].

Several glycosyltransferases have been associated as cancer biomarkers [4,9,11]. A glycosyltransferase used as a biomarker is UDP-N-acetyl-D-glucosamine: N-acetylglucosamine transferase V (GlcNAcT-V), which catalyzes the β1-6 branching of N-glycans. Increased β1-6 branching, due to GlcNAcT-V overexpression, has been observed in breast carcinoma [14]. Sialyltransferases are glycosyltransferases that are abnormally expressed in cancers and are involved in carcinogenesis, progression, and metastasis [15–17]. An overexpression of α2-3 sialyltransferase III (ST3Gal-III) in pancreatic cancer has been implicated in pancreatic tumor progression. The overexpression of α2-6 sialyltransferase I (ST6GalNAc-I) was related to poor patient survival in colorectal carcinoma patients [18]. As such, the glycosyltransferases and glycosidases are underexploited drug targets for cancer therapeutics and there is a relative lack of small molecule inhibitors of these enzymes with drug-like properties. The glycosylation inhibitors that were previously reported include metabolic inhibitors, which target the formation of nucleotide sugars; tunicamycin which targets dolichol-PP-GlcNAc formation (biosynthesis of N-glycans); plant alkaloids that inhibit the processing of glycosidases; substrate analogs which are specific towards glycosyltransferases and glycosidases; glycoside primers which divert the assembly of glycans from endogenous acceptors towards exogenous primers; and tagged monosaccharides which target several different biosynthesis pathways [6]. Synthetic compounds, such as 2-deoxy-2-fluorofucose and 2,4,7,8,9-Penta-O-acetyl-N-acetyl-3-fluoro-b-D-neuraminic acid ester, were previously shown to inhibit the expression of fucosylated and sialylated N-glycan structures, respectively, in the glycocalyx of Caco-2, A549, and PNT2 cells, using LC–MS/MS methods [19]. The natural product, tunicamycin, induces the inhibition of the protein N-glycosylation by blocking the GlcNAc phosphotransferase-catalyzed transfer of N-acetylglucosamine-1-phosphate from UDP-GlcNAc to dolichol-P, which results in the decreased production of dolichol-PP-GlcNAc. In combination with anticancer drugs, tunicamycin has also been shown to be cytotoxic against multidrug-resistant human ovarian cystadenocarcinoma cells, by inhibiting protein and glycoprotein syntheses [20]. However, the effects of these glycan inhibitors on cancer-associated pathways, and in correlation with protein glycosylation, have not been explored before.

In this study, we integrated LC–MS/MS methods—glycomics, proteomics, and glycoproteomics—to characterize the effect on protein glycosylation of a small-molecule inhibitor, pictilisib, identified through our computational docking predictions. In conjunction with the LC–MS/MS methods, we developed molecular and bioinformatics models to understand how pictilisib affects the protein glycosylation, and subsequently affects the cancer-associated biological pathways.

2. Results

2.1. Pictilisib Was Validated to Reduce the Relative Abundance of Fucosylated and Sialylated N-glycans

We previously discovered, through in silico screening, that pictilisib is able to bind and inhibit several glycosyltransferases [21]. To determine the effect of changes in the protein glycosylation after pictilisib treatment, in vitro assays were performed using an A549 non-small cell lung cancer (NSCLC) cell line as a model for lung cancer. Dose–response cytotoxic assay and preliminary drug-titration assay after 24 h of treatment were conducted to determine the nontoxic drug concentrations that could still affect the protein glycosylation (Figures S1 and S2, Supplementary Materials). After optimization of the assay conditions, the A549 cells were treated with pictilisib (4 μM) for 24 h, then subjected to glycomics profiling using an established mass spectrometric method [22]. Glycomics profiling with mass spectrometry allows for a comprehensive and reproducible analysis of the glycan composition of the cell's glycocalyx, after treatment with pictilisib or vehicle control (Figures 1a and S3; Table S1, Supplementary Materials). Comparing the sum of the relative abundances of the primary N-glycan types—high-mannose, undecorated, fucosylated, sialylated, and sialo-fucosylated—shows that pictilisib treatment significantly reduced the total relative abundances of the fucosylated and sialylated N-glycans (Figure 1b). A total of 138 glycans were profiled, of which 36 were found to be significant ($p < 0.05$). A closer inspection of these N-glycans showed that the total fucosylated complex- and high-mannose-type N-glycans, as well as both the total sialylated complex- and hybrid-type N-glycans were significantly reduced by the pictilisib treatment (Figure 1b). Specifically, the N-glycan compositions $Hex_6HexNAc_4NeuAc_1$, $Hex_6HexNAc_4NeuAc_2$, $Hex_3HexNAc_2Fuc_1$, $Hex_5HexNAc_4Fuc_2$, $Hex_7HexNAc_6Fuc_1$, $Hex_8HexNAc_7Fuc_6$, $Hex_8HexNAc_7Sia_2$, $Hex_9HexNAc_8Fuc_1$, and $Hex_9HexNAc_8Fuc_1NeuAc_2$ were found to be very significantly underexpressed in the pictilisib-treated cells (Table 1). Mapping these N-glycan compositions to the known N-glycan biosynthetic pathway shows potential glycosylation enzyme reactions that could be inhibited from the pictilisib treatment (Figure 1c), specifically those glycosylation reactions involving the addition of fucose and sialic acid residues [23,24]. These significantly underexpressed N-glycans represent several known cancer-related N-glycan epitopes, such as Lewis and Sialyl Lewis antigens, core fucosylation, and α2,6-sialylated lactosamine [13].

2.2. Proteomic Analysis Shows Upregulated Pathways Involving Apoptosis and Cell Adhesion, and Downregulated Pathways Involving Cell Cycle Process, mRNA Processing, and Protein Translation

To validate the bioactivity effects of pictilisib on A549, we conducted in vitro assays coupled with label-free quantitative proteomics, to identify the specific pathways targeted by pictilisib. Proteins were filtered by setting the Byologic score higher than or equal to 100 and having two unique peptides per protein. The protein intensities were reported as the sum of the top two peptides for each protein, normalized to the total intensity per sample. The dataset was further filtered based on the presence of a specific protein in at least two replicates per group and then analyzed using multiple *t*-tests ($\alpha = 0.05$) (Figure S4, Supplementary Materials). Based on the proteomics data (Table S2, Supplementary Materials), 1518 proteins were quantified, with 380 proteins identified as significantly different (p-value < 0.05, Figure 2a). A gene set enrichment analysis of these significantly different proteins showed interesting biological processes affected by the pictilisib treatment, such as the upregulation of apoptosis and biological adhesion processes and the downregulation of cell cycle processes (Figures 2b,c, S5 and S6, Supplementary Materials). In vitro apoptosis and cell cycle assays verify that, indeed, pictilisib treatment induced apoptosis (Figures 2d and S7, Supplementary Materials) and G0/G1 cell cycle arrest (Figures 2e and S8, Supplementary Materials) in A549 cells. Correspondingly, the quantification of specific apoptosis, cell cycle, and DNA damage-related proteins show significant differences in key proteins, such as HELLS [25], TOP2A [26,27], MCM6 [28], PSMB4 [29], EIF4G1 [30–32], EIF5A [33], DDX5 [34], and RACK1 [35,36], involved in these pathways (Figure 2f–h). Likewise, the effect of pictilisib on cell migration was verified using both scratch assay and transwell migration assay, with pictilisib treatment causing a significant reduction in cell migration (Figures 2i,j, S9 and S10, Supplementary Materials). Proteomics analysis shows that the mechanism affecting the cell migration was observed by overexpressing adhesion proteins and upregulating cell adhesion pathways (Figure 2k). Interestingly, the pictilisib treatment also significantly downregulated the proteins involved in mRNA processing (Figure 3a,b), and the protein translation (Figure 3c) processes.

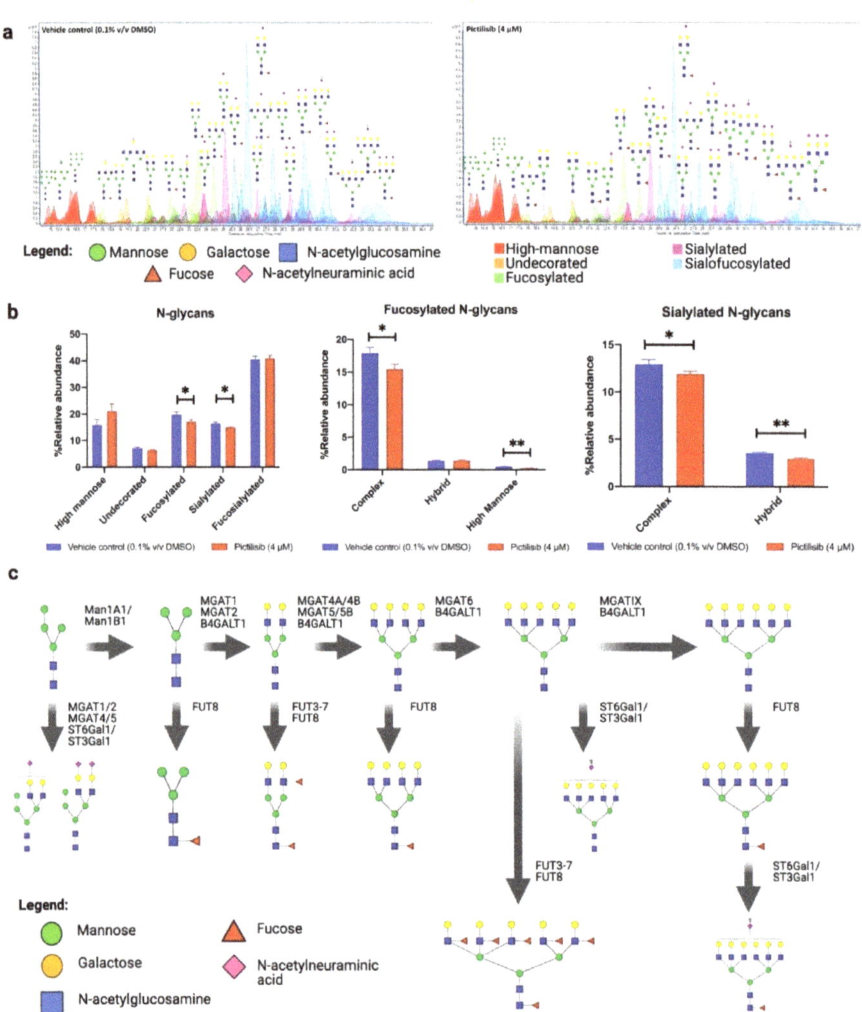

Figure 1. Pictilisib significantly reduced the relative abundance of fucosylated and sialylated N-glycans. (**a**) Glycan-annotated extracted ion chromatograms (EICs) of N-glycomes of vehicle control and pictilisib-treated A549 cells; (**b**) Relative abundances of N-glycan types in vehicle control and pictilisib-treated A549 cells; (**c**) Biosynthetic map showing the abundance of each significantly different N-glycan (Table 1), N-glycan precursor, and known enzymes catalyzing the glycosylation reaction. The data represent mean ± SD; $n = 3$. The statistical differences were detected utilizing multiple t-test with FDR correction, * $q < 0.05$; ** $q < 0.01$.

Table 1. Glycan composition, putative structures, and log 2 fold-change (and *q*-values) of highly significantly different glycans in pictilisib-treated A549 cells compared to vehicle control.

Glycan Composition	Putative Structure	$\log_2$ Fold-Change	$-\log_{10}$ *q*-Value
$Hex_8HexNAc_7NeuAc_2$	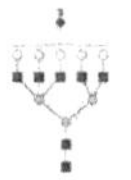	−2.4214	1.6249
$Hex_8HexNAc_7Fuc_6$	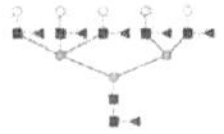	−1.6756	1.6249
$Hex_9HexNAc_8Fuc_1NeuAc_2$	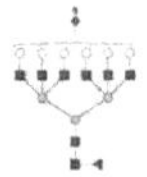	−1.3536	1.5544
$Hex_9HexNAc_8Fuc_1$	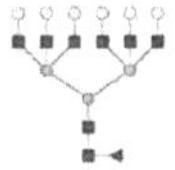	−1.3269	1.5403
$Hex_6HexNAc_4NeuAc_2$		−1.0587	1.6249
$Hex_3HexNAc_2Fuc_1$		−1.0000	1.4854
$Hex_5HexNAc_4Fuc_2$		−0.7675	1.6249
$Hex_6HexNAc_4NeuAc_1$		−0.5385	1.6249
$Hex_7HexNAc_6Fuc_1$		−0.3078	1.6249

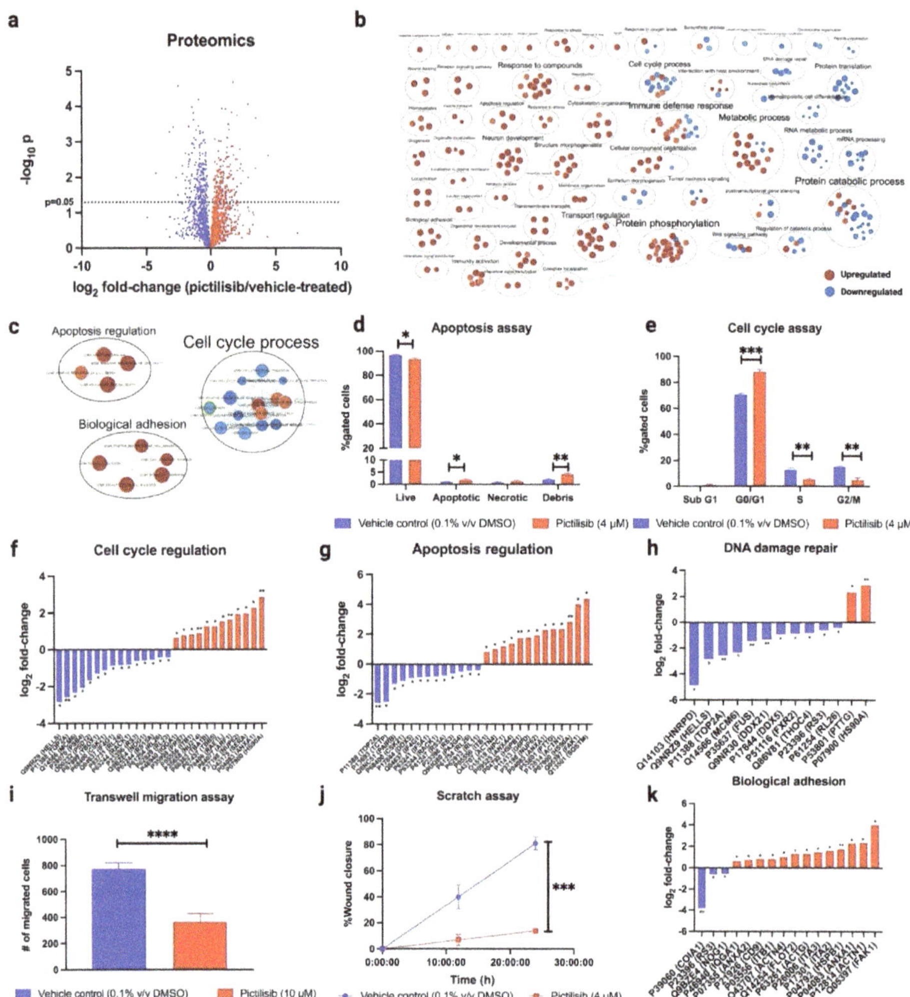

Figure 2. Pictilisib treatment significantly affected the pathways involving ECM interactions and migration, and cell death and proliferation, in A549. (**a**) Volcano plot of differentially expressed proteins in pictilisib-treated A549 cells; (**b**) Gene-set enrichment analysis of pre-ranked protein expression profiles of pictilisib- vs. vehicle control-treated cells; (**c**) Processes involved in apoptosis regulation and biological adhesion were upregulated, while processes involved in cell cycle regulation were downregulated; (**d**,**e**) In vitro assays of pictilisib-treated cells show significantly increased apoptosis and G0/G1 cell cycle arrest; (**f**–**h**) Quantification of proteins related to cell cycle regulation, apoptosis regulation, and DNA damage repair show significant differences (q-value < 0.05); (**i**,**j**) In vitro scratch and transwell migration assays show significantly reduced migration activity of pictilisib-treated cells; (**k**) Quantification of proteins related to biological adhesion shows significant differences (q-value < 0.05). The data represent mean ± SD; n = 3. The statistical differences were detected utilizing multiple t-test with FDR correction, * $q < 0.05$; ** $q < 0.01$; *** $q < 0.001$; **** $q < 0.0001$.

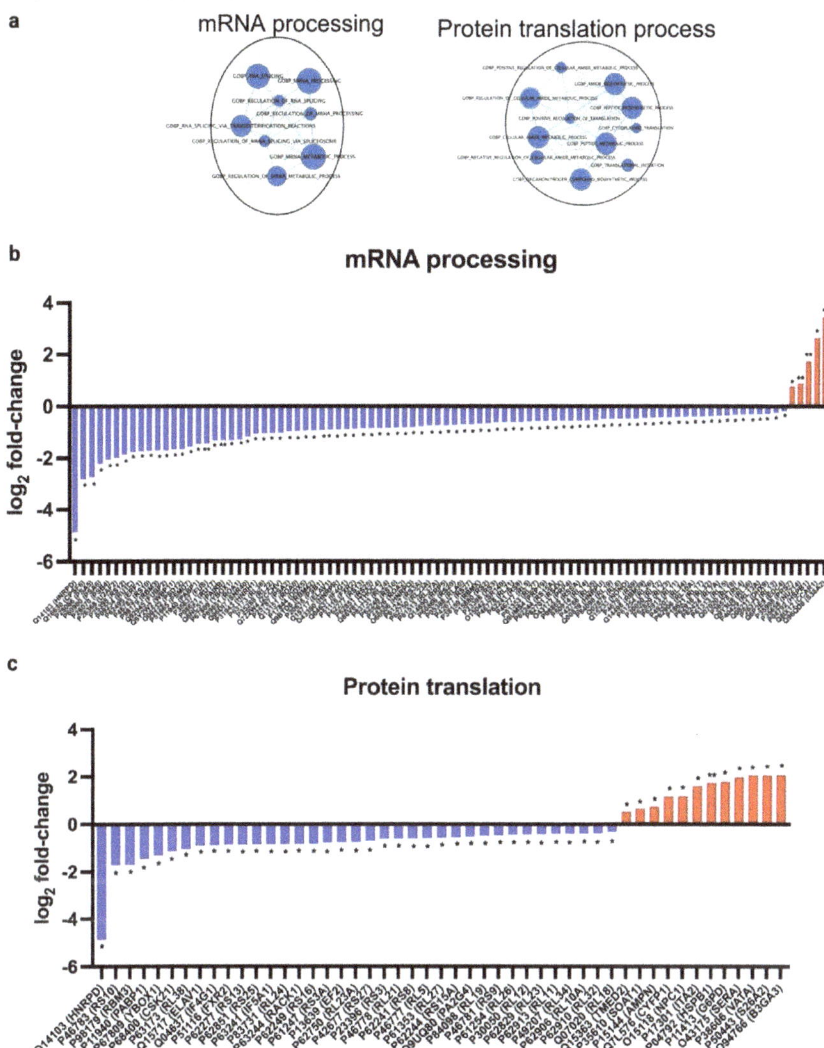

Figure 3. Pictilisib treatment also significantly affected pathways involving mRNA processing and protein translation in A549. (**a**) Based on the GSEA analysis, clusters of pathways involved in mRNA processing and protein translation are downregulated. Most genes involved in mRNA processing (**b**) and protein translation (**c**) are significantly underexpressed following pictilisib treatment. The data represent mean ± SD; n = 3. The statistical differences were detected utilizing multiple t-test with FDR correction, * $q < 0.05$; ** $q < 0.01$.

2.3. Glycoproteins with Reduced Fucosylation and Sialylation Were Involved in Apoptosis, DNA Damage Repair, and Cell Adhesion

Aberrant glycosylation has been well-documented in cancer, with fundamental changes in the glycosylation patterns of cell surface and secreted proteins during cancer progression. Growing evidence supported the role of glycosylation during multiple steps in tumor progression, cancer cell proliferation, invasion, metastasis, and angiogenesis [5,11,13,37,38].

Our glycomics results show that pictilisib treatment significantly reduced global fucosylation and sialylation of the cell membrane N-glycans. Likewise, our proteomics results show that pictilisib treatment significantly affected adhesion, apoptotic, and cell cycle pathways. To identify which of the glycoproteins have reduced fucosylation and sialylation and their involvement in these pathways, we performed quantitative site-specific glycoproteomics coupled with gene ontology analysis of the pictilisib-treated cells. Glycoforms were identified after score-filtering, replicate-filtering (present in at least two of the replicates), and normalized glycopeptides per protein glycosite (Figures 4a, S11 and S12; Table S3, Supplementary Materials). Normalized glycoforms were categorized based on the N-glycan type—high-mannose, undecorated, fucosylated, sialylated, and sialofucosylated—and then summed up for each glycosite. For example, the changes in glycoform occupancy in ANPEP, ADA10, ITGB1, and ITGA3 following pictilisib treatment reduced the fucosylation and sialylation or sialofucosylation in specific glycosites (Figure 4b). The glycosites with reduced fucosylation, sialylation, and sialofucosylation were represented as heat maps annotated with gene ontologies of the corresponding glycoproteins. Indeed, the glycoproteins associated with biological adhesion and locomotion, and apoptosis, had reduced fucosylation, sialylation, and sialofucosylation (Figure 4c). Increasing sialylation leads to a buildup of negative charges, physically disrupting cell-cell adhesion and promotes detachment through electrostatic repulsion [35]. Overexpression of the enzyme ST6GAL1 and its glycan product, sTn, leads to an increased migration and invasion in carcinoma [39].

Interestingly, the pictilisib treatment also reduced fucosylation, sialylation, and sialofucosylation of glycoproteins involved in chemical stimulus–response (Figure 4c). On the other hand, the glycoproteins involved in the immune system response have only reduced sialylation and sialofucosylation after pictilisib treatment. Looking specifically at the pathway effects, such as the integrin pathway, the integrins showed reduced fucosylation, sialylation, and sialofucosylation of several of their glycosites. While the epidermal growth factor receptor (EGFR) pathway and the ubiquitin-proteasome pathway glycoproteins had reduced sialofucosylation following pictilisib treatment (Figures S13 and S14, Supplementary Materials), these glycoproteins are also shown to perform functions in binding, catalysis, regulation, signal transduction, transport, and structural support. When mapped to show the protein–protein interaction network using STRING (Figure 4d), the enrichment analysis further confirms that these glycoproteins are significantly enriched in pathways related to cell adhesion, apoptotic process, and signaling pathways, DNA damage responses, and cellular responses to chemical stimuli (Figure 4e).

Site-specific protein glycoproteomics also allowed us to investigate deeper into the molecular interactions between glycoproteins, such as integrins. Integrin α-5 (ITGA5) and integrin β-1 (ITGB1) are integrins involved in several biological processes, including cell adhesion and survival. Following the pictilisib treatment of the A549 cells, we found several glycosites in both of the glycoproteins that had either reduced fucosylation, sialylation, or sialofucosylation (Figure 5a,c; Table S4, Supplementary Materials). This site-specific glycosylation information was overlaid with protein domain annotations from PFAM (http://pfam.xfam.org/, accessed on 15 May 2021) [40], to show how certain glycosites could potentially contribute to the protein interactions (Figure 5b,d). Further analysis through molecular dynamics showed that specific glycosites are significantly affected by the type of glycan decorations (Table S5, Supplementary Materials). $Hex_6HexNAc_7Fuc_4NeuAc_2$ was found to be downregulated, while $Hex_7HexNAc_2$ was upregulated in ITGB1 glycosite Asn269, following pictilisib treatment. Glycosite Asn269 was deemed important, due to its proximity to the interaction interface of ITGB1 and ITGA5. These glycans were modeled into ITGA5 and ITGB1 complex (PDB ID: 3vi4) using CHARMM-GUI [41], then simulated over 45 ns using NAMD [42] (Tables S6 and S7; Videos S1 and S2, Supplementary Materials). Hydrogen-bonding interactions were then monitored in VMD to show additional residue contacts by $Hex_6HexNAc_7Fuc_4NeuAc_2$ in the negative control compared to $Hex_7HexNAc_2$ in the pictilisib-treated cells (Figure 5e).

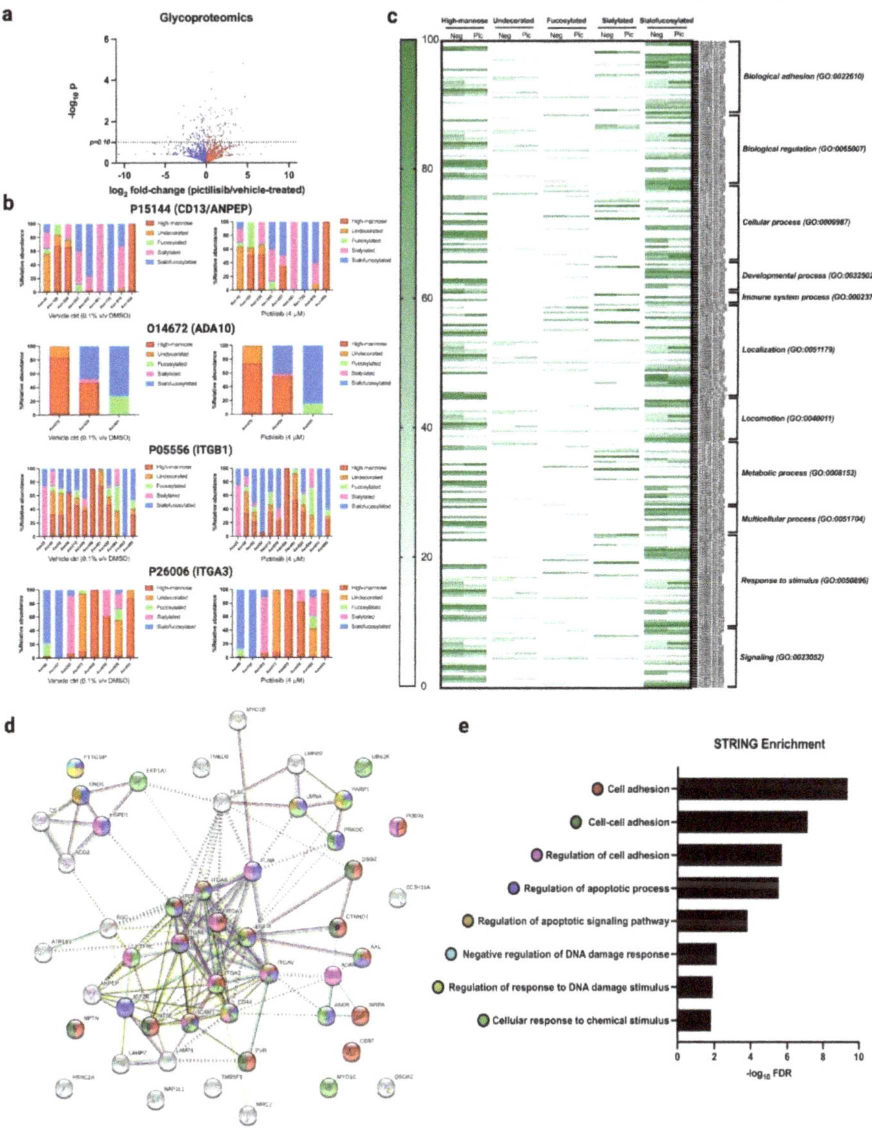

Figure 4. Pictilisib treatment reduced sialylation and fucosylation in specific glycoproteins. (**a**) Volcano plot of differentially abundant glycopeptides; (**b**) Site-specific glycosylation—high-mannose, undecorated, fucosylated, sialylated, and sialofucosylated—of several glycoproteins shown to have reduced fucosylation, sialylation, or sialofucosylation following pictilisib treatment; (**c**) Gene ontology analysis of proteins with reduced fucosylation, sialylation, and fucosylation shows glycoproteins involved in several biological processes; (**d**) STRING interaction analysis shows the interaction of the glycoproteins with reduced fucosylation, sialylation, or sialofucosylation; (**e**) Subsequent STRING enrichment analysis shows a significant enrichment of biological processes involved in adhesion, apoptosis, response to chemicals, and DNA damage. The data represent mean ± SD; n = 3. The statistical differences were detected utilizing multiple t-test with FDR correction.

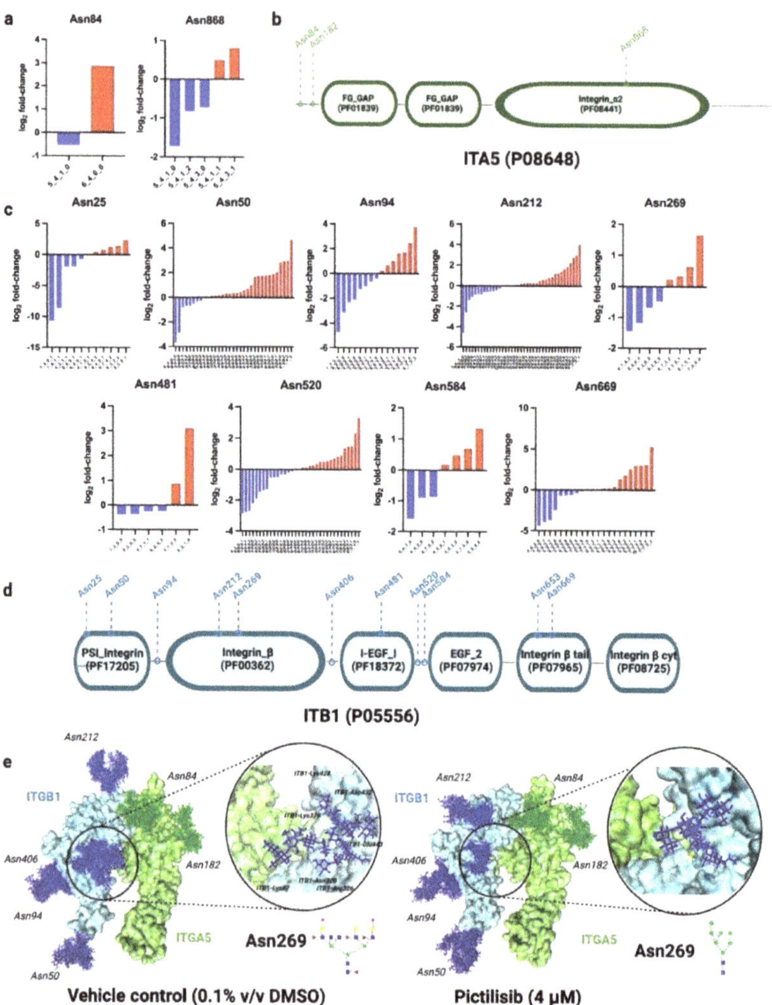

Figure 5. Site-specific glycosylation analysis of ITGA5-ITGB1 illustrates how specific glycosites potentially contribute to protein interactions. (**a**) Site-specific glycosylation of ITGA5 glycoprotein. Bar graphs represent log2 fold-changes in glycoform abundance following pictilisib treatment. Glycosite Asn182 did not change glycosylation; (**b**) Site-specific glycosylation overlaid with protein domain information of ITGA5, annotated using PFAM (http://pfam.xfam.org/, accessed on 15 May 2021); (**c**) Site-specific glycosylation of ITGB1 glycoprotein. Bar graphs represent log2 fold-changes in glycoform abundance following pictilisib treatment. Glycosites Asn406 and Asn653 did not change glycosylation. The *X*-axis represents glycoforms, annotated as $Hex_aHexNAc_bFuc_cSia_d$; (**d**) Site-specific glycosylation overlaid with protein domain information of ITGB1, annotated using PFAM (http://pfam.xfam.org/, accessed on 15 May 2021); (**e**) 3D trajectories of ITGA5-ITGB1 glycoprotein complexes following treatment with pictilisib. Specific glycoform structures can be seen in Table S7 (Supplementary Materials). Dynamics simulation of negative control (Video S1, Supplementary Materials) and pictilisib-treated (Video S2, Supplementary Materials) are also available.

2.4. Pictilisib Was Predicted to Interact and Inhibit Glycosylation Enzymes Using In Silico Docking and Network Pharmacology

Combined network pharmacology and in silico docking approaches were previously used to identify potential interactors with N-glycosylation-related proteins [21]. Several gene–drug interaction databases were surveyed, resulting in 185 predicted glycosylation interactors that were mapped against 356 glycosylation-related proteins and enzymes (Figure 6a; Table S8, Supplementary Materials). From this set of compounds, pictilisib was selected due to its high degree of interactions (Figure 6b). Specifically, pictilisib was predicted to lower the expression of the glycosyltransferase genes B3GALNT1, B4GALT2, and glycosidase MAN1A1 through interactions with PIK3CA [43].

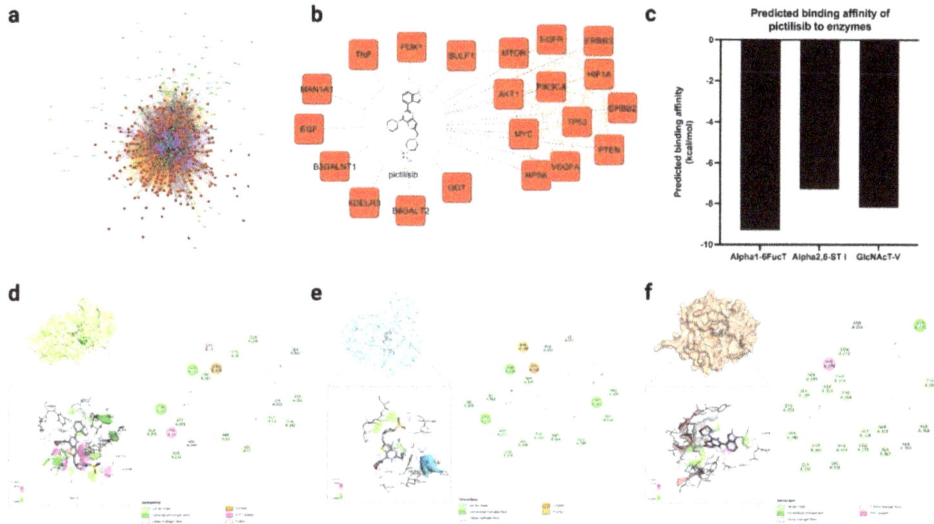

Figure 6. Pictilisib was predicted to interact and inhibit several glycosylation-related genes through network pharmacology and in silico binding approach. (**a**) Drug–gene interaction network of selected glycosylation targeting compounds; (**b**) Sub-network of pictilisib drug–gene interactions; (**c**) The binding affinity of pictilisib against Alpha1-6FucT, GlcNAcT-V, and Alpha2,6-ST I; (**d**–**f**) Docking conformation and residue interactions of pictilisib with Alpha1-6FucT, GlcNAcT-V, and Alpha2,6-ST I, respectively.

Additionally, the compounds were screened and docked onto the available crystal structures of three glycosylation proteins—Alpha1-6FucT, Alpha2,6-ST I, and GlcNAcT-V—to predict the binding affinities. Here, pictilisib was predicted to bind to the active sites of Alpha1-6FucT, Alpha2,6-ST I, and GlcNAcT-V with a higher binding affinity than the natural substrate (Figure 6c; Table S9, Supplementary Materials). Our analysis suggests potential pictilisib interactions with critical amino acid residues in each of the enzyme's active sites. Against Alpha1-6FucT (Alpha1-6FucT), pictilisib formed pi–cation interactions with Arg365, hydrogen-bonding interactions with His363, and pi–pi T-shaped interactions with Tyr220 (Figure 6d). In a similar docking experiment by Manabe et al., the diphosphate group of GDP-fucose was predicted to form hydrogen bonds with Gly221, Arg365, Ser469, and Gln470 [44]. Additionally, the His363 sidechain and Tyr250 backbone residues were shown to tether the guanosine moiety with hydrogen bonds [45]. Pictilisib also formed hydrogen bonding interactions with Gln235, pi–pi T-shaped interactions with His370, and Van der Waals interactions with Ala363 in Alpha2,6-ST I (Figure 6e). It is noteworthy that a proposed reaction mechanism of Alpha2,6-ST I showed His370 acting as the catalytic base for the deprotonation of the 6′-hydroxyl group of the N-glycan acceptor, leading to a S_N2 attack at the C2 position of Neu5Ac [46]. Due to these predicted interactions, pictilisib was chosen for further in vitro studies, to determine if it can modulate glycosylation. On the other hand, pictilisib formed hydrogen bonding interactions with Trp401, Asp378, and Leu372 of GlcNAcT-V (Figure 6f). The sulfur atom in pictilisib also formed pi–sulfur interactions with Phe380 and Lys554. The same aromatic amino acid residues, Phe380, and Trp401 also interacted with the sugar acceptor, with Trp401 further restraining the conformation of the α1,6-branch. Most of the top ligands also formed hydrogen bonding interactions with Lys554, a residue with known interaction with the sugar acceptor. The residues—Phe390, Trp401, and Lys554—are also found in the acceptor substrate binding site for MGAT-IX, suggesting that these residues are relevant in acceptor sugar recognition.

3. Discussion

Glycans also play a role in regulating the processes that lead to cell death, such as controlling intra- and extracellular pathways that promote the initiation and execution of apoptosis [47]. Cancer cells resist apoptosis through the modification of glycans presented on cell death receptors. Glycosylation can also modulate the function of the death receptors of the extrinsic apoptotic pathway, Fas (CD95), and TNFR1 (tumor necrosis factor receptor 1) [48]. These glycosylations may positively regulate the apoptotic machinery. Galectin-3 binding to β1,6 branched glycans regulate the tumor cell motility by stimulating focal adhesion modeling, FAK and PI3K activation, local F-actin instability, and α5β1 integrin translocation to fibrillar adhesions [49]. Lewis a and Lewis b antigens originated from the mono- or di-fucosyl substitution of type 1 chains, while Lewis x and Lewis y are derived from the mono- or di-fucosyl substitution of type 2 chains. The mono-fucosyl substitution of the α2,3-sialylated type 1 or type 2 chains leads to the formation of Sialyl Lewis a (sLe^a) and Sialyl Lewis × (sLe^x), respectively. Core fucosylation is also observed in several cancers [5]. This involves the addition of α1,6-fucose to the innermost GlcNAc residue of N-glycans through Fuc-TVIII (FUT8), and overexpression is additionally observed in several cancers, including lung cancer [50]. In breast cancer, the increased core fucosylation of EGFR was associated with an increased dimerization and phosphorylation, resulting in increased EGFR-mediated signaling, promoting tumor growth [37]. The expression of β-galactoside α2,6-sialyltransferase (ST6Gal1) is altered in several cancers, including colon, stomach, and ovarian [51]. The Ras pathway regulates the transcription and expression of ST6Gal1, and transfectants containing ST6Gal1-expressing cells indicate an increased adhesion to the extracellular matrix molecules in colon [52] and breast cancer [18].

Glycans associated with aberrant fucosylation and sialylation are implicated in cancer progression and metastasis [39,53]. Thus, the inhibition of enzymes involved in the biosynthesis of fucosylated and sialylated glycans has attracted interest as a novel strategy in the development of potential anti-cancer therapeutics. Despite advances made in the development of fucosylation and sialylation inhibitors [54], our use of an in silico, mass spectrometry- and bioinformatics-based approach in the identification of a small molecule inhibitor, pictilisib, serves as an alternative and complementary method for rapid identification of potential glycosyltransferase inhibitors.

We previously discovered, through in silico screening, that pictilisib gave the highest binding affinity, among more than 14,000 compounds and drugs, towards the homology-modeled Alpha1-6FucT at −9.3 kcal/mol [21]. Pictilisib was previously identified as a clinically well-tolerated chemotherapeutic agent that specifically targets the enzyme PI3 kinase among patients with advanced solid tumors, mostly arising from colorectal and breast cancers [55]. To the best of our knowledge, this is the first study to investigate the effect of this chemotherapeutic drug on protein glycosylation, specifically with respect to glycosyltransferase inhibition. However, one aspect of protein glycosylation that was not explored in this study was the possible effects of the drugs on the expression of sugar transporters [7,8], glycosyltransferase retention in the Golgi apparatus [56,57], and the overall integrity of the N-glycan biosynthesis machinery. Future experiments utilizing mass spectrometry to characterize these processes can be developed and incorporated into the glycomics-driven glycoproteomics mass spectrometry workflow, such as this.

4. Materials and Methods

4.1. Biochemical Assays

4.1.1. Cell Culture

The cell line A549 (CCL-185TM) was obtained from the American Type Culture Collection (ATCC). The A549 cells were grown in RPMI 1640 medium (Gibco), supplemented with 10% fetal bovine serum (FBS) and 1% penicillin-streptomycin (Thermo Scientific, Waltham, MA, USA) in T75 flasks. The media was changed every other day. For each assay, the cells were grown in at least three biological replicates and maintained in a humidified incubator at 37 °C and in an atmosphere of 5% CO_2.

4.1.2. Dose–Response Assay

The A549 cells were seeded into 96-well plates at 3000 cells/well. The plates were incubated at 37 °C, 5% CO_2, for 24 h to allow attachment and proliferation. After which, the cells were treated with half-log dilutions of pictilisib (SelleckChem, Houston, TX, USA) and negative control (1% *v*/*v* DMSO). The cells were incubated with the test compounds at 37 °C, 5% CO_2, for 24 h. The cell viability was detected using CellTiter^{96} AQueous MTS assay reagent (Promega), following the manufacturer's instructions. The IC_{50} value was calculated using GraphPad Prism (version 9.3.1 for Windows, GraphPad Software, San Diego, CA, USA) using % viability as input values for each log (pictilisib) concentration. The assays were completed in triplicate.

4.1.3. Cell Cycle Assay

The A549 cells were seeded into 100 mm plates. Upon reaching approximately 80% confluency, the cells were treated with 4 μM pictilisib (final concentration), 0.1% *v*/*v* DMSO (negative control), or 100 μM docetaxel (positive control) for 24 h at 37 °C, 5% CO_2. The cell cycle assay was performed using Cellometer™ PI Cell cycle kit (Nexcelom), according to the manufacturer instructions. The data were acquired using Cellometer Vision CBA (version 5, Nexcelom, Lawrence, MA, USA) using the protocol CBA_Cell Cycle-PI660 nm, with an exposure time of 15,000 ms. The acquired image cytometry data were analyzed using FCS Express 7.0 (version 7.12.0007, De Novo Software, Pasadena, CA, USA). Cell gating was adjusted based on negative control. Assays were completed in triplicate.

4.1.4. Apoptosis Assay

The A549 cells were seeded into 100 mm plates. Upon reaching approximately 80% confluency, the cells were treated with 4 μM pictilisib (final concentration), 0.1% *v*/*v* DMSO (negative control), or 100 μM docetaxel (positive control) for 24 h at 37 °C, 5% CO_2. The cell cycle assay was performed using Annexin V-FITC/PITM Apoptosis kit (Nexcelom), according to the manufacturer's instructions. Data were acquired using Cellometer Vision CBA (version 5, Nexcelom, Lawrence, MA, USA), using the protocol CBA_Annexin V + PI assay, with an F1 exposure time of 8000 ms and F2 exposure time of 20,000 ms. Acquired data were analyzed using FCS Express 7.0. Cell gating was adjusted based on negative and positive controls. Assays were completed in triplicate.

4.1.5. Scratch Assay

The A549 cells were seeded into 6-well plates and allowed to grow to confluency at 37 °C, 5% CO_2. Cell surface scratches were made using P200 pipette tips, then washed twice with PBS to remove the debris. The plates were supplemented with RPMI media (2% FBS, 1% penicillin-streptomycin) to reduce the effects of cell proliferation. The cells were treated with a final concentration of 4 μM pictilisib or with 0.1% *v*/*v* DMSO (negative control) for 48 h. Micrographs were taken starting from 0 h and every 12 h thereafter. The wound size areas were measured using ImageJ software [58] and reported relative to initial wound size. Assays were completed in triplicate.

4.1.6. Transwell Migration Assay

The A549 cells were grown in 100 mm plates until reaching approximately 80% confluency. Then, the cells were treated to a final concentration of 4 µM pictilisib or with 0.1% *v*/*v* DMSO (negative control) for 24 h. After which, the cells were harvested using trypsin and adjusted to 50,000 cells per mL in RPMI media (1% penicillin-streptomycin) without FBS. One (1) mL of the resulting suspension was pipetted into the top compartment of a transwell plate. The bottom chamber was filled with complete media to establish chemotaxis. The plates were subsequently incubated for 3 h at 37 °C, 5% CO_2, to allow cell migration. The plates were washed twice with HBSS (Hank's Balanced Salt Solution), fixed with 1% formaldehyde for 5 min, and the bottom compartment stained with 0.1% crystal violet. The cells that migrated through the transwell membrane were visualized in micrographs and manually counted using ImageJ analysis software [58]. Assays were completed in triplicate.

4.2. Mass Spectrometry Assay and Data Processing

4.2.1. Cell Treatment and Glycan, Protein, and Glycoprotein Enrichment

The cells were grown in T75 flasks until reaching approximately 80% confluency. The cells were treated to a final concentration of 4 µM pictilisib or with 0.1% *v*/*v* DMSO (negative control) for 24 h. For the glycomic and proteomic mass spectrometric assay, the cells were grown in triplicate T75 flasks for each group (n = 3). For the glycoproteomic mass spectrometric assay, the cells were grown in 15 replicate T75 flasks for each group (n = 15). After culturing, the established general protocol for all of the mass spectrometric analyses was used [22,24].

4.2.2. Glycomics Assay

The N-Glycan profiling was performed using an Agilent 6200 series nanoHPLC-Chip-QTOF-MS (Agilent) with an Agilent 6210 Time-of-Flight mass spectrometer. The chip (glycan chip II, Agilent) contained a 9 mm × 0.075 mm i.d. enrichment column coupled to a 43 mm × 0.075 mm i.d. analytical column; both are packed with 5-µm porous graphitized carbon (PGC). The N-glycan samples were reconstituted in 40 µL of water, and 5 µL of the resulting solution was used for injection into the LC–MS/MS system. Alongside the samples, blanks and quality controls (N-glycans released from a 1:1 mixture of RNAse B and commercially available serum) were included to ensure data reproducibility and quality. Upon injection, the sample was loaded onto the enrichment column using 3% ACN containing 0.1% formic acid (FA, Fluka, St. Louis, MO, USA). After the analytical column was switched in-line, the nano pump delivered a gradient of 3% ACN with 0.1% FA (solvent A) and 90% ACN with 1% FA (solvent B). The sample was delivered by the capillary pump to the enrichment column at a flow rate of 3 µL/min and separated in the analytical column by the nano pump at a flow rate of 0.3 µL/min using a gradient optimized for N-glycans (0% B, 0–2.5 min; 0 to 16% B, 2.5–20 min; 16 to 44% B, 20–30 min; 44 to 100% B, 30–35 min; and 100% B, 35–45 min) followed by 20-min equilibration for pure A. The tandem MS spectra were acquired via collision-induced dissociation (CID).

Analysis of the N-glycan data was performed using MassHunter Qualitative Analysis Software B.07.00 (Agilent Technologies, Santa Clara, CA, USA). Matching of the monoisotopic masses obtained was completed against our in-house database for glycan composition identification and subsequently verified through their corresponding MS/MS spectra. The relative abundance of each glycan in a sample was determined using the peak area of all glycans from extracted ion chromatograms. The comparison between the relative abundances of primary N-glycan types—high-mannose, undecorated, fucosylated, sialylated, and sialofucosylated glycans—was completed by adding the relative abundances of each glycoform belonging to a specific glycan type. Further comparison of each glycoform was completed using multiple *t*-tests (GraphPad version 9.3.1 for Windows, GraphPad Software,

San Diego, CA, USA) at a significance level of $\alpha = 0.05$. Significantly different N-glycans were mapped on the N-glycan biosynthesis pathway, based on the known biosynthetic sequence [22,23].

4.2.3. Proteomics and Glycoproteomics Assay

The pellets containing membrane proteins were reconstituted with 60 μL of 8 M urea and sonicated for 20 min for denaturation. Two microliters (2 μL) Dithiothreitol (DTT, 550 mM in 50 mM NH_4HCO_3) were then added to the samples and incubated for 50 min at 55 °C. The free cysteine was alkylated with 4 μL of iodoacetamide (450 mM) for 20 min in the dark at an ambient temperature. To quench the reaction, 420 μL of a protein digestion buffer (50 mM NH_4HCO_3) was added. Trypsin (10 μL, 0.1 mg/mL) was then added to the mixture and tryptic digestion was performed at 37 °C for 18 h. The resulting peptides were purified using a C18 solid-phase extraction cartridge and dried before LC–MS/MS analysis. To enrich the glycopeptides, the tryptic digests were cleaned up using HILIC solid-phase extraction and dried before LC–MS/MS analysis. The purified peptides were adjusted to 0.5 μg/μL while the glycopeptides were adjusted to 0.2 μg/μL before injection using Pierce BCA assay kit, following the manufacturer's instructions (ThermoFisher, Waltham, MA, USA).

The proteomics and glycoproteomics samples were characterized using an UltiMate™ WPS-3000RS nanoLC system coupled with an Orbitrap Fusion Lumos MS system (ThermoFisher Scientific). One (1) μL of each sample was injected, alongside blanks and quality controls (tryptic digest of commercially available HeLa proteins), which were included to ensure data reproducibility and quality. The analytes were separated using an Acclaim™ PepMap™ 100 C18 LC Column (3 mm, 0.075 mm × 250 mm, ThermoFisher Scientific) at a flow rate of 300 nL/min. Water containing 0.1% formic acid and 80% acetonitrile containing 0.1% formic acid were used as solvents A and B, respectively. The MS spectra were collected with a mass range of m/z 600–2000 at a rate of 1.5 s per spectrum in positive ionization mode. The filtered precursor ions in each MS spectrum were subjected to fragmentation through 30% higher-energy C-trap dissociation (HCD) using nitrogen gas as carrier.

The mass spectrometry data were analyzed using Byos workflow (Protein Metrics, Cupertino, CA, USA). For the qualitative analysis in Byonic (version 4.3.4, Protein Metrics, Cupertino, CA, USA), the proteins were identified against the human proteome database [59] using a precursor mass tolerance of 20 ppm and fragment mass tolerance of 10 ppm. The digestion parameters used included C-terminal cleavage by trypsin (K and R cleavage sites), with at most two missed cleavages. The following peptide modifications were included: carbamidomethyl @ C; oxidation @ M; deamidation @ N and Q; acetylation at protein N-terminal; Gln to pyro-Glu at N-terminal Q; Glu to pyro-Glu at N-terminal E. The protein IDs were filtered at 1% FDR. To identify the glycoproteins and glycoforms, an additional search was performed in Byonic, using an in-house N-glycan database. Quantification for each protein was completed in Byologic (version 4.3-117, Protein Metrics, Cupertino, CA, USA) by quantifying the XIC area sum of the top three most abundant peptides. The XICs were then normalized to sum total before statistical analysis. On the other hand, glycoform quantification was normalized to each protein's glycosite to yield the percentage occupancy of a particular glycoform.

4.2.4. Gene Ontology Analysis

To identify significantly different proteins and glycopeptides, multiple t-tests were conducted in GraphPad Prism (version 9.3.1 for Windows, GraphPad Software, San Diego, CA, USA), using an FDR approach (FDR = 5%). Significantly over- and under-expressed protein IDs were annotated, using Gene Set Enrichment Analysis [60]. Similarly, the glycopeptides were annotated using g: Profiler [61] to yield significantly enriched pathways and then plotted as a heatmap in GraphPad Prism.

4.3. Computational Methods

4.3.1. Network Pharmacology

A ligand database was prepared by downloading the structure data files (.sdf) from several online databases, such as the Comparative Toxicogenomics database (http://ctdbase.org/, accessed on 21 August 2021) [43], STITCH database (http://stitch.embl.de/, accessed on 21 August 2021) [62], GeneCards (https://www.genecards.org/, accessed on 21 August 2021) [63], the Drug Gene Interaction database (http://www.dgidb.org/, accessed on 21 August 2021) [64], and the Protein Databank (https://www.rcsb.org/, accessed on 21 August 2021) [65], and 185 compounds predicted to bind or interact with glycosylation enzymes from the DrugBank (https://go.drugbank.com/, accessed on 21 August 2021) [66]. The drug–gene interactions were predicted using the STITCH database and the Comparative Toxicogenomics database and then visualized using Cytoscape [67].

4.3.2. In Silico Docking

All compounds from the ligand database were loaded onto PyRx [68] and minimized using the Universal Force Field [69], as implemented in Open Babel [70]. The enzymes GlcNAcT-V, Alpha2,6-ST I, and Alpha1-6FucT were selected as the drug target for this study due to the availability of their 3D crystal structures in PDB. The enzyme GlcNAcT-V (PDB ID: 5ZIC, 2.10 Å) [71] was downloaded as a complex with its acceptor sugar, 2-acetamido-2-deoxy-beta-D-glucopyranose-(1-2)-6-thio-alpha-D-mannopyranose-(1-6)-beta-d-mannopyranose. The Alpha2,6-ST I (PDB ID: 4JS2, 2.30 Å) [46] was downloaded as a complex with cytidine monophosphate. The human Alpha1-6FucT (PDB ID: 2de0) [72] was homology-modeled from *Caenorhabditis elegans* POFUT1 (PDB ID: 3ZY6, 1.91 Å) [73] in complex with GDP-fucose, using SWISS-MODELLER [74]. These protein structures were prepared for docking by using the Dock Prep protocol in Chimera [75]. The prepared protein structures were loaded in PyRx as macromolecule receptors.

In silico screening methods were performed in PyRx [68] using the AutoDock VINA docking protocol [76], at exhaustiveness level 8. Validation of the docking protocol was completed by redocking the ligands, complexed with their respective enzymes, using precise grid box parameters. After docking validation, all of the compounds in the ligand database were screened against each of the three enzymes. The compounds were ranked according to VINA-predicted binding energy (kcal/mol). The top binding molecules against each enzyme were visualized for residue interactions with the target enzyme, using Discovery Studio™ (version 21.1.0.20298, Dassault Systemes, BIOVIA Corp, San Diego, CA, USA). From the screening, pictilisib was found to have high compound cross-reactivity (binding to multiple enzyme targets) and a high number of network interactions, and was selected for further in vitro studies.

4.3.3. Glycoprotein Molecular Modeling and Molecular Dynamics

The top glycoproteins were modeled to visualize the effects of changes in glycosylation on protein dynamics and interactions. Interesting glycoproteins, such as ITGA5 and ITGB1, were modeled for visualization of the changes in glycosylation to protein dynamics and interactions. The glycoforms in each glycosite were selected, based on the highest fold-change between the pictilisib- and negative control-treated cells from the glycoproteomics results. The crystal structures of selected proteins were downloaded from PDB, and then the glycans were attached to these proteins using CHARMM-GUI Glycan modeler [77]. The system was solvated using the TIP3P model, and 150 mM KCl was added. The CHARMM36 force field was used for both proteins and carbohydrates [78]. The resulting molecular dynamics input files were used to simulate the glycoprotein dynamics for one ns (10,000 fs/timestep) using the DOST-ASTI High-Performance Computing (HPC; Cluster, Quezon City, Philippines). Simulations were visualized, and the number of interacting hydrogen bonds between the glycans and proteins was analyzed using VMD [79].

5. Conclusions

This study highlights the use of combined mass spectrometric (glycomics, proteomics, and glycoproteomics) and bioinformatics approaches that resulted in the identification of an inhibitor, pictilisib, affecting protein N-glycosylation and the associated glycoprotein pathways. By integrating network pharmacology and in silico docking approaches, we were able to identify a glycosylation inhibitor, and subsequently utilized integrated glycomic-driven glycoproteomic and proteomic mass spectrometric analyses to validate its effect in lung cancer cells. The compound was validated to inhibit the formation of fucosylated and sialylated N-glycans attached to glycoproteins involved in apoptosis, cell adhesion, DNA damage repair, and chemical response processes. Furthermore, the compound also significantly affected the cellular processes involved in cell cycle, apoptosis, cell adhesion, transcription, and translation, which we validated using in vitro biochemical assays. Finally, we simulated the differences in interactions of a model glycoprotein complex, ITGA5-ITGB1, due to glycosylation alteration after pictilisib treatment of A549 cells. To further validate the efficacy of the compound in affecting protein glycosylation, it is recommended to conduct future in vivo studies with animal models, coupled with our glycomics-based glycoproteomics mass spectrometric method. These combined methods may be used as tools to reveal potential untapped glycosylation inhibitors.

Supplementary Materials: The following supporting information can be downloaded at: https://www.mdpi.com/article/10.3390/molecules27123834/s1: Figure S1: 24- and 48-h dose-response toxicity of pictilisib; Figure S2: Dose-response glycomics assay of pictilisib; Figure S3: Reproducibility of glycomics results between groups; Figure S4: Reproducibility of proteomics results between groups; Figure S5: $-\log_{10}$ FDR value results of under-expressed proteins by GSEA using Biological Process Gene Ontology set; Figure S6: $-$log10 FDR value results of overexpressed proteins by GSEA using Biological Process Gene Ontology set; Figure S7: Apoptosis assay results of pictilisib treatment; Figure S8: Cell cycle assay results; Figure S9: Scratch assay results of pictilisib-treatment; Figure S10: Transwell migration assay results of pictilisib-treatment; Figure S11: Reproducibility of glycoproteomics results between groups; Figure S12: Variation of protein glycosylation across all glycosites; Figure S13: Pathways of proteins with reduced fucosylation, sialylation, or sialofucosylation; Figure S14: Molecular function of proteins with reduced fucosylation, sialylation, or sialofucosylation; Table S1. Mass spectrometry glycomics results showing relative abundances of each N-glycan; Table S2. Mass spectrometry proteomics results showing relative abundances of each protein; Table S3. Mass spectrometry glycoproteomics results showing relative abundances of each glycopeptide; Table S4. Mass spectrometry glycoproteomics results showing relative abundances of each protein glycosite; Table S5. Mass spectrometry glycoproteomics results showing relative abundances of each integrin alpha-5 (ITGA5) and integrin beta-1 (ITGB1) glycosylation; Table S6. Molecular dynamics analysis of ITGA5-ITGB1 complex in vehicle control-treated cells; Table S7. Molecular dynamics analysis of ITGA5-ITGB1 complex in pictilisib-treated cells; Table S8. Predicted drug–gene interaction network of compound library against glycosylation genes; Table S9. Predicted in silico binding energies of compound library against glycosylation enzymes; Video S1. Molecular dynamics trajectories of ITGA5-ITGB1 complex in vehicle control-treated cells; Video S2. Molecular dynamics trajectories of ITGA5-ITGB1 complex in pictilisib-treated cells.

Author Contributions: Conceptualization, M.R.S.A. and G.C.C.; methodology, M.R.S.A. and Q.Z.; software, M.R.S.A. and J.F.R.; validation, Q.Z. and J.F.R.; formal analysis, M.R.S.A., Q.Z. and S.J.B.G.; investigation, Q.Z., C.B.L. and F.M.H.III; resources, C.B.L. and F.M.H.III; data curation, Q.Z. and F.M.H.III; writing—original draft preparation, M.R.S.A., S.J.B.G. and Q.Z.; writing—review and editing, M.R.S.A., G.C.C., R.C.N. and C.B.L.; visualization, M.R.S.A., Q.Z. and S.J.B.G.; supervision, G.C.C., R.C.N. and C.B.L.; project administration, S.J.B.G. and R.C.N.; funding acquisition, G.C.C., R.C.N. and C.B.L. All authors have read and agreed to the published version of the manuscript.

Funding: This research was funded by the Philippine Commission on Higher Education—Philippine-California Advanced Research Institutes, grant number PCARI-IHITM 2017-18.

Institutional Review Board Statement: Not Applicable.

Informed Consent Statement: Not Applicable.

Data Availability Statement: The data presented in this study are available on request from the corresponding author.

Acknowledgments: The authors would like to acknowledge the COARE facility of the Department of Science and Technology, Philippines, for their assistance in performing the molecular dynamics simulations.

Conflicts of Interest: The authors declare no conflict of interest.

References

1. Cheng, T.-Y.D.; Cramb, S.M.; Baade, P.D.; Youlden, D.R.; Nwogu, C.; Reid, M.E. The International Epidemiology of Lung Cancer: Latest Trends, Disparities, and Tumor Characteristics. *J. Thorac. Oncol. Off. Publ. Int. Assoc. Study Lung Cancer* **2016**, *11*, 1653–1671. [CrossRef] [PubMed]
2. Bray, F.; Ferlay, J.; Soerjomataram, I.; Siegel, R.L.; Torre, L.A.; Jemal, A. Global Cancer Statistics 2018: GLOBOCAN Estimates of Incidence and Mortality Worldwide for 36 Cancers in 185 Countries. CA. *Cancer J. Clin.* **2018**, *68*, 394–424. [CrossRef] [PubMed]
3. Sung, H.; Ferlay, J.; Siegel, R.L.; Laversanne, M.; Soerjomataram, I.; Jemal, A.; Bray, F. Global Cancer Statistics 2020: GLOBOCAN Estimates of Incidence and Mortality Worldwide for 36 Cancers in 185 Countries. CA. *Cancer J. Clin.* **2021**, *71*, 209–249. [CrossRef] [PubMed]
4. Meany, D.L.; Chan, D.W. Aberrant Glycosylation Associated with Enzymes as Cancer Biomarkers. *Clin. Proteomics* **2011**, *8*, 7. [CrossRef]
5. Pinho, S.S.; Reis, C.A. Glycosylation in Cancer: Mechanisms and Clinical Implications. *Nat. Rev. Cancer* **2015**, *15*, 540–555. [CrossRef]
6. Esko, J.D.; Bertozzi, C.R. Chapter 50: Chemical Tools for Inhibiting Glycosylation. In *Essentials of Glycobiology*, 2nd ed.; Varki, A., Cummings, R.D., Esko, J.D., Freeze, H.H., Stanley, P., Bertozzi, C.R., Hart, G.W., Etzler, M.E., Eds.; Cold Spring Harbor Laboratory Press: Cold Spring Harbor, NY, USA, 2009; ISBN 978-0-87969-770-9.
7. Viinikangas, T.; Khosrowabadi, E.; Kellokumpu, S. N-Glycan Biosynthesis: Basic Principles and Factors Affecting Its Outcome. In *Antibody Glycosylation*; Pezer, M., Ed.; Experientia Supplementum; Springer International Publishing: Cham, Switzerland, 2021; Volume 112, pp. 237–257.
8. Kellokumpu, S.; Hassinen, A.; Glumoff, T. Glycosyltransferase Complexes in Eukaryotes: Long-Known, Prevalent but Still Unrecognized. *Cell. Mol. Life Sci.* **2016**, *73*, 305–325. [CrossRef]
9. Landi, M.T.; Dracheva, T.; Rotunno, M.; Figueroa, J.D.; Liu, H.; Dasgupta, A.; Mann, F.E.; Fukuoka, J.; Hames, M.; Bergen, A.W.; et al. Gene Expression Signature of Cigarette Smoking and Its Role in Lung Adenocarcinoma Development and Survival. *PLoS ONE* **2008**, *3*, e1651. [CrossRef]
10. Hong, Q.; Ruhaak, L.R.; Stroble, C.; Parker, E.; Huang, J.; Maverakis, E.; Lebrilla, C.B. A Method for Comprehensive Glycosite-Mapping and Direct Quantitation of Serum Glycoproteins. *J. Proteome Res.* **2015**, *14*, 5179–5192. [CrossRef]
11. Ruhaak, L.R.; Taylor, S.L.; Stroble, C.; Nguyen, U.T.; Parker, E.A.; Song, T.; Lebrilla, C.B.; Rom, W.N.; Pass, H.; Kim, K.; et al. Differential N-Glycosylation Patterns in Lung Adenocarcinoma Tissue. *J. Proteome Res.* **2015**, *14*, 4538–4549. [CrossRef]
12. Vajaria, B.N.; Patel, P.S. Glycosylation: A Hallmark of Cancer? *Glycoconj. J.* **2017**, *34*, 147–156. [CrossRef]
13. Dall'Olio, F.; Malagolini, N.; Trinchera, M.; Chiricolo, M. Mechanisms of Cancer-Associated Glycosylation Changes. *Front. Biosci. Landmark Ed.* **2012**, *17*, 670–699. [CrossRef] [PubMed]
14. Handerson, T.; Camp, R.; Harigopal, M.; Rimm, D.; Pawelek, J. Beta1,6-Branched Oligosaccharides Are Increased in Lymph Node Metastases and Predict Poor Outcome in Breast Carcinoma. *Clin. Cancer Res. Off. J. Am. Assoc. Cancer Res.* **2005**, *11*, 2969–2973. [CrossRef] [PubMed]
15. Recchi, M.A.; Hebbar, M.; Hornez, L.; Harduin-Lepers, A.; Peyrat, J.P.; Delannoy, P. Multiplex Reverse Transcription Polymerase Chain Reaction Assessment of Sialyltransferase Expression in Human Breast Cancer. *Cancer Res.* **1998**, *58*, 4066–4070. [PubMed]
16. Burchell, J.; Poulsom, R.; Hanby, A.; Whitehouse, C.; Cooper, L.; Clausen, H.; Miles, D.; Taylor-Papadimitriou, J. An Alpha2,3 Sialyltransferase (ST3Gal I) Is Elevated in Primary Breast Carcinomas. *Glycobiology* **1999**, *9*, 1307–1311. [CrossRef] [PubMed]
17. Picco, G.; Julien, S.; Brockhausen, I.; Beatson, R.; Antonopoulos, A.; Haslam, S.; Mandel, U.; Dell, A.; Pinder, S.; Taylor-Papadimitriou, J.; et al. Over-Expression of ST3Gal-I Promotes Mammary Tumorigenesis. *Glycobiology* **2010**, *20*, 1241–1250. [CrossRef] [PubMed]
18. Schneider, F.; Kemmner, W.; Haensch, W.; Franke, G.; Gretschel, S.; Karsten, U.; Schlag, P.M. Overexpression of Sialyltransferase CMP-Sialic Acid: Galbeta1,3GalNAc-R Alpha6-Sialyltransferase Is Related to Poor Patient Survival in Human Colorectal Carcinomas. *Cancer Res.* **2001**, *61*, 4605–4611.
19. Zhou, Q.; Xie, Y.; Lam, M.; Lebrilla, C.B. N-Glycomic Analysis of the Cell Shows Specific Effects of Glycosyl Transferase Inhibitors. *Cells* **2021**, *10*, 2318. [CrossRef]
20. Hiss, D.C.; Gabriels, G.A.; Folb, P.I. Combination of Tunicamycin with Anticancer Drugs Synergistically Enhances Their Toxicity in Multidrug-Resistant Human Ovarian Cystadenocarcinoma Cells. *Cancer Cell Int.* **2007**, *7*, 5. [CrossRef]
21. Alvarez, M.R.S.; Grijaldo, S.J.B.; Nacario, R.C.; Rabajante, J.F.; Heralde, F.M.; Lebrilla, C.B.; Completo, G.C. In Silico Screening-Based Discovery of Inhibitors against Glycosylation Proteins Dysregulated in Cancer. *J. Biomol. Struct. Dyn.* **2022**, 1–13. [CrossRef]

22. Li, Q.; Xie, Y.; Wong, M.; Barboza, M.; Lebrilla, C.B. Comprehensive Structural Glycomic Characterization of the Glycocalyxes of Cells and Tissues. *Nat. Protoc.* **2020**, *15*, 2668–2704. [CrossRef]
23. Kronewitter, S.R.; An, H.J.; de Leoz, M.L.; Lebrilla, C.B.; Miyamoto, S.; Leiserowitz, G.S. The Development of Retrosynthetic Glycan Libraries to Profile and Classify the Human Serum N-Linked Glycome. *Proteomics* **2009**, *9*, 2986–2994. [CrossRef] [PubMed]
24. Li, Q.; Xie, Y.; Wong, M.; Lebrilla, C. Characterization of Cell Glycocalyx with Mass Spectrometry Methods. *Cells* **2019**, *8*, 882. [CrossRef] [PubMed]
25. Zhu, W.; Li, L.L.; Songyang, Y.; Shi, Z.; Li, D. Identification and Validation of HELLS (Helicase, Lymphoid-Specific) and ICAM1 (Intercellular Adhesion Molecule 1) as Potential Diagnostic Biomarkers of Lung Cancer. *PeerJ.* **2020**, *8*, e8731. [CrossRef] [PubMed]
26. Pabla, S.; Conroy, J.M.; Nesline, M.K.; Glenn, S.T.; Papanicolau-Sengos, A.; Burgher, B.; Hagen, J.; Giamo, V.; Andreas, J.; Lenzo, F.L.; et al. Proliferative Potential and Resistance to Immune Checkpoint Blockade in Lung Cancer Patients. *J. Immunother. Cancer* **2019**, *7*, 27. [CrossRef] [PubMed]
27. Du, X.; Xue, Z.; Lv, J.; Wang, H. Expression of the Topoisomerase II Alpha (TOP2A) Gene in Lung Adenocarcinoma Cells and the Association with Patient Outcomes. *Med. Sci. Monit. Int. Med. J. Exp. Clin. Res.* **2020**, *26*, e929120. [CrossRef] [PubMed]
28. Misono, S.; Mizuno, K.; Suetsugu, T.; Tanigawa, K.; Nohata, N.; Uchida, A.; Sanada, H.; Okada, R.; Moriya, S.; Inoue, H.; et al. Molecular Signature of Small Cell Lung Cancer after Treatment Failure: The MCM Complex as Therapeutic Target. *Cancers* **2021**, *13*, 1187. [CrossRef]
29. Lee, G.Y.; Haverty, P.M.; Li, L.; Kljavin, N.M.; Bourgon, R.; Lee, J.; Stern, H.; Modrusan, Z.; Seshagiri, S.; Zhang, Z.; et al. Comparative Oncogenomics Identifies PSMB4 and SHMT2 as Potential Cancer Driver Genes. *Cancer Res.* **2014**, *74*, 3114LP–3126LP. [CrossRef]
30. Cao, Y.; Wei, M.; Li, B.; Liu, Y.; Lu, Y.; Tang, Z.; Lu, T.; Yin, Y.; Qin, Z.; Xu, Z. Functional Role of Eukaryotic Translation Initiation Factor 4 Gamma 1 (EIF4G1) in NSCLC. *Oncotarget* **2016**, *7*, 24242–24251. [CrossRef]
31. Jaiswal, P.K.; Koul, S.; Palanisamy, N.; Koul, H.K. Eukaryotic Translation Initiation Factor 4 Gamma 1 (EIF4G1): A Target for Cancer Therapeutic Intervention? *Cancer Cell Int.* **2019**, *19*, 224. [CrossRef]
32. Lu, Y.; Yu, S.; Wang, G.; Ma, Z.; Fu, X.; Cao, Y.; Li, Q.; Xu, Z. Elevation of EIF4G1 Promotes Non-Small Cell Lung Cancer Progression by Activating MTOR Signalling. *J. Cell. Mol. Med.* **2021**, *25*, 2994–3005. [CrossRef]
33. He, L.-R.; Zhao, H.-Y.; Li, B.-K.; Liu, Y.-H.; Liu, M.-Z.; Guan, X.-Y.; Bian, X.-W.; Zeng, Y.-X.; Xie, D. Overexpression of EIF5A-2 Is an Adverse Prognostic Marker of Survival in Stage I Non–Small Cell Lung Cancer Patients. *Int. J. Cancer* **2011**, *129*, 143–150. [CrossRef] [PubMed]
34. Xing, Z.; Russon, M.P.; Utturkar, S.M.; Tran, E.J. The RNA Helicase DDX5 Supports Mitochondrial Function in Small Cell Lung Cancer. *J. Biol. Chem.* **2020**, *295*, 8988–8998. [CrossRef] [PubMed]
35. Nagashio, R.; Sato, Y.; Matsumoto, T.; Kageyama, T.; Satoh, Y.; Shinichiro, R.; Masuda, N.; Goshima, N.; Jiang, S.-X.; Okayasu, I. Expression of RACK1 Is a Novel Biomarker in Pulmonary Adenocarcinomas. *Lung Cancer* **2010**, *69*, 54–59. [CrossRef] [PubMed]
36. Peng, H.; Gong, P.-G.; Li, J.-B.; Cai, L.-M.; Yang, L.; Liu, Y.; Yao, K.; Li, X. The Important Role of the Receptor for Activated C Kinase 1 (RACK1) in Nasopharyngeal Carcinoma Progression. *J. Transl. Med.* **2016**, *14*, 131. [CrossRef]
37. Potapenko, I.O.; Haakensen, V.D.; Lüders, T.; Helland, A.; Bukholm, I.; Sørlie, T.; Kristensen, V.N.; Lingjaerde, O.C.; Børresen-Dale, A.-L. Glycan Gene Expression Signatures in Normal and Malignant Breast Tissue; Possible Role in Diagnosis and Progression. *Mol. Oncol.* **2010**, *4*, 98–118. [CrossRef]
38. Ruhaak, L.R.; Stroble, C.; Dai, J.; Barnett, M.; Taguchi, A.; Goodman, G.E.; Miyamoto, S.; Gandara, D.; Feng, Z.; Lebrilla, C.B.; et al. Serum Glycans as Risk Markers for Non-Small Cell Lung Cancer. *Cancer Prev. Res.* **2016**, *9*, 317–323. [CrossRef]
39. Munkley, J.; Elliott, D.J. Hallmarks of Glycosylation in Cancer. *Oncotarget* **2016**, *7*, 35478–35489. [CrossRef]
40. El-Gebali, S.; Mistry, J.; Bateman, A.; Eddy, S.R.; Luciani, A.; Potter, S.C.; Qureshi, M.; Richardson, L.J.; Salazar, G.A.; Smart, A.; et al. The Pfam Protein Families Database in 2019. *Nucleic Acids Res.* **2019**, *47*, D427–D432. [CrossRef]
41. Mallajosyula, S.S.; Jo, S.; Im, W.; MacKerell Jr, A.D. Molecular Dynamics Simulations of Glycoproteins Using CHARMM. *Methods Mol. Biol.* **2015**, *1273*, 407–429. [CrossRef]
42. Acun, B.; Hardy, D.J.; Kale, L.V.; Li, K.; Phillips, J.C.; Stone, J.E. Scalable Molecular Dynamics with NAMD on the Summit System. *IBM J. Res. Dev.* **2018**, *62*, 4:1–4:9. [CrossRef]
43. Davis, A.P.; Grondin, C.J.; Johnson, R.J.; Sciaky, D.; McMorran, R.; Wiegers, J.; Wiegers, T.C.; Mattingly, C.J. The Comparative Toxicogenomics Database: Update 2019. *Nucleic Acids Res.* **2019**, *47*, D948–D954. [CrossRef] [PubMed]
44. Manabe, Y.; Kasahara, S.; Takakura, Y.; Yang, X.; Takamatsu, S.; Kamada, Y.; Miyoshi, E.; Yoshidome, D.; Fukase, K. Development of A1,6-Fucosyltransferase Inhibitors through the Diversity-Oriented Syntheses of GDP-Fucose Mimics Using the Coupling between Alkyne and Sulfonyl Azide. *Bioorg. Med. Chem.* **2017**, *25*, 2844–2850. [CrossRef] [PubMed]
45. García-García, A.; Ceballos-Laita, L.; Serna, S.; Artschwager, R.; Reichardt, N.C.; Corzana, F.; Hurtado-Guerrero, R. Structural Basis for Substrate Specificity and Catalysis of A1,6-Fucosyltransferase. *Nat. Commun.* **2020**, *11*, 973. [CrossRef] [PubMed]
46. Kuhn, B.; Benz, J.; Greif, M.; Engel, A.M.; Sobek, H.; Rudolph, M.G. The Structure of Human α-2,6-Sialyltransferase Reveals the Binding Mode of Complex Glycans. *Acta Crystallogr. D Biol. Crystallogr.* **2013**, *69*, 1826–1838. [CrossRef]
47. Lichtenstein, R.G.; Rabinovich, G.A. Glycobiology of Cell Death: When Glycans and Lectins Govern Cell Fate. *Cell Death Differ.* **2013**, *20*, 976–986. [CrossRef]
48. Shatnyeva, O.M.; Kubarenko, A.V.; Weber, C.E.M.; Pappa, A.; Schwartz-Albiez, R.; Weber, A.N.R.; Krammer, P.H.; Lavrik, I.N. Modulation of the CD95-Induced Apoptosis: The Role of CD95 N-Glycosylation. *PLoS ONE* **2011**, *6*, e19927. [CrossRef]

49. Lagana, A.; Goetz, J.G.; Cheung, P.; Raz, A.; Dennis, J.W.; Nabi, I.R. Galectin Binding to Mgat5-Modified N-Glycans Regulates Fibronectin Matrix Remodeling in Tumor Cells. *Mol. Cell. Biol.* **2006**, *26*, 3181–3193. [CrossRef]
50. Liu, Y.-C.; Yen, H.-Y.; Chen, C.-Y.; Chen, C.-H.; Cheng, P.-F.; Juan, Y.-H.; Chen, C.-H.; Khoo, K.-H.; Yu, C.-J.; Yang, P.-C.; et al. Sialylation and Fucosylation of Epidermal Growth Factor Receptor Suppress Its Dimerization and Activation in Lung Cancer Cells. *Proc. Natl. Acad. Sci. USA* **2011**, *108*, 11332–11337. [CrossRef]
51. Dall'Olio, F.; Chiricolo, M. Sialyltransferases in Cancer. *Glycoconj. J.* **2001**, *18*, 841–850. [CrossRef]
52. Seales, E.C.; Shaikh, F.M.; Woodard-Grice, A.V.; Aggarwal, P.; McBrayer, A.C.; Hennessy, K.M.; Bellis, S.L. A Protein Kinase C/Ras/ERK Signaling Pathway Activates Myeloid Fibronectin Receptors by Altering Beta1 Integrin Sialylation. *J. Biol. Chem.* **2005**, *280*, 37610–37615. [CrossRef]
53. Lee, J.; Ha, S.; Kim, M.; Kim, S.-W.; Yun, J.; Ozcan, S.; Hwang, H.; Ji, I.J.; Yin, D.; Webster, M.J.; et al. Spatial and Temporal Diversity of Glycome Expression in Mammalian Brain. *Proc. Natl. Acad. Sci. USA* **2020**, *117*, 28743–28753. [CrossRef] [PubMed]
54. Chen, X.; Wu, J.; Huang, H.; Ding, Q.; Liu, X.; Chen, L.; Zha, X.; Liang, M.; He, J.; Zhu, Q.; et al. Comparative Profiling of Triple-Negative Breast Carcinomas Tissue Glycoproteome by Sequential Purification of Glycoproteins and Stable Isotope Labeling. *Cell Physiol. Biochem.* **2016**, *38*, 110–121. [CrossRef] [PubMed]
55. Sarker, U.; Oba, S. Drought Stress Enhances Nutritional and Bioactive Compounds, Phenolic Acids and Antioxidant Capacity of Amaranthus Leafy Vegetable. *BMC Plant. Biol.* **2018**, *18*, 258. [CrossRef] [PubMed]
56. Rizzo, R.; Russo, D.; Kurokawa, K.; Sahu, P.; Lombardi, B.; Supino, D.; Zhukovsky, M.; Vocat, A.; Pothukuchi, P.; Kunnathully, V.; et al. The Glyco-Enzyme Adaptor GOLPH3 Links Intra-Golgi Transport. Dynamics to Glycosylation Patterns and Cell Proliferation. *bioRxiv* **2019**. *preprint*. [CrossRef]
57. Rizzo, R.; Russo, D.; Kurokawa, K.; Sahu, P.; Lombardi, B.; Supino, D.; Zhukovsky, M.A.; Vocat, A.; Pothukuchi, P.; Kunnathully, V.; et al. Golgi Maturation-dependent Glycoenzyme Recycling Controls Glycosphingolipid Biosynthesis and Cell Growth via GOLPH3. *EMBO J.* **2021**, *40*, 107238. [CrossRef]
58. Schneider, C.A.; Rasband, W.S.; Eliceiri, K.W. NIH Image to ImageJ: 25 Years of Image Analysis. *Nat. Methods* **2012**, *9*, 671–675. [CrossRef]
59. The Uniprot Consortium UniProt: The Universal Protein Knowledgebase in 2021. *Nucleic Acids Res.* **2021**, *49*, D480–D489. [CrossRef]
60. Subramanian, A.; Tamayo, P.; Mootha, V.K.; Mukherjee, S.; Ebert, B.L.; Gillette, M.A.; Paulovich, A.; Pomeroy, S.L.; Golub, T.R.; Lander, E.S.; et al. Gene Set Enrichment Analysis: A Knowledge-Based Approach for Interpreting Genome-Wide Expression Profiles. *Proc. Natl. Acad. Sci. USA* **2005**, *102*, 15545–15550. [CrossRef]
61. Raudvere, U.; Kolberg, L.; Kuzmin, I.; Arak, T.; Adler, P.; Peterson, H.; Vilo, J. G:Profiler: A Web Server for Functional Enrichment Analysis and Conversions of Gene Lists (2019 Update). *Nucleic Acids Res.* **2019**, *47*, W191–W198. [CrossRef]
62. Szklarczyk, D.; Franceschini, A.; Wyder, S.; Forslund, K.; Heller, D.; Huerta-Cepas, J.; Simonovic, M.; Roth, A.; Santos, A.; Tsafou, K.P.; et al. STRING V10: Protein-Protein Interaction Networks, Integrated over the Tree of Life. *Nucleic Acids Res.* **2015**, *43*, D447–D452. [CrossRef]
63. Stelzer, G.; Rosen, N.; Plaschkes, I.; Zimmerman, S.; Twik, M.; Fishilevich, S.; Stein, T.I.; Nudel, R.; Lieder, I.; Mazor, Y.; et al. The GeneCards Suite: From Gene Data Mining to Disease Genome Sequence Analyses. *Curr. Protoc. Bioinform.* **2016**, *54*, 1.30.1–1.30.33. [CrossRef] [PubMed]
64. Freshour, S.L.; Kiwala, S.; Cotto, K.C.; Coffman, A.C.; McMichael, J.F.; Song, J.J.; Griffith, M.; Griffith, O.L.; Wagner, A.H. Integration of the Drug-Gene Interaction Database (DGIdb 4.0) with Open Crowdsource Efforts. *Nucleic Acids Res.* **2021**, *49*, D1144–D1151. [CrossRef] [PubMed]
65. Burley, S.K.; Bhikadiya, C.; Bi, C.; Bittrich, S.; Chen, L.; Crichlow, G.V.; Christie, C.H.; Dalenberg, K.; Di Costanzo, L.; Duarte, J.M.; et al. RCSB Protein Data Bank: Powerful New Tools for Exploring 3D Structures of Biological Macromolecules for Basic and Applied Research and Education in Fundamental Biology, Biomedicine, Biotechnology, Bioengineering and Energy Sciences. *Nucleic Acids Res.* **2021**, *49*, D437–D451. [CrossRef] [PubMed]
66. Wishart, D.S.; Feunang, Y.D.; Guo, A.C.; Lo, E.J.; Marcu, A.; Grant, J.R.; Sajed, T.; Johnson, D.; Li, C.; Sayeeda, Z.; et al. DrugBank 5.0: A Major Update to the DrugBank Database for 2018. *Nucleic Acids Res.* **2018**, *46*, D1074–D1082. [CrossRef] [PubMed]
67. Shannon, P.; Markiel, A.; Ozier, O.; Baliga, N.S.; Wang, J.T.; Ramage, D.; Amin, N.; Schwikowski, B.; Ideker, T. Cytoscape: A Software Environment for Integrated Models of Biomolecular Interaction Networks. *Genome Res.* **2003**, *13*, 2498–2504. [CrossRef]
68. Dallakyan, S.; Olson, A.J. Small-Molecule Library Screening by Docking with PyRx BT - Chemical Biology: Methods and Protocols. In *Chemical Biology*; Hempel, J.E., Williams, C.H., Hong, C.C., Eds.; Springer: New York, NY, USA, 2015; pp. 243–250, ISBN 978-1-4939-2269-7.
69. Rappe, A.K.; Casewit, C.J.; Colwell, K.S.; Goddard, W.A.; Skiff, W.M. UFF, a Full Periodic Table Force Field for Molecular Mechanics and Molecular Dynamics Simulations. *J. Am. Chem. Soc.* **1992**, *114*, 10024–10035. [CrossRef]
70. O'Boyle, N.M.; Banck, M.; James, C.A.; Morley, C.; Vandermeersch, T.; Hutchison, G.R. Open Babel: An Open Chemical Toolbox. *J. Cheminform.* **2011**, *3*, 33. [CrossRef]
71. Nagae, M.; Kizuka, Y.; Mihara, E.; Kitago, Y.; Hanashima, S.; Ito, Y.; Takagi, J.; Taniguchi, N.; Yamaguchi, Y. Structure and Mechanism of Cancer-Associated N-Acetylglucosaminyltransferase-V. *Nat. Commun.* **2018**, *9*, 3380. [CrossRef]
72. Ihara, H.; Ikeda, Y.; Toma, S.; Wang, X.; Suzuki, T.; Gu, J.; Miyoshi, E.; Tsukihara, T.; Honke, K.; Matsumoto, A.; et al. Crystal Structure of Mammalian A1,6-Fucosyltransferase, FUT8. *Glycobiology* **2007**, *17*, 455–466. [CrossRef]

73. Lira-Navarrete, E.; Valero-González, J.; Villanueva, R.; Martínez-Júlvez, M.; Tejero, T.; Merino, P.; Panjikar, S.; Hurtado-Guerrero, R. Structural Insights into the Mechanism of Protein O-Fucosylation. *PLoS ONE* **2011**, *6*, e25365. [CrossRef]
74. Waterhouse, A.; Bertoni, M.; Bienert, S.; Studer, G.; Tauriello, G.; Gumienny, R.; Heer, F.T.; de Beer, T.A.P.; Rempfer, C.; Bordoli, L.; et al. SWISS-MODEL: Homology Modelling of Protein Structures and Complexes. *Nucleic Acids Res.* **2018**, *46*, W296–W303. [CrossRef] [PubMed]
75. Pettersen, E.F.; Goddard, T.D.; Huang, C.C.; Couch, G.S.; Greenblatt, D.M.; Meng, E.C.; Ferrin, T.E. UCSF Chimera–a Visualization System for Exploratory Research and Analysis. *J. Comput. Chem.* **2004**, *25*, 1605–1612. [CrossRef] [PubMed]
76. Trott, O.; Olson, A.J. AutoDock Vina: Improving the Speed and Accuracy of Docking with a New Scoring Function, Efficient Optimization, and Multithreading. *J. Comput. Chem.* **2010**, *31*, 455–461. [CrossRef] [PubMed]
77. Park, S.-J.; Lee, J.; Qi, Y.; Kern, N.R.; Lee, H.S.; Jo, S.; Joung, I.; Joo, K.; Lee, J.; Im, W. CHARMM-GUI Glycan Modeler for Modeling and Simulation of Carbohydrates and Glycoconjugates. *Glycobiology* **2019**, *29*, 320–331. [CrossRef]
78. Huang, J.; Rauscher, S.; Nawrocki, G.; Ran, T.; Feig, M.; de Groot, B.L.; Grubmüller, H.; MacKerell, A.D. CHARMM36m: An Improved Force Field for Folded and Intrinsically Disordered Proteins. *Nat. Methods* **2017**, *14*, 71–73. [CrossRef]
79. Humphrey, W.; Dalke, A.; Schulten, K. VMD: Visual Molecular Dynamics. *J. Mol. Graph.* **1996**, *14*, 33–38. [CrossRef]

MDPI
St. Alban-Anlage 66
4052 Basel
Switzerland
Tel. +41 61 683 77 34
Fax +41 61 302 89 18
www.mdpi.com

Molecules Editorial Office
E-mail: molecules@mdpi.com
www.mdpi.com/journal/molecules

www.ingramcontent.com/pod-product-compliance
Lightning Source LLC
LaVergne TN
LVHW070121170726
843515LV00007B/1729

* 9 7 8 3 0 3 6 5 6 1 2 5 7 *